Key Topics in Quantum and Statistical Mechanics

Key Topics in Quantum and Statistical Mechanics

Editor: Mac Shaw

NY RESEARCH PRESS
New York

Published by NY Research Press
118-35 Queens Blvd., Suite 400,
Forest Hills, NY 11375, USA
www.nyresearchpress.com

Key Topics in Quantum and Statistical Mechanics
Edited by Mac Shaw

International Standard Book Number: 978-1-64725-466-7 (Hardback)

Cataloging-in-Publication Data

Key topics in quantum and statistical mechanics / edited by Mac Shaw.
p. cm.
Includes bibliographical references and index.
ISBN 978-1-64725-466-7
1. Quantum theory. 2. Statistical mechanics. 3. Statistical physics. I. Shaw, Mac.
QC174.12 .K49 2023
530.12--dc23

Contents

Preface

Quantum mechanics refers to a branch of physics concerned with the behavior of matter and light at the atomic and subatomic levels. It tries to explain the properties of molecules and atoms. It also seeks to describe the properties of the fundamental particles of atoms and molecules, which include neutrons, gluons, protons, quarks and electrons. Statistical mechanics is a branch of physics that uses probability theory to study the thermodynamic behavior of systems composed of several particles. It is helpful in the fundamental study of physical systems with multiple degrees of freedom. Statistical mechanics provides a framework for connecting microscopic properties of individual molecules and atoms to macroscopic bulk properties of materials that can be seen in everyday life. This book includes some of the vital pieces of works being conducted across the world, on key topics related to quantum and statistical mechanics. It consists of contributions made by international experts. A number of latest researches have been included to keep the readers up-to-date with the global concepts in this area of study.

The researches compiled throughout the book are authentic and of high quality, combining several disciplines and from very diverse regions from around the world. Drawing on the contributions of many researchers from diverse countries, the book's objective is to provide the readers with the latest achievements in the area of research. This book will surely be a source of knowledge to all interested and researching the field.

In the end, I would like to express my deep sense of gratitude to all the authors for meeting the set deadlines in completing and submitting their research chapters. I would also like to thank the publisher for the support offered to us throughout the course of the book. Finally, I extend my sincere thanks to my family for being a constant source of inspiration and encouragement.

Editor

1

Algebraic DVR Approaches applied to Describe the Stark Effect

Marisol Bermúdez-Montaña [1,2], Marisol Rodríguez-Arcos [1], Renato Lemus [1,*], José M. Arias [3,4], Joaquín Gómez-Camacho [3,5] and Emilio Orgaz [2]

1 Instituto de Ciencias Nucleares, Universidad Nacional Autónoma de México, Apartado Postal 70-543, CDMX 04510, Mexico; marisol.bermudez@nucleares.unam.mx (M.B.-M.); marisol_ra@ciencias.unam.mx (M.R.-A.)
2 Facultad de Química, Universidad Nacional Autónoma de México, Apartado Postal 70-543, CDMX 04510, Mexico; emilio.Orgaz@unam.mx
3 Departamento de Física Atómica, Molecular y Nuclear, Facultad de Física, Universidad de Sevilla, Apartado 1065, 41080 Sevilla, Spain; ariasc@us.es (J.M.A.); gomez@us.es (J.G.-C.)
4 Instituto Carlos I (iCI) de Física Teórica y Computacional, Universidad de Sevilla, Apartado 1065, 41080 Sevilla, Spain
5 CN de Aceleradores (U. Sevilla, J. Andalucía, CSIC), 41092 Sevilla, Spain
* Correspondence: renato@nucleares.unam.mx

Abstract: Two algebraic approaches based on a discrete variable representation are introduced and applied to describe the Stark effect in the non-relativistic Hydrogen atom. One approach consists of considering an algebraic representation of a cutoff 3D harmonic oscillator where the matrix representation of the operators r^2 and p^2 are diagonalized to define useful bases to obtain the matrix representation of the Hamiltonian in a simple form in terms of diagonal matrices. The second approach is based on the $U(4)$ dynamical algebra which consists of the addition of a scalar boson to the 3D harmonic oscillator space keeping constant the total number of bosons. This allows the kets associated with the different subgroup chains to be linked to energy, coordinate and momentum representations, whose involved branching rules define the discrete variable representation. Both methods, although originating from the harmonic oscillator basis, provide different convergence tests due to the fact that the associated discrete bases turn out to be different. These approaches provide powerful tools to obtain the matrix representation of 3D general Hamiltonians in a simple form. In particular, the Hydrogen atom interacting with a static electric field is described. To accomplish this task, the diagonalization of the exact matrix representation of the Hamiltonian is carried out. Particular attention is paid to the subspaces associated with the quantum numbers $n = 2, 3$ with $m = 0$.

Keywords: algebraic approach; DVR method; coulombic potential; stark effect

---

1. Introduction

A discrete variable representation approach is based on the search of a discrete basis in terms of which any function of the coordinates is diagonal. The discrete variable representation methods (DVR) in configuration space were developed with some variants since the 1960s, but systematically widely used during the 1980s with different names: discrete-variable representation method [1–4], quadrature discretization method [5,6], configuration localized states (CLS) approach [7], and Lagrange mesh method (LMM) [8–11], whose similarities and differences are discussed in Ref. [11]. The basic ingredient of these methods is the use of orthogonal polynomials and their associated grids from Gaussian quadratures.

In this contribution, we introduce two algebraic DVR methods where the discrete bases are obtained using purely algebraic means without any explicit reference to polynomials. This is accomplished through the diagonalization of the matrix representations associated with the coordinates and momenta. From this perspective, the recently-proposed unitary group approach ($U(4)$-UGA) belongs to an algebraic DVR method [12–16]. In this approach, the discretization provided by the zeros of the polynomials in configuration space is substituted by the branching rules involved in the subgroup chains embedded in the dynamical group. An alternative approach to establish an algebraic DVR method consists of taking a cutoff 3D harmonic oscillator in Fock space and diagonalize the matrix representation of the squares of the coordinates and momenta to obtain two discrete bases (HO-DVR approach), which in turn are used to diagonalize the matrix representation of the Hamiltonian in a simple manner. Both methods are based on a harmonic oscillator basis, albeit they provide different convergence behavior due to the fact that the provided discrete bases are distinct.

In this contribution we present the HO-DVR approach as well as the salient features of the $U(4)$-UGA, both applied to the Coulombic potential. We first show the confidence of our approaches to reproduce the analytical wave functions for the non relativistic Hydrogen atom. Both approaches are able to arrive at the same level of accuracy with the appropriate basis dimension. In order to explore the capabilities and differences between these approaches, the Stark effect is studied by adding the potential energy for the electron in an homogeneous electric field. Our treatment involves a complete dipole matrix representation contemplating the possibility of the action of fields beyond the perturbation region. Since the conservation of the angular momentum is broken, the calculations become particularly heavy. This fact forces us to constraint the basis to a limited angular momentum. With the proposal of these DVR algebraic approaches, we do not intend to improve the variational approaches used to describe the Stark effect, but rather for showing a simple way to deal with this problem as well as to evaluate the differences between them.

Our contribution is organized as follows. In Section 2 the salient features of both the algebraic HO-DVR method and the $U(4)$-UGA are presented. Section 3 is devoted to applying the methods to describe the isolated non-relativistic Hydrogen atom. To this end, both energies and wave functions are tested. In Section 4 the Stark effect is described, paying attention to the bound states originated from the subspaces with $n = 2, 3$. Finally, in Section 5 the summary and our conclusions are presented.

2. Algebraic DVR Methods in 1D Systems

In this section, we present the salient features of the HO-DVR method as well as of the $U(4)$-UGA. Since both approaches are based on the 3D harmonic oscillator we start presenting the main features associated with its algebraic solutions.

2.1. 3D Harmonic Oscillator

We start considering the isotropic 3D harmonic oscillator with reduced mass m and frequency ω. The Hamiltonian is given by

$$\hat{H}^{\text{HO}}_{\text{CS}} = \frac{1}{2m}p^2 + \frac{m\omega^2}{2}r^2. \tag{1}$$

The solutions in configuration space are well known, but this is not the case in Fock space [17,18]. The Hamiltonian, as well as the eigenfunctions can be translated into the Fock space through the introduction of the bosonic creation $a^\dagger_\mu$ and annihilation a_μ operators with commutation relations

$$[a_\mu, a^\dagger_\nu] = \delta_{\mu\nu}; \quad [a^\dagger_\mu, a^\dagger_\nu] = [a_\mu, a_\nu] = 0. \tag{2}$$

The reflection and rotation properties of $a^\dagger_\mu$ are summarized by their transformation under a rotation $\hat{R}(\alpha, \beta, \gamma)$:

$$\hat{R}^{-1}a^\dagger_\mu\hat{R} = \sum_{\mu'} D^{(1)}_{\mu'\mu}(\alpha, \beta, \gamma)a^\dagger_{\mu'}, \tag{3}$$

characterized by the three Euler angles (α, β, γ). The annihilation operators a_μ, however, do not transform as the components of a spherical tensor operator, but the modified operator

$$\tilde{a}_\mu = (-1)^{1-\mu} a_{-\mu} \tag{4}$$

do satisfy the transformation property (3). In this way the the operators $a^\dagger_\mu$ and $\tilde{a}_\mu$ can be coupled to definite angular momentum λ and projection m

$$[a^\dagger \times \tilde{a}]^{(\lambda)}_m = \sum_{\mu',\mu} \langle 1\mu 1\mu' | \lambda m \rangle a^\dagger_\mu \tilde{a}_{\mu'}, \tag{5}$$

using the standard notations $\times$ for the angular momentum coupling and the Clebsh–Gordan coefficients. These operators in terms of the spherical coordinate and momentum components are given by

$$a^\dagger_\mu = \frac{1}{\sqrt{2}} \left(\frac{1}{d} q_\mu - i\frac{d}{\hbar} (-1)^\mu p_{-\mu} \right); \tag{6}$$

$$\tilde{a}_\mu = - \frac{1}{\sqrt{2}} \left(\frac{1}{d} q_\mu + i\frac{d}{\hbar} (-1)^\mu p_{-\mu} \right), \tag{7}$$

with commutation relations

$$[\tilde{a}_\mu, a^\dagger_\nu] = (-1)^{1-\mu} \delta_{-\mu\nu}; \ \ [a^\dagger_\mu, a^\dagger_\nu] = [\tilde{a}_\mu, \tilde{a}_\nu] = 0, \tag{8}$$

where $d = \sqrt{\hbar/m\omega}$ has been introduced as unit length, $q_\mu (\mu = 1, 0, -1)$ are the coordinate components with $r^2 = \sum_\mu (-1)^\mu q_\mu q_{-\mu}$ and corresponding momenta $p_\mu = -i\hbar \frac{\partial}{\partial q_\mu}$. The solutions of the eigensystem associated with (1) in Fock space are given by the kets [17,18]

$$|nlm\rangle = B_{nl} \, (\mathbf{a}^\dagger \cdot \mathbf{a}^\dagger)^{(n-l)/2} \mathcal{Y}^l_m(\mathbf{a}^\dagger)|0\rangle, \tag{9}$$

characterized by the eigensystem

$$\hat{n}|nlm\rangle = n|nlm\rangle, \tag{10}$$

$$\hat{L}^2|nlm\rangle = l(l+1)|nlm\rangle, \tag{11}$$

$$\hat{L}_z|nlm\rangle = m|nlm\rangle, \tag{12}$$

where $\mathcal{Y}^l_m(\mathbf{a}^\dagger)$ are the solid spherical harmonics defined by

$$\mathcal{Y}^l_m(\mathbf{a}^\dagger) = 2^{-l/2} (\mathbf{a}^\dagger \cdot \mathbf{a}^\dagger)^{l/2} Y^l_m(\mathbf{a}^\dagger) \tag{13}$$

in terms of the spherical harmonics $Y^l_m(\mathbf{a}^\dagger)$ with normalization [17,18]

$$B_{nl} = (-1)^{(n-l)/2} \sqrt{\frac{4\pi}{(n-l)!!(n+l+1)!!}}, \tag{14}$$

and eigenvalues

$$E^{\mathrm{HO}}_n = \hbar\omega(n + 3/2), \tag{15}$$

with frequency $\omega = \sqrt{k/m}$ in terms of the force constant k and number operator

$$\hat{n} = \sum_\mu a^\dagger_\mu a_\mu = \sqrt{3} [\mathbf{a}^\dagger \times \tilde{\mathbf{a}}]^{(\mathbf{0})}_{\mathbf{0}}, \tag{16}$$

while the notation $\mathbf{a}^\dagger \cdot \mathbf{a}^\dagger$ stands for the dot product defined by

$$\mathbf{a}^\dagger \cdot \mathbf{a}^\dagger = \sqrt{3}[\mathbf{a}^\dagger \times \mathbf{a}^\dagger]^{(0)}_0 = \sum_\mu (-1)^\mu a^\dagger_\mu a^\dagger_{-\mu}, \tag{17}$$

with the realization of the angular momentum components

$$\hat{L}_\mu = \sqrt{2}[a^\dagger \times \tilde{a}]^{(1)}_\mu. \tag{18}$$

This summary for the harmonic oscillator paves the way to introduce the algebraic DVR methods.

2.2. HO-DVR Approach

The HO-DVR approach consists of establishing discrete bases where functions of the radial coordinate and the corresponding momentum are diagonal. To accomplish this task we should consider the algebraic realization of the variables involved in Equation (1). Based on the relations (6) and (7), the squared of the radius and momenta in Fock space takes the form

$$r^2 = (\hat{n} + 3/2) + \frac{1}{2}(\mathbf{a}^\dagger \cdot \mathbf{a}^\dagger + \mathbf{a} \cdot \mathbf{a}), \tag{19}$$

$$p^2 = (\hat{n} + 3/2) - \frac{1}{2}(\mathbf{a}^\dagger \cdot \mathbf{a}^\dagger + \mathbf{a} \cdot \mathbf{a}), \tag{20}$$

Since both operators (19) and (20) preserve the angular momentum we can obtain their matrix representation for a given l and m in the N-dimensional space defined by

$$\mathcal{L}^l_N = \{|nlm\rangle, n = 0, 1, \ldots, N-1\}, \tag{21}$$

with the restrictions $l = n, n-2, \ldots, 1$ or 0, and the usual reduction $SO(3) \supset SO(2)$ given by $-l \leq m \leq l$. It is clear that the N-dimensional space (21) is not complete, but it establishes a space to carry our the calculations in an approximate way. Explicitly the matrix elements of the operators (19) and (20) in dimensionless units turn out to be

$$\begin{aligned} \langle nlm|r^2|nlm\rangle &= \langle nlm|p^2|nlm\rangle, \\ &= n + \frac{3}{2}, \end{aligned} \tag{22}$$

$$\begin{aligned} \langle n+2, lm|r^2|nlm\rangle &= -\langle n+2, lm|p^2|nlm\rangle \\ &= \frac{1}{2}\sqrt{(n+l+2)(n-l+2)}. \end{aligned} \tag{23}$$

The numerical diagonalizations of the representation matrix of these operators provide a discrete set of eigenvectors

$$|lm, r^2_i\rangle = \sum_{n=0}^{N-1} \langle nlm|lm, r^2_i\rangle |nlm\rangle, \tag{24}$$

$$|lm, p^2_i\rangle = \sum_{n=0}^{N-1} \langle nlm|lm, p^2_i\rangle |nlm\rangle, \tag{25}$$

which provide the radial coordinate and momentum representations characterized by

$$r|lm, r^2_i\rangle = \sqrt{r^2_i}\, |lm, r^2_i\rangle, \tag{26}$$

$$p|lm, p^2_i\rangle = \sqrt{p^2_i} |lm, p^2_i\rangle. \tag{27}$$

Because of these properties, the basis $|lm, r_i^2\rangle$ is identified with the position representation, while $|lm, p_i^2\rangle$ stands for the momentum representation. In addition since the Hamiltonian (15) is diagonal in the basis (9) we refer the $U(3)$ basis as the energy representation. As a consequence, for any functions $V(r)$ and $G(p)$ we have for their matrix elements

$$\langle lm, r_j^2|V(r)|lm, r_i^2\rangle = V\left(\sqrt{r_i^2}\right)\delta_{ij}, \tag{28}$$

$$\langle lm, p_j^2|G(p)|lm, p_i^2\rangle = G\left(\sqrt{p_i^2}\right)\delta_{ij}. \tag{29}$$

The eigenvectors (24) and (25) define the transformation coefficients

$$\mathbf{W} = ||\langle nlm|lm, p_i^2\rangle||, \tag{30}$$
$$\mathbf{T} = ||\langle nlm|lm, r_i^2\rangle||, \tag{31}$$

while the properties (28) and (29) lead to a DVR method where the matrix representation of the general 3D Hamiltonian depending on the radial coordinate

$$\hat{H} = \frac{p^2}{2m} + V(r), \tag{32}$$

takes the following matrix form in the energy representation:

$$\mathbf{H} = \mathbf{W}^\dagger \mathbf{\Lambda}^{(p)} \mathbf{W} + \mathbf{T}^\dagger \mathbf{\Lambda}^{(r)} \mathbf{T}, \tag{33}$$

where $\mathbf{\Lambda}^{(p)}$ is the diagonal contribution of the kinetic energy in the momentum representation (25):

$$||\mathbf{\Lambda}^{(p)}|| = ||\langle lm, p_j^2|\frac{p^2}{2m}|lm, p_i^2\rangle|| = \frac{p_i^2}{2m}\delta_{ij}, \tag{34}$$

where the notation for the mass m should not be confused with the angular momentum projection m. In addition the matrix $\mathbf{\Lambda}^{(r)}$ corresponds to the diagonal matrix of the potential $V(r)$ in the position representation:

$$||\mathbf{\Lambda}^{(r)}|| = ||\langle lm, r_j^2|\left[V\left(\sqrt{r^2}\right)\right]|lm, r_i^2\rangle|| = \left[V\left(\sqrt{r_i^2}\right)\right]\delta_{ij}. \tag{35}$$

Both transformation coefficients **W** and **T** are numerically calculated only once for a given l and m and space dimension N, providing the methodology for any potential in the framework of the HO-DVR method. However since the basis corresponds to the harmonic oscillator functions, adding and subtracting the quadratic potential makes possible to reformulate the Equation (33) in terms of only the transformation coefficients **T** in the form:

$$\mathbf{H} = \mathbf{\Lambda}^{(E)} + \kappa \mathbf{T}^\dagger \mathbf{\Lambda}'^{(r)} \mathbf{T}, \tag{36}$$

where for a given l and m, $\mathbf{\Lambda}^{(E)}$ is the diagonal matrix in the energy representation

$$||\Lambda^{(n)}|| = ||\langle n'lm|\hbar\omega(\hat{n} + 3/2)|nlm\rangle|| = \hbar\omega(n + 3/2)\delta_{n'n}, \tag{37}$$

while in the position representation

$$||\Lambda'^{(r)}|| = -\frac{m\omega^2}{2} r_i^2 \delta_{ij} + V\left(\sqrt{r_i^2}\right)\delta_{ij}. \tag{38}$$

In Equation (36), κ stands for a control parameter in the interval $[0,1]$, whose variation allows the potential to be transformed from a harmonic oscillator ($\kappa = 0$) to the arbitrary potential $V(\sqrt{r^2})$ ($\kappa = 1$).

The methodology of this approach is simple. We diagonalize both matrix representations (22) and (23) to obtain the coefficients (30) and (31). Then we proceed to diagonalize in the energy representation the Hamiltonian (36) associated with the potential we are interested in for a given angular momentum and projection.

2.3. *SU*(4)-*UGA*

This approach starts by establishing the $U(4)$ dynamical group for 3D systems by adding a scalar boson $s^\dagger(s)$ to the space of the same bosonic operators $a^\dagger_\mu(\tilde{a}_\mu)$ previously introduced carrying angular momentum $l = 1$ in such a way that the bilinear products satisfy the commutation relations associated with the generators of the $U(4)$ group [19]:

$$\hat{n}_a = \sqrt{3}[a \times \tilde{a}]^{(0)}_0; \qquad \hat{n}_s = s^\dagger s, \tag{39}$$

$$\hat{L}_\mu = \sqrt{2}[a \times \tilde{a}]^{(1)}_\mu; \qquad \hat{D}_\mu = a^\dagger_\mu s - s^\dagger \tilde{a}_\mu, \tag{40}$$

$$\hat{Q}_\mu = [a \times \tilde{a}]^{(2)}_\mu; \qquad \hat{R}_\mu = i(a^\dagger_\mu s + s^\dagger \tilde{a}_\mu), \tag{41}$$

with the constraint that the total number of bosons $\hat{N} = \hat{n}_a + s^\dagger s$ is constant. Hence, the associated quantum number N establishes the dimension of the dynamical space. From the dynamical group $U(4)$, three subgroup chains can be obtained, preserving the angular momentum [18,19]:

$$U(4) \supset U(3) \supset SO(3) \supset SO(2), \tag{42}$$

$$U(4) \supset SO(4) \supset SO(3) \supset SO(2), \tag{43}$$

$$U(4) \supset \overline{SO}(4) \supset SO(3) \supset SO(2). \tag{44}$$

Each chain provides a basis with a definitive meaning, as will become clear shortly. A particular chain, (42), leads to the basis $|[N]n_aLM\rangle$ defined by the operator equations

$$\hat{N}|[N]n_aLM\rangle = N|[N]n_aLM\rangle, \tag{45}$$

$$\hat{n}_a|[N]n_aLM\rangle = n_a|[N]n_aLM\rangle, \tag{46}$$

$$\hat{L}^2|[N]n_aLM\rangle = L(L+1)|[N]n_aLM\rangle, \tag{47}$$

$$\hat{L}_z|[N]n_aLM\rangle = M|[N]n_aLM\rangle. \tag{48}$$

We now turn our attention to the connection with configuration space. Using the approach recently proposed [20] with a mapping to the 3D harmonic oscillator basis

$$|[N]n_aLM\rangle \to |nlm\rangle. \tag{49}$$

It is possible to show that the $U(4)$ algebraic realization $\hat{\mathcal{Q}}_\mu$ and $\hat{\mathcal{P}}_{-\mu}$ for the coordinates and momenta, respectively, take the form [16]

$$\hat{\mathcal{Q}}_\mu = \frac{d}{\sqrt{2N}}\hat{D}_\mu, \tag{50}$$

$$\hat{\mathcal{P}}_{-\mu} = -(-)^{1-\mu}\frac{\hbar}{d}\frac{1}{\sqrt{2N}}\hat{R}_\mu. \tag{51}$$

The mapping

$$q_\mu \to \hat{\mathcal{Q}}_\mu \qquad p_\mu \to \hat{\mathcal{P}}_\mu \tag{52}$$

allows the algebraic Hamiltonian (1) to be obtained in the $U(4)$ space:

$$\hat{H}^{U(4)}_{alg} = \frac{1}{2m}\hat{\mathcal{P}}^2 + \frac{m\omega^2}{2}\hat{\mathcal{Q}}^2. \tag{53}$$

Notice that the operators $\hat{\mathcal{P}}^2$ and $\hat{\mathcal{Q}}^2$ are not related in a straightforward way with the physical momentum p^2 and coordinate q^2, but their matrix elements coincide in the N large limit:

$$\lim_{N\to\infty} ||\langle Nn'_a LM|\hat{\mathcal{P}}^2|Nn_a LM\rangle|| = ||\langle n'lm|p^2|nlm\rangle||, \tag{54}$$

with similar equation for the coordinate.

Taking into account (50) and (51), the Hamiltonian (53) is recast in the form:

$$\hat{H}^{U(4)}_{alg} = \frac{\hbar\omega}{4}\left[\frac{\hat{R}^2 + \hat{D}^2}{N}\right]. \tag{55}$$

The second order Casimir operators of $U(4)$ and $U(3)$ are given by [18]

$$\hat{C}_2[U(4)] = \frac{1}{3}\hat{n}_a^2 + \frac{1}{2}(\hat{L}^2 + \hat{D}^2 + \hat{R}^2) + \hat{Q}^2 + (\hat{N} - \hat{n}_a)^2, \tag{56}$$

$$\hat{C}_2[U(3)] = \frac{1}{3}\hat{n}_a^2 + \frac{1}{2}\hat{L}^2 + \hat{Q}^2, \tag{57}$$

and in the symmetrical space of identical bosons the following replacements are valid [18]

$$C_2[U(4)] \to \hat{N}(\hat{N}+3) \tag{58}$$

$$C_2[U(3)] \to \hat{n}_a(\hat{n}_a+2), \tag{59}$$

which allows (55) to be transformed into

$$\hat{H}^{U(4)}_{alg} = \hbar\omega\left[\left(1 - \frac{1}{N}\right)\hat{n}_a + \frac{3}{2} - \frac{\hat{n}_a^2}{N}\right]. \tag{60}$$

Here we have taken into account that the operator $\hat{N}$ is diagonal in any basis and consequently it can be substituted by its eigenvalue N. This result leads to the identification of the $U(3)$ chain (42) as the energy representation with eigenkets

$$\hat{n}_a|[N]n_a LM\rangle = n_a|[N]n_a LM\rangle, \tag{61}$$

and branching rules $n_a = 0, 1, \ldots, N$ and $L = n_a, n_a - 2, \ldots, 1$ or 0.

Let us now come back to chains (43, 44). Since the Casimir operators characterizing the $SO(4)$ and $\overline{SO}(4)$ groups are $\hat{W}^2 = \hat{D}^2 + \hat{L}^2$ and $\hat{\bar{W}}^2 = \hat{R}^2 + \hat{L}^2$ respectively, Equation (50) and (51) implies that chain (43) provides the coordinate representation while chain (44) gives the momentum representation. The corresponding eigenvectors satisfy

$$\hat{W}^2|[N]\zeta LM\rangle = \zeta(\zeta+2)|[N]\zeta LM\rangle, \tag{62}$$

$$\hat{\bar{W}}^2|[N]\bar{\zeta} LM\rangle = \bar{\zeta}(\bar{\zeta}+2)|[N]\bar{\zeta} LM\rangle, \tag{63}$$

with branching rules $\zeta, \bar{\zeta} = N, N-2, \ldots, 1$ or 0, and $L = 0, 1, \ldots, \zeta(\bar{\zeta})$. The importance of these transformations is twofold. On one hand, we have the mapping (49) and consequently it is possible to project any state to the position representation as an expansion of harmonic oscillator functions. On the other hand, an arbitrary potential depending on $\sqrt{\hat{\mathcal{Q}}^2}$ takes a diagonal form in the basis (62), providing a direct way to obtain its matrix representation in the energy representation through a similarity transformation.

Let us now consider a general situation of a 3D Hamiltonian of the form (32). The corresponding algebraic Hamiltonian is obtained by applying the anharmonization procedure (52):

$$\hat{H}^{U(4)}_{alg} = \hat{H}\Big|_{q\to\hat{\mathcal{Q}},p\to\hat{\mathcal{P}}} = \frac{1}{2m}\hat{\mathcal{P}}^2 + \hat{V}(\sqrt{\hat{\mathcal{Q}}^2}). \tag{64}$$

A practical convenient form for this Hamiltonian is obtained by adding and subtracting a quadratic term in such a way that the harmonic oscillator term (1) is identified:

$$\hat{H}^{U(4)}_{alg} = \hbar\omega\left[\left(1-\frac{1}{N}\right)\hat{n}_a + \frac{3}{2} - \frac{\hat{n}_a^2}{N}\right] + \kappa\hat{V}'(\sqrt{\hat{\mathcal{Q}}^2}). \tag{65}$$

where

$$V'(\hat{\mathcal{Q}}^2) = -\frac{m\omega^2}{2}\hat{\mathcal{Q}}^2 + \hat{V}(\sqrt{\hat{\mathcal{Q}}^2}). \tag{66}$$

Taking into account the scalar character of $\hat{\mathcal{Q}}^2$, we have its following matrix elements in the coordinate representation:

$$\langle [N]\zeta' LM|\hat{\mathcal{Q}}^2|[N]\zeta LM\rangle = \frac{d^2}{2}\frac{[\zeta(\zeta+2)-L(L+1)]}{N}\,\delta_{\zeta',\zeta}. \tag{67}$$

Hence, the matrix elements of the Hamiltonian (for a given L and M) in the energy representation take the general form

$$\mathbf{H}^{(E)} = \mathbf{\Lambda}^{(E)} + \kappa\mathbf{T}^{\dagger}\mathbf{\Lambda}^{(\mathcal{Q})}\mathbf{T}, \tag{68}$$

where $\mathbf{\Lambda}^{(E)}$ is the diagonal contribution of the deformed harmonic oscillator

$$||\mathbf{\Lambda}^{(E)}|| = \hbar\omega\left[\left(1-\frac{1}{N}\right)n_a + \frac{3}{2} - \frac{n_a^2}{N}\right]\delta_{\hat{n}'_a,n_a}, \tag{69}$$

while $\mathbf{\Lambda}^{(\mathcal{Q})}$ is the diagonal matrix of the potential $V'(\sqrt{\hat{\mathcal{Q}}^2})$ in the position representation:

$$||\mathbf{\Lambda}^{(\mathcal{Q})}|| = \hbar\omega\left[-\frac{1}{2}\frac{\xi(\zeta,L)^2}{2N} + \frac{1}{\hbar\omega}V\left(d\frac{\xi(\zeta,L)}{\sqrt{2N}}\right)\right]\delta_{\zeta,\zeta'}, \tag{70}$$

with $\xi(\zeta,L) = \sqrt{\zeta(\zeta+2)-L(L+1)}$. The $\mathbf{T}$ matrix stands for the transformation brackets: $\mathbf{T} = ||\langle [N]\zeta LM|[N]n_a LM\rangle||$, which are obtained by diagonalizing the matrix representation of the Casimir operators $\hat{W}^2$ given by

$$\begin{aligned}
&\langle [N]n_a LM|\hat{W}^2|[N];n_a LM\rangle = \\
&\qquad\qquad N(2n_a+3) - 2n_a(n_a+1) + L(L+1), \\
&\langle [N]n_a+2, LM|\hat{W}^2|[N];n_a LM\rangle = \\
&-\sqrt{(N-n_a-1)(N-n_a)(n_a-L+2)(n_a+L+3)},
\end{aligned}$$

or in analytic form [18,21]. This matrix representation of $\hat{W}^2$ is tridiagonal and its diagonalization to provide $\mathbf{T}$ is calculated once and for all for a given N, while the diagonal matrix $\mathbf{\Lambda}^{(\mathcal{Q})}$ is modified in accordance with the specific potential. The contribution (69) involves accidental degeneracy due to the quadratic contribution $\hat{n}_a^2$. This degeneracy manifests in the full Hamiltonian (68) by the appearance of degeneracy for the bound states in the N large limit. This degeneracy is removed by substituting the diagonal term (69) by [16]

$$\left[\left(1-\frac{1}{N}\right)\hat{n}_a+\frac{3}{2}-\delta\frac{\hat{n}_a^2}{N}\right], \tag{71}$$

with $\delta = 1$ for $n_a < N/2$ and $\delta = 0$ for $n_a \geq N/2$.

The expression (68) contains the main features of our approach. The physical information characterizing the system is given by the diagonal matrix $\mathbf{\Lambda}^{(\mathcal{Q})}$ easily generated and given by (70), while the calculation of the transformation brackets allows any system to be analyzed in simple form using the standard tools of matrix diagonalization. It is important to notice that this approach is intended to be considered in the N large limit [16]. This is in contrast to the previous analysis of the algebraic models where $N \to \infty$ represents the classical limit [22].

3. Coulombic Potential: Hydrogen Atom

Because of its importance in dealing with molecular systems the Coulombic potential deserves special attention. For this reason we shall consider the non-relativistic Hydrogen atom together with the effect of an external electric field to show the advantages of the proposed approaches. We start considering the Hamiltonian of the Hydrogen atom.

Following M. Moshinsky, [17], in energy units of the first Bohr orbit $E_B = \frac{1}{2}me^4/\hbar^2$, the Hamiltonian takes the form

$$\frac{\hat{H}}{E_B} = \frac{1}{2mE_B}p^2 - \sqrt{\frac{2\hbar^2}{mE_B}}\frac{1}{r}. \tag{72}$$

Here, the mass associated with the harmonic oscillator is identified with the reduced mass of the Hydrogen system. Adding and subtracting the harmonic oscillator potential we obtain

$$\frac{\hat{H}}{E_B} = \epsilon^2\hat{H}^{\mathrm{HO}} - \left[\frac{\epsilon^2}{2}\bar{r}^2 + \epsilon\sqrt{2}\frac{1}{\bar{r}}\right], \tag{73}$$

where $\bar{r}$ is assumed to be in units of $d = \sqrt{\hbar/m\omega}$, the harmonic oscillator Hamiltonian $\hat{H}^{\mathrm{HO}}$ in units of $\hbar\omega$ and ϵ a dimensionless parameter given by

$$\epsilon = \sqrt{\frac{\hbar\omega}{E_B}}, \tag{74}$$

with ω a variational parameter to be determined.

Let us now establish, on one hand, the application of the HO-DVR approach to this problem. The matrix representation of the Hamiltonian in units of $\hbar\omega$ takes the form

$$\mathbf{H} = \epsilon^2\mathbf{\Lambda}^{(n)} - \kappa\mathbf{T}^{\dagger}\mathbf{\Lambda}'^{(r)}\mathbf{T}, \tag{75}$$

where the diagonal matrix $\mathbf{\Lambda}^{(n)}$ in the energy representation is given by

$$||\mathbf{\Lambda}^{(n)}|| = (n+3/2)\delta_{n'n}, \tag{76}$$

while for the diagonal matrix $\Lambda'^{(r)}$ in the position representation we have

$$||\Lambda'^{(r)}|| = \left[\frac{\epsilon^2}{2}\bar{r}_i^2 + \epsilon\sqrt{2}\frac{1}{\sqrt{\bar{r}_i^2}}\right]\delta_{ij}. \tag{77}$$

On the other hand, in the framework of the $U(4)$-DVR, we introduce the realization (50) and (51) into (73) to obtain

$$\frac{\hat{H}}{E_B} = \epsilon^2\left[\left(1-\frac{1}{N}\right)\hat{n}_a + \frac{3}{2} - \frac{\hat{n}_a^2}{N}\right] - \left[\frac{1}{2}\epsilon^2\left(\frac{\hat{D}^2}{2N}\right) + \epsilon\frac{2\sqrt{N}}{\sqrt{\hat{D}^2}}\right]. \tag{78}$$

The corresponding matrix representation takes the form

$$\mathbf{H}^{(E)} = \mathbf{\Lambda}^{(E)} - \kappa\mathbf{T}^{\dagger}\mathbf{\Lambda}^{(\mathcal{Q})}\mathbf{T}, \tag{79}$$

where $\mathbf{\Lambda}^{(E)}$ is the diagonal contribution of the deformed harmonic oscillator

$$||\mathbf{\Lambda}^{(E)}|| = \epsilon^2\left[\left(1-\frac{1}{N}\right)n_a + \frac{3}{2} - \frac{n_a^2}{N}\right]\delta_{n_a',n_a}, \tag{80}$$

while $\mathbf{\Lambda}^{(\mathcal{Q})}$ is the diagonal matrix containing the potential with the form

$$||\mathbf{\Lambda}^{(\mathcal{Q})}|| = \left[\frac{1}{2}\epsilon^2\left(\frac{1}{2N}\right)\xi(\zeta,L)^2 + \frac{\epsilon\, 2\sqrt{N}}{\xi(\zeta,L)}\right]\delta_{\zeta,\zeta'}, \tag{81}$$

where the preservation of the angular momentum has been considered implicitly. It is importance to notice that because of the branching rule $\zeta = N, N-2, \ldots, 1$ or 0, the number of bosons N must be taken to be odd in order to avoid a singularity (obtained when $L = \zeta = 0$).

The diagonalization of the Hamiltonian matrices (75) and (79) lead to the correlation diagram displayed in Figure 1 as a function of κ. The parameter (74) turns out to be $\epsilon = 1$. Although the correlations diagram is the same for both methods, the corresponding basis dimension is different in order to reach convergence. As noticed, the levels corresponding to the principal quantum number in the Hydrogen atom converge to the well known n^2 degeneracy, although to simplify the figure we have only included angular momenta up to 2. The degeneracy $2L+1 = 2l+1$ corresponding to the projection (M, m) should be assumed but not shown because of the (M, m)-independence of the Hamiltonian. The states above the zero energy represent a discretization of the continuum. This kind of the description of the continuum is also present in the Morse and Pöeschl–Teller potentials [23–25].

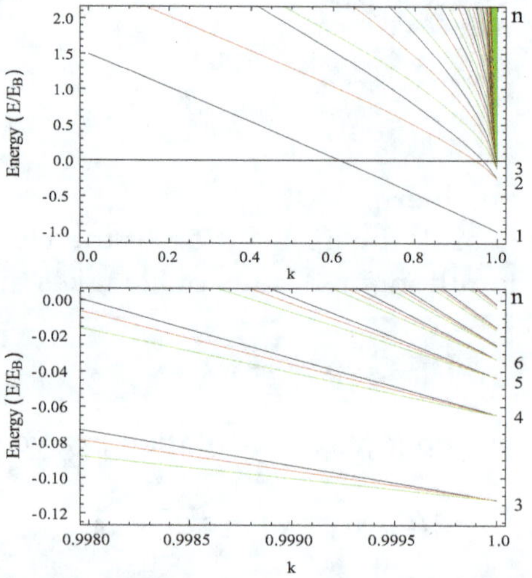

Figure 1. Correlation diagram from the harmonic oscillator ($\kappa = 0$) to the Coulomb potential ($\kappa = 1$) corresponding to the Hamiltonians (36) and (68), with (77) and (81), respectively. At the right of the figures the principal quantum number for the electron in the H atom, n, is indicated. To simplify only the levels of the Hydrogen system with $L, l = 0, 1, 2$ are included. In the bottom panel a zoom is displayed in order to show the convergence of the levels with different angular momenta. The top level in each group corresponds to angular momentum $L, l = 0$. The parameters were taken to be $\epsilon = 1$ and $N = 4001$ for the $U(4)$-unitary group approach (UGA), while $N = 2001$ for the HO-DVR method.

In order to explicitly show the energy convergence, the comparison between the exact energy levels $E_n = -1/n^2$ and the calculated ones are displayed in Table 1 for the HO-DVR method and in Table 2 for the $U(4)$-UGA. The convergence is similar, but while in the HO-DVR it is reached for $N = 2001$, the $U(4)$-UGA needed a number of bosons $N = 4001$. This difference may be explained by the accidental degeneracy involved in the $U(4)$-UGA but also because the bases are different. In both cases, the space dimension is large enough to obtain a reasonable convergence to the exact values. In general, the convergence can be measured through the fidelity F or overlap of eigenstates involving different parameters N [26]. Along this venue we have calculated the energies for number of bosons $N \pm 50$, registering the energy difference as the error displayed in Table 3 for the HO-DVR method and in Table 4 for the $U(4)$-UGA. It is interesting that the uniformity of the errors in the latter case. On the other hand, we know the analytical solutions and consequently the wave functions can be directly compared.

Table 1. Exact bond energies compared with the energies provided by the HO-DVR method, the basis dimension was taken to be $N = 2001$.

	Basis Dimension N		
	2001		
E_n	$L = 0$	$L = 1$	$L = 2$
−1.	−0.9969		
−0.25	−0.2496	−0.2500	
−0.1111	−0.1111	−0.1111	−0.1111
−0.0625	−0.0625	−0.0625	−0.0625
−0.04	−0.0400	−0.0400	−0.0400
−0.0278	−0.0275	−0.0276	−0.0276
−0.0204	−0.0173	−0.0175	−0.0180

Table 2. Exact bond energies compared with the energies provided by the diagonalization of (68) taking a total number of bosons $N = 4001$ with the $U(4)$-UGA method.

	Total Number of Bosons N		
	4001		
E_n	$L = 0$	$L = 1$	$L = 2$
−1.	−1.00330		
−0.25	−0.25088	−0.25039	
−0.1111	−0.11237	−0.11222	−0.11220
−0.0625	−0.06463	−0.06457	−0.06455
−0.0400	−0.04344	−0.04340	−0.04336
−0.0278	−0.0330	−0.0330	−0.0330
−0.0204	−0.02541	−0.02555	−0.02586

Table 3. Errors of the energies provided by the HO-DVR method. Errors were calculated considering the difference of energies $|E(N = 2051) - E(N = 1951)|$.

Error for $L = 0$	Error for $L = 1$	Error for $L = 2$
0.0001		
0.00002	9×10^{-9}	
5×10^{-6}	3×10^{-9}	8×10^{-13}
2×10^{-6}	1×10^{-9}	1×10^{-11}
2×10^{-6}	5×10^{-7}	3×10^{-7}
0.00008	0.00007	0.00005
0.0006	0.0005	0.0005

Table 4. Errors of the energies provided by the $U(4)$-UGA method. Errors were calculated considering the difference of energies $|E(N = 4051) - E(N = 3951)|$.

Error for $L = 0$	Error for $L = 1$	Error for $L = 2$
0.00007		
0.00002	0.00001	
0.00003	0.00003	0.00003
0.00005	0.00005	0.00005
0.00009	0.00009	0.00009
0.0001	0.0001	0.0001
0.00008	0.00006	0.00003

In our approach the α-th eigenstate in the HO-DVR method takes the form

$$|\psi^N_{\alpha,lm}\rangle = \sum_{n=0}^{N} \langle nlm|\psi^N_{\alpha,lm}\rangle \; |nlm\rangle, \tag{82}$$

while for the $U(4)$-UGA

$$|\psi^N_{\alpha,LM}\rangle = \sum_{n_a=0}^{N} \langle [N]n_a LM|\psi^N_{\alpha,LM}\rangle \; |[N]n_a LM\rangle. \tag{83}$$

From these expressions we obtain the corresponding position representation for the radial contribution of the wave function. Integrating over the angular coordinates and applying the coordinate projection we obtain $\langle r|\psi^N_{\alpha L}\rangle$ or $\langle r|\psi^N_{\alpha l}\rangle$ with

$$\langle r|nl\rangle = \langle r|[N]n_a L\rangle = A_{nl} r^l e^{r^2/2} L^{l+1/2}_{(n-l)/2}(r^2), \tag{84}$$

where $A_{nl} = \sqrt{(2^{l+2}(n-l)!!)/(\sqrt{\pi}(n+l+1)!!)}$. The projections $\langle r|\psi^N_{\alpha L}\rangle$ or $\langle r|\psi^N_{\alpha l}\rangle$ can be compared to the exact analytical functions $R_{nL}(r)$. Here the variable r is given in the distance unit d. Since $\epsilon = 1$ we have that $\hbar\omega = E_B$, a result that determines the distance units in the MKS units system. In Figure 2 the position projection of the eigenstates (82) and (83) together with the exact solutions for the $1s, 2s$ and $3s$ orbitals are displayed, while in Figure 3 the corresponding $2p, 3p$ and $3d$ orbitals are shown. Beyond the intrinsic convergence ($N = 2001$ for the HO-DVR method and $N = 4001$ for the U(4)-UGA) provided by the harmonic oscillator basis, we notice that the wave functions are very well described, a feature that proves the validity of both approaches for the free Hydrogen atom.

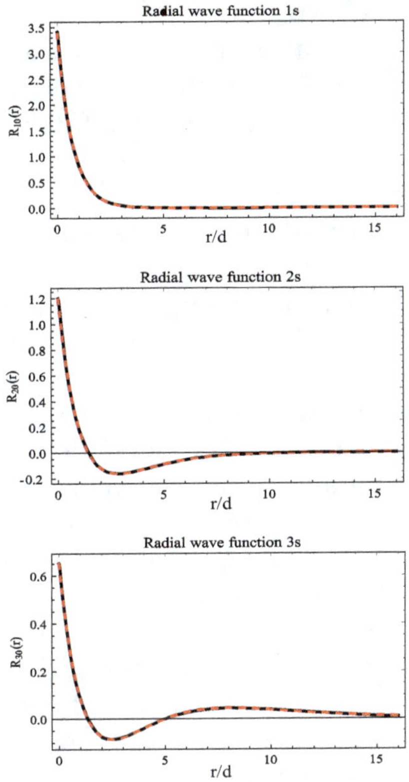

Figure 2. Comparison between exact and calculated radial wave functions $R_{n0}(r); n = 1, 2, 3$ for the Hydrogen atom using (75) and (79) with parameters $N = 2001$ and $N = 4001$ respectively, with $\epsilon = 1$. The dash lines correspond to the exact wave functions.

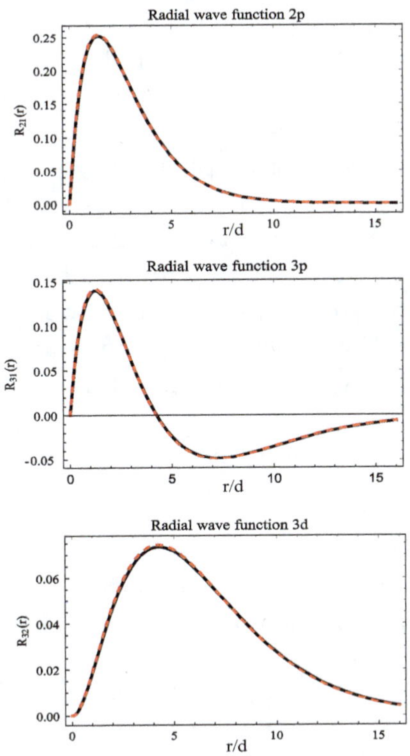

Figure 3. Comparison between exact and calculated radial wave functions $R_{21}(r), R_{31}(r), R_{32}(r)$ for the Hydrogen atom using (75) and (79) with parameters $N = 2001$ and $N = 4001$ respectively, with $\epsilon = 1$. The dash lines correspond to the exact wave functions.

4. Stark Effect

In this section, we study the effect of an static external electric field on the spectrum of the Hydrogen atom. The electric field interacts through the dipole producing a splitting of the spectral lines called Stark effect, since this phenomenon was experimentally shown by Stark in 1913 [27]. There is a vast amount of literature on the Stark effect in the Hydrogen atom. This problem has been treated in configuration space using semi-classical methods [28–30], as well as variational approaches [31,32], but also algebraically in an elegant way. Since it is impossible to mention all the contributions to this problem we confine our discussion to some relevant works to put in context our algebraic method. We do not pretend that our method represents an improvement to the previous descriptions. Instead we are interested in showing that it is possible to tackle this problem along the venue worked out by *M. Moshinsky*, where the harmonic oscillator was the basis to solve problems from atoms to quarks [17].

Algebraic methods applied to the Hydrogen atom may be considered to have originated from the identification of the $SO(4)$ group as the symmetry group of the non-relativistic Hydrogen atom for the bound states and $SO(3,1)$ for the continuum [33–39]. Based upon the $SO(4)$ symmetry group the Stark effect for both static and time-dependent electric fields has been described from different points of view, including semiclassical methods taking into account the crucial result that the Runge–Lenz vector is proportional to the coordinate and the energy [40–43]. In any case, these treatments are confined to the subspace characterized for a given principal quantum number n since the generators of the group commute with the Hamiltonian, a fact that constraints the descriptions to perturbative schemes. In addition the natural basis to deal with this problem, called the Stark basis $|n_1n_2m\rangle$, is based on the parabolic coordinates where the Schrödinger equation is separable. In this context an important role is played by the transformation brackets $\langle n_1n_2m|nlm\rangle$ connecting to the states $|nlm\rangle$ with good angular momentum [44].

A full algebraic treatment for the Stark effect needs the introduction of the $SO(4,2)$ dynamical group [38,45–48] or alternative groups [49]. The $SO(4,2)$ group can be identified by the merging of the $SO(2,1)$ and $SO(4)$ groups [50]. The $SO(2,1)$ group, or alternatively, the $SU(1,1)$ group deals with the radial contributions that shift the principal quantum numbers [50–52]. Even though the $SO(4,2)$ group allows the dipole matrix elements to be calculated, the use of the dynamical group tends to perturbation treatments [53,54]. In contrast, our algebraic approach is not based on the use of the dynamical group, instead, our approach falls in the framework of a discrete variable representation approach, whose salient feature consists in its simplicity as we next present.

In the previous section, we have shown that both algebraic DVR approaches are able to describe the free Hydrogen atom, although with a different convergence degree. The result is that the HO-DVR method needs half of the number of the basis functions, albeit with somewhat less uniformity of errors. In this section, the interaction with a static electric field is introduced whose better description is tempting to be given by the HO-DVR approach. The result, however, is that the $SU(4)$-UGA is more suitable to describe the Stark Effect given the dimensions taken into account for the free Hydrogen atom, as we show next.

The Hydrogen atom under the action of the electric field takes the form

$$\hat{H}^S = \hat{H} + \hat{V}(E), \tag{85}$$

where $\hat{H}$ represents the unperturbed Hydrogen atom (72) and $\hat{V}(E)$ is the potential energy associated with the interaction of the electron with the external electric field $|\mathbf{E}|$ along the z direction

$$\hat{V}(E) = -e|\mathbf{E}|z, \tag{86}$$

and in the convenient units previously used

$$\frac{\hat{V}(E)}{E_B} = -\frac{\bar{z}}{\epsilon}2\sqrt{2}|\bar{\mathbf{E}}|, \tag{87}$$

with $\bar{z}$ in units of d and $\bar{\mathbf{E}}$ in atomic units defined by the electric field of the nucleus given by $e/a_0^2 \approx 5 \times 10^9$ V/cm.

Let us now consider the realization of the potential $\hat{V}(E)$ in accordance with the two algebraic DVR methods. According to the relations (6) and (7) the z component is given by

$$\bar{z} = \frac{1}{2}(a_0^\dagger + a_0). \tag{88}$$

Hence, in the HO-DVR method we just calculate the matrix elements of the operator $a_0^\dagger$ in the basis (9). We obtain

$$\langle n+1, l+1, m|a_0^\dagger|nlm\rangle = \sqrt{(n+l+3)}\sqrt{\frac{(l+1)}{(2l+3)}}\sqrt{\frac{(l+m+1)(l-m+1)}{(2l+1)(l+1)}}, \tag{89}$$

$$\langle n+1, l-1, m|a_0^\dagger|nlm\rangle = -\sqrt{(n-l+2)}\sqrt{\frac{l}{(2l-1)}}\sqrt{\frac{(l+m)(l-m)}{l(2l+1)}}. \tag{90}$$

On the other hand, in the U(4)-UGA method, taking into account (50) and (51) with the correspondence $\bar{z} \to \hat{Q}/d$, we have

$$\bar{z} = \frac{1}{\sqrt{2N}}\hat{D}_0. \tag{91}$$

Using the Wigner–Eckart theorem, its matrix elements take the form

$$\langle [N]n'_a L'M|\hat{D}|[N]n_a LM\rangle = \langle LM;10|L'M\rangle\langle [N]n'_a L'||\hat{D}||[N]n_a L\rangle, \tag{92}$$

with the reduced matrix elements

$$\langle [N]n_a - 1L - 1\rangle||\hat{D}||[N]n_a L\rangle = -\sqrt{\frac{(N-n_a+1)(n_a+L+1)L}{2L-1}}, \tag{93}$$

$$\langle [N]n_a - 1L + 1\rangle||\hat{D}||[N]n_a L\rangle = -\sqrt{\frac{(N-n_a+1)(n_a-L)(L+1)}{2L+3}}, \tag{94}$$

$$\langle [N]n_a + 1L - 1\rangle||\hat{D}||[N]n_a L\rangle = \sqrt{\frac{(N-n_a)(n_a-L+2)L}{2L-1}}, \tag{95}$$

$$\langle [N]n_a + 1L + 1\rangle||\hat{D}||[N]n_a L\rangle = \sqrt{\frac{(N-n_a)(n_a+L+3)(L+1)}{2L+3}}. \tag{96}$$

Taking into account these technical ingredients, we carried out the diagonalization of the Hamiltonian (85) in the framework of both methods.

In Figure 4 the energy levels associated with the subspace $n = 2$ as a function of the electrical field are depicted. In order to obtain these results, a cutoff of the dimension space determined by the feasibility of the calculations was necessary as well as a constraint in the involved angular momenta. The black solid circles correspond to the spectrum obtained using the $U(4)$-UGA, where the number of boson was taken to be $N = 3501$ up to $L = 16$, while the blue triangles correspond to HO-DVR method taking $N = 3001$ up to $l = 11$. The continuous lines correspond to the energy-splitting obtained by the variational approach described in Ref. [32], where an accurate variational method is presented for the $n = 2$–5 shells for different $m's$ including a display of the wave functions. Since, for each approach, the convergence of the energy levels for the null field is different the continuous lines were appropriately located to follow the splitting trend. There are two remarkable facts to be highlighted. First, the splitting is correctly described by the $SU(4)$-UGA, while the HO-DVR method tends to overestimate the splitting as the field intensity increases. This behavior is explained by the different basis dimensions. As a second point, we notice that for small-field intensity the splitting is rather flattened, showing an evident deviation from the linearity predicted by perturbation theory. In order to clearly see this effect a zoom of Figure 4 is depicted in Figure 5 for low field intensities in the interval $[0, 10^{-4}]$. Both approaches present this anomalous behavior, which we believe corresponds to a lack of convergence. To make clear that this is in fact the case, a similar plot is depicted in Figure 6 for the HO-DVR approach for different boson numbers. The black triangles correspond to $N = 3001$, while the red ones to $N = 1501$. From these plots, it becomes clear that as the basis dimension increases the splittings approach to the correct results.

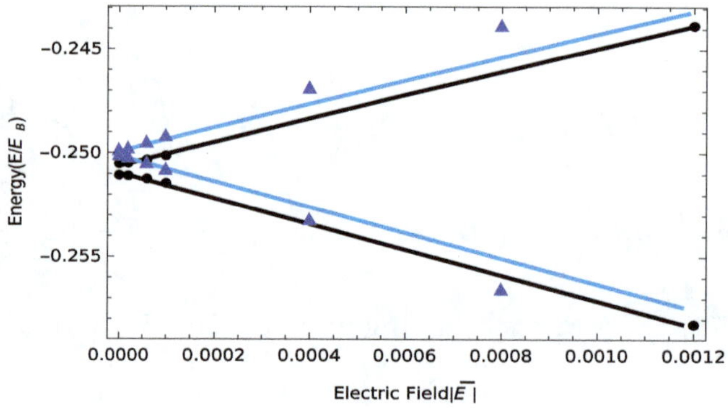

Figure 4. Effect of the electric field effect over the subspace $n = 2$ with $m = M = 0$ in the Hydrogen atom. The black solid circles correspond to spectrum obtained with the $U(4)$-UGA taking $N = 3501$ up to $L = 16$, while the blue triangles correspond to HO-DVR method taking $N = 3001$ up to $l = 11$. Since the calculations were carried out independently, the plotted points do not coincide. The continuous line correspond to the variational approach described in Ref. [32]. $|\bar{E}|$ is the electric field in atomic units.

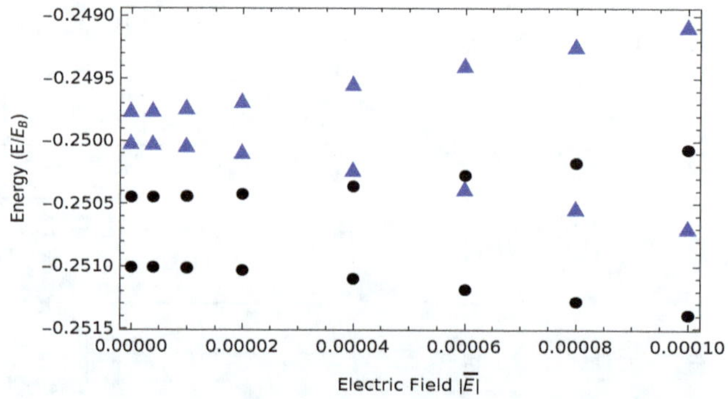

Figure 5. Electric field effect over the $n = 2$ levels of Hydrogen atom for $M = m = 0$. The black solid circles correspond to the results using the $U(4)$-UGA method taking $N = 3501$ and $L = 16$. The blue triangles correspond to HO-DVR method taking $N = 3001$ and $l = 11$.

It is worth mentioning the complexity of these calculations due to the spherical symmetry breaking. Due to our computational limitations, we are forced to fix a maximum value for the angular momentum. Without this approximation, we would be dealing with a $(3,067,752)^2$ dimension matrices, which are impossible to calculate given our computational capacity. However, the restriction of the angular momentum components has physical justification in the sense that not all the functions with a given projection are expected to be mixed. Indeed for the $n = 2$ and $n = 3$ levels the mixing of states with different angular momenta are not expected to be so large (may be up to $L, l = 5$), but in this work we have included the maximum angular momenta that allows our calculations to be carried out.

In order to see whether the previous behavior for the p-space is reproduced for other subspaces, in Figure 7 we display the effect of the electric field provided by both algebraic DVR methods over the subspace with $n = 3$ for $M = m = 0$. The same trend is obtained given the comparison with the exact result given in Ref. [32]. While the $SU(4)$-UGA reproduces the correct splitting, the HO-DVR method overestimates it.

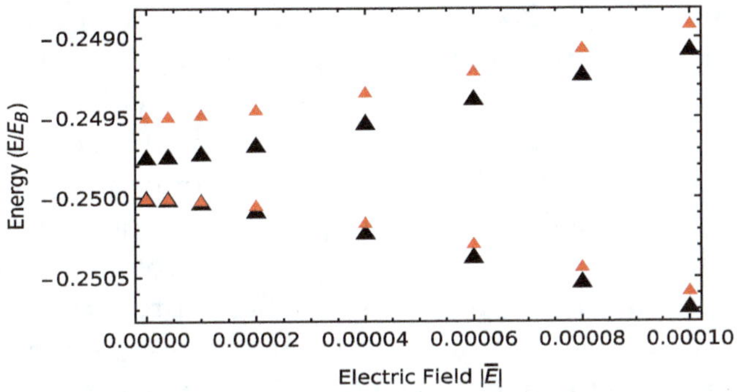

Figure 6. Zoom of the electric field effect over the subspace $n = 2$ with $m = M = 0$ in the Hydrogen atom provided by the HO-DVR method for $N = 1501$ (red triangles) and $N = 3001$ (black triangles).

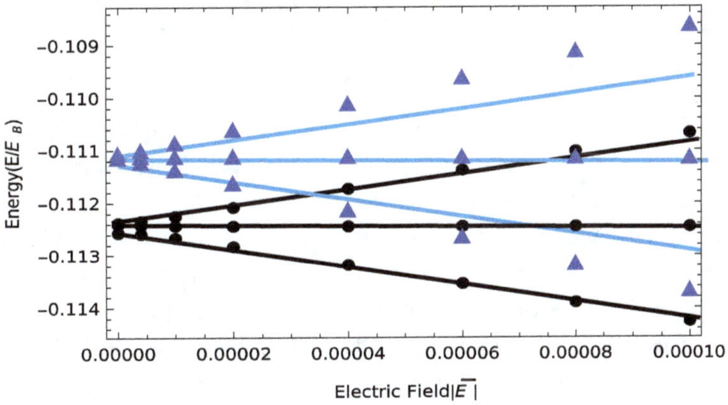

Figure 7. Electric field effect over the subspace characterized by the principal quantum number $n = 3$ with $M = m = 0$ in the Hydrogen atom. The solid black dots correspond to the $U(4)$-UGA method taking $N = 3501$ up to $L = 16$, while the blue triangles correspond to the HO-DVR method taking $N = 3001$ up to $l = 11$. The continuous lines correspond to the exact results provided by Ref. [32].

Finally it is convenient to recall the usual treatment of the Stark effect using perturbation theory. For the first excited state ($n = 2$) the fourfold degeneracy must be born in mind. In this context, since the angular momentum projection is preserved and z is odd, only the states $|\Psi_{200}\rangle$ and $|\Psi_{210}\rangle$ interact through (86). The rest of the matrix element vanish leading to the eigenvalues for $M = m = 0$ [55,56]:

$$E_{\pm} = E_2 \pm 6|\bar{\mathbf{E}}|, \tag{97}$$

where the energy is intended to be in units of E_B and the electric field in atomic units. The magnitude of the splitting of the levels is proportional to the electric field strength, giving rise to the linear Stark effect. The straight lines provided by the equation (97) coincide with the exact values displayed in Figure 4. Deviation from linearity manifests around field intensities of 0.004 au.

5. Conclusions

In this work, we have applied two algebraic discrete variable representation approaches to describe the non-relativistic Hydrogen atom together with the Sark effect. The selection of the Coulombic potential to test our approaches is twofold. On one hand, its study is compulsory to consider more complex situations like molecular systems. On the other hand it is reasonable to evaluate the feasibility of these approaches to take into account additional interactions breaking the symmetry, in particular for the $SU(4)$-UGA where the condition (71) is imposed to remove the intrinsic degeneracy associated with the deformed harmonic oscillator (60).

The two approaches we have proposed may be, in principle, applied to general potentials to obtain the solutions of the time-independent Schrödinger equation. In both approaches, the matrix representation of the Hamiltonian given by either (36) or (79) is given in terms of two diagonal matrices together with the transformation brackets. The calculation of the transformation brackets is independent of the potential and hence they are calculated only once for a given space dimension. The diagonal matrices $\mathbf{\Lambda}^{(E)}$ corresponding to the simple or deformed harmonic oscillator is trivially generated and fixed for any system, while $\mathbf{\Lambda}^{(r)}$ and $\mathbf{\Lambda}^{(\mathcal{Q})}$ are diagonal in the coordinate representation which are also easily constructed. Hence, these approaches represent powerful simple alternatives to study systems where the analytical solutions of the Schrödinger equations are either not known or difficult to obtain by integration methods. Even considering moderate values of space dimension, the approach provides solutions reflecting the main salient features of the system, a situation very useful when the solution cannot be obtained by conventional methods. Hence, this approach may be useful in catastrophe theory where a small change in the original stable potential changes the topology of the problem [57], as well as in the study of quantum phase transitions [58], but also provides new ways to obtain Franck–Condon factors between different potentials since in our approach a common basis of harmonic oscillators are used [59].

We have shown the feasibility of both approaches, the HO-DVR method and U(4)-UGA to obtain the solutions of the 3D Coulombic potential. Although both algebraic methods provide the solutions, the HO-DVR approach proves to show a faster convergence, a fact that makes it the best candidate to study this system. A quite different situation is manifested when the dipole interaction is taken into account to describe the Stark effect, which is particularly difficult due to the spherical symmetry breaking. The full dimension of the matrices to be diagonalized is too large to be consider in realistic terms and consequently, we were forced to cutoff the included angular momenta. With this approximation, the calculations become feasible with reasonable convergence. Fixing the total number of bosons obtained to describe the free Hydrogen atom, the $SU(4)$-UGA turns out to be the suitable method to describe the energy splittings, while the HO-DVR method overestimates the Stark effect. Although both approaches are based on the harmonic oscillator basis, the localized functions associated with the discrete variable representation basis are different, a fact that makes the difference between these approaches. We have thus obtained the unexpected remarkable result that a preference of a model based on the convergence in the zeroth-order Hamiltonian does not imply the same preference when a symmetry breaking interaction is included.

The two presented algebraic approaches provide alternatives to obtain the solutions of 3D systems for any potential where the harmonic oscillator functions represent a good basis. Even though for the Coulombic potential the harmonic oscillator is not by far the best basis, our methods provide reasonable results in a simple way. This opens the possibility of studying systems with different algebraic DVR methods where the exact solutions are not known, providing the possibility of extracting properties by purely algebraic means using the standard tools of matrix diagonalization.

Author Contributions: M.R.-A., M.B.-M.: investigation, software, methodology, reviewing, discussion. R.L.: conceptualization, supervision, investigation, original draft preparation, discussion. J.M.A., J.G.-C.: conceptualization, investigation, writing, discussion. E.O.: investigation, discussion. All authors have read and agreed to the published version of the manuscript.

Acknowledgments: This work is partially supported by DGAPA-UNAM, Mexico, under project IN-212020, by the Consejería de Economía, Conocimiento, Empresas y Universidad de la Junta de Andalucía (Spain) under Group FQM-160, by the Spanish Ministerio de Ciencia e Innovaciíon, ref. FIS2017-88410-P, RTI2018-098117-B-C21 and PID2019-104002GB-C22, and by the European Commission, ref. H2020-INFRAIA-2014-2015 (ENSAR2). First author is grateful to DGAPA-UNAM for the postdoctoral scholarship (Facultad de Química). Second author is also grateful for the scholarship (Posgrado en Ciencia e Ingeniería de Materiales) provided by CONACyT, México.

References

1. Light, J.C.; Hamilton, I.P.; Lill, J.V. Generalized discrete variable approximation in quantum mechanics. *J. Chem. Phys.* **1985**, *82*, 1400. [CrossRef]
2. Light, J.C.; Carrington, T., Jr. Discrete-variable representations and their utilization. *Adv. Chem. Phys.* **2000**, *114*, 263.
3. Littlejohn, R.G.; Cargo, M.; Carrington; Mitchell, K.A.; Poirier, B. A general framework for discrete variable representation basis sets. *J. Chem. Phys.* **2002**, *116*, 8691. [CrossRef]
4. Wang, X.-G.; Carrington, T. J. A discrete variable representation method for studying the rovibrational quantum dynamics of molecules with more than three atoms. *Chem. Phys.* **2009**, *130*, 094101. [CrossRef] [PubMed]
5. Shizgal, B.; Blackmore, R. A discrete ordinate method of solution of linear boundary value and eigenvalue problems. *J. Comput. Phys.* **1984**, *55*, 313. [CrossRef]
6. Shizgal, B.D.; Chen, H. The quadrature discretization method (QDM) in the solution of the Schrödinger equation with nonclassical basis functions. *J. Chem. Phys.* **1996**, *104*, 4137. [CrossRef]
7. Pérez-Bernal, F.; Arias, J.M.; Carvajal, M.; Gómez-Camacho, J. Configuration localized wave functions: General formalism and applications to vibrational spectroscopy of diatomic molecules. *Phys. Rev. A* **2000**, *61*, 042504. [CrossRef]
8. Baye, D.; Heenen, P.-H. Generalised meshes for quantum mechanical problems. *J. Phys. A* **1986**, *19*, 2041. [CrossRef]
9. Vincke, M.; Malegat, L.; Baye, D. Regularization of singularities in Lagrange-mesh calculations. *J. Phys. B At. Mol. Opt. Phys.* **1993**, *26*, 811. [CrossRef]
10. Baye, D. Lagrange-mesh method for quantum-mechanical problems. *Phys. Status Solidi B* **2006**, *243*, 1095. [CrossRef]
11. Baye, D. The Lagrange-mesh method. *Phys. Rep.* **2015**, *565*, 1. [CrossRef]
12. Estévez-Fregoso, M.M.; Lemus, R. Connection between the $su(3)$ algebraic and configuration spaces: Bending modes of linear molecules. *Mol. Phys.* **2018**, *116*, 2374. [CrossRef]
13. Rodríguez-Arcos, M.; Lemus, R. Unitary group approach for effective potentials in 2D systems: Application to carbon suboxide C_3O_2. *Chem. Phys. Lett.* **2018**, *713*, 266. [CrossRef]
14. Lemus, R. Unitary group approach for effective molecular potentials: 1D systems. *Mol. Phys.* **2019**, *117*, 167. [CrossRef]
15. Lemus, R. A simple approach to solve the time independent Schrödinger equation for 1D systems. *J. Phys. Commun.* **2019**, *3*, 025012. [CrossRef]
16. Rodríguez-Arcos, M.; Lemus, R.; Arias, J.M.; Gómez-Camacho, J. Unitary group approach to describe interatomic potentials in 3D systems. *Mol. Phys.* **2019**, *118*, e1662959. [CrossRef]
17. Moshinsky, M. *The Harmonic Oscillator in Modern Physics: From Atoms to Quarks*; Gordon and Breach: New York, NY, USA, 1969.
18. Frank, A.; Van Isacker, P. *Algebraic Methods in Molecular and Nuclear Structure Physics*; Wiley and Sons: New York, NY, USA, 1994.
19. Iachello, F.; Levine, R.D. *Algebraic Theory of Molecules*; Oxford University Press: Oxford, UK, 1995.
20. Santiago, R.D.; Arias, J.M.; Gómez-Camacho, J.; Lemus, R. An approach to establish the connection between configuration and $su(n+1)$ algebraic spaces in molecular physics: application to ammonia. *Mol. Phys.* **2017**, *115*, 3206. [CrossRef]
21. Santopinto, E.; Bijker, R.; Iachello, F. Transformation brackets between $U(\nu+1) \supset U(\nu) \supset SO(\nu)$ and $U(\nu+1) \supset SO(\nu+1) \supset SO(\nu)$. *J. Math. Phys.* **1996**, *37*, 2674. [CrossRef]
22. Cejnar, P.; Jolie, J. Quantum phase transitions in the interacting boson model. *Prog. Part. Nucl. Phys.* **2009**, *62*, 210. [CrossRef]
23. Arias, J.M.; Camacho, J.G.; Lemus, R. An $su(1,1)$ dynamical algebra for the Pöschl–Teller potential. *J. Phys. A Math. Gen.* **2004**, *37*, 877. [CrossRef]
24. Lemus, R.; Arias, J.M.; Gómez-Camacho, J. An $su(1,1)$ dynamical algebra for the Morse potential. *J. Phys. A Math. Gen.* **2004**, *37*, 1805. [CrossRef]
25. Álvarez-Bajo, O.; Arias, J.M.; Gómez-Camacho, J.; Lemus, R. An approach to the study of the continuum effects in systems of interacting Morse oscillators. *Mol. Phys.* **2008**, *106*, 1275. [CrossRef]

26. Bermúdez-Montaña, M.; Lemus, R.; Castaños, O. Polyad breaking phenomenon associated with a local-to-normal mode transition and suitability to estimate force constants. *Mol. Phys.* **2017**, *115*, 3076. [CrossRef]
27. Condon, E.; Shortley, G. *The Theory of Atomic Spectra*; Cambridge University Press: Cambridge, UK, 1951.
28. Harmin, D. Hydrogenic Stark effect: Properties of the wave functions. *Phys. Rev. A* **1981**, *24*, 2491. [CrossRef]
29. Harmin, D. Theory of the Stark effect. *Phys. Rev. A* **1982**, *26*, 2656. [CrossRef]
30. Rice, M.H.; Good, R.H., Jr. Stark effect in Hydrogen. *J. Opt. Soc. Am.* **1962**, *52*, 239. [CrossRef]
31. Fernández, F.M. Direct calculation of Stark resonances in Hydrogen. *Phys. Rev. A* **1996**, *54*, 1206. [CrossRef]
32. Fernández-Menchero, L.; Summers, H.P. Stark effect in neutral Hydrogen by direct integration of the Hamiltonian in parabolic coordinates. *Phys. Rev. A.* **2013**, *88*, 022509. [CrossRef]
33. Lenz, W. Über den Bewegungsverlauf und die Quantenzustände der gestörten Keplerbewegung. *Z. Phys.* **1924**, *24*, 197. [CrossRef]
34. Pauli, W. On the spectrum of the Hydrogen from the standpoint of the new Quantum Mechanics. *Z. Phys.* **1926**, *36*, 336. [CrossRef]
35. Fock, V. 2. Phys. 98 145 Bargmann V 1936. *Z. Phys.* **1935**, *98*, 145. [CrossRef]
36. Bargman, V. Zur theorie des Wasserstoffatoms. *Z. Phys.* **1936**, *99*, 576. [CrossRef]
37. Schiff, L.I. *Quantum Mechanics*; McGraw-Hi11: New York, NY, USA,1968.
38. Wybourne, B.G. *Classical Groups for Physicists*; John Wiley: New York, NY, USA, 1974.
39. Majumbdar, S.D.; Basu, D. $O(3,1)$ symmetry of the Hydrogen atom. *J. Phys. A* **1974**, *7*, 787. [CrossRef]
40. Flamand, G. The Solution of a Stark-Effect Model as a Dynamical Group Calculation. *J. Math. Phys.* **1966**, *7*, 1924. [CrossRef]
41. Bakshi, P.; Kalman, G.; Cohn, A. hydrogenic Stark-Zeeman Spectra for Combined Static and Dynamic Fields. *Phys. Rev. Lett.* **1973**, *31*, 1576. [CrossRef]
42. Gigosos, M.A.; Fraile, J.; Torres, F. Hydrogen Stark profiles: A simulation-oriented mathematical simplification. *Phys. Rev. A* **1985**, *31*, 3509. [CrossRef]
43. Demkov, Y.N.; Monozon, B.S.; Ostrovskii, V.N. Energy levels of a Hydrogen atom in crossed electric and magnetic fields. *Sov. Phys. JETP* **1970**, *30*, 775.
44. Hughes, J.W.B. Stark states and $O(4)$ symmetry of hydrogenic atoms. *Proc. Phys. Soc.* **1967**, *91*, 810. [CrossRef]
45. Barut, A.O.; Kleinert, H. Transition probabilities of the Hydrogen atom from noncompact dynamical groups. *Phys. Rev. A* **1967**, *156*, 1541. [CrossRef]
46. Fronsdal, C. Infinite multiplets and the Hydrogen atom. *Phys. Rev.* **1967**, *156*, 1665. [CrossRef]
47. Kleinert, H. Group Dynamics of Elementary Particles. *Fortsch. Phys.* **1968**, *16*, 1. [CrossRef]
48. Bednar, M. Algebraic treatment of quantum-mechanical models with modified Coulomb potentials. *Ann. Phys.* **1973**, *75*, 305. [CrossRef]
49. Hughes, J.W.B.; Yadagar, J. Theory of laser-induced inelastic collisions. *J. Phys. A.* **1976**, *9*, 1569. [CrossRef]
50. Adams, B.G.; Cizek, J.; Paldus, J. Representation theory of $so(4,2)$ for the perturbation treatment of Hydrogenic-type hamiltonians by algebraic methods. *Int. J. Quantum. Chem.* **1982**, *23*, 153. [CrossRef]
51. Dimitriev, V.F.; Rumer, Y.B. $O(2,1)$ Algebra and the Hydrogen atom. *Theor. Math. Phys.* **1970**, *5*, 1146. [CrossRef]
52. Martínez-Yépez, R.P.; Salas-Brito, A.L. An $su(1,1)$ algebraic method for the Hydrogen atom. *J. Phys. A* **2005**, *38*, 8579. [CrossRef]
53. Tsai, W.-Y. Third-order Stark effect: An operator approach. *Phys. Rev. A.* **1974**, *9*, 1081. [CrossRef]
54. Romero-Lópes, A.; Leal-Ferreira, P. Algebraic treatment of the Stark effect for Hydrogen. *Il Nuovo C* **1971**, *3*, 23. [CrossRef]
55. Greiner, W. *Quantum Mechanics: An Introduction*; Springer Science & Business Media: Berlin, Germany, 2011.
56. Davydov, A.S. *Quantum Mechanics*, 2nd ed.; Pergamon Press: New York, NY, USA, 1965.
57. Gilmore, R. *Catastrophe Theory for Scientists and Engineers*; Wiley: New York, NY, USA, 1981.
58. Castaños, O.; Calixto, M.; Pérez-Bernal, F.;Romera, E. Identifying the order of a quantum phase transition by means of Wehrl entropy in phase space. *Phys. Rev. E* **2015**, *92*, 052106.
59. Lemus, R. An algebraic approach to calculate Franck–Condon factors. *J. Math. Chem.* **2020**, *58*, 29. [CrossRef]

2

Permutation Symmetry in Coherent Electrons Scattering by Disordered Media

Elena V. Orlenko [1,*] and Fedor E. Orlenko [2]

1 Higher School of Engineering Physics, Institute of Physics, Nanotechnology and Telecommunication, Peter the Great Saint Petersburg Polytechnic University, 195251 St. Petersburg, Russia
2 Scientific and Educational Center for Biophysical Research in the Field of Pharmaceuticals, Saint Petersburg State Chemical Pharmaceutical University (SPCPA), 197376 St. Petersburg, Russia; elena.orlenko@phmf.spbstu.ru
* Correspondence: eorlenko@mail.ru

Abstract: A non-Anderson weak localization of an electron beam scattered from disordered matter is considered with respect to the principle of electron indistinguishability. A weak localization of electrons of a new type is essentially associated with inelastic processing. The origin of inelasticity is not essential. We take into account the identity principle for electron beam and electrons of the atom of the scatterer with an open shell. In spite of isotropic scattering by each individual scatterer, the electron exchange contribution has a hidden parameters effect on the resulting angular dependence of the scattering cross-section. In this case, the electrons of the open shell of an atomic scatterer can be in the s-state, that is, the atomic shell remains spherically symmetric. The methods of an invariant time-dependent exchange perturbation theory and a Green functions with exchange were applied. An additional angular dependence of the scattering cross-section appears during the coherent scattering process. It is shown exactly for the helium scatterer that the role of exchange effects in the case of a singlet is negligible, while for the triplet state, it is decisive, especially for those values of the energy of incident electrons when de Broglie's waves are commensurate with the atomic.

Keywords: identity principle; exchange contribution; new type weak localization; inelastic coherent scattering

---

1. Introduction

The phenomenon of weak localization of conduction electrons, which manifests itself in the enhancement of backscattering of classical waves in disordered media, has attracted scientific interest in recent decades [1–9]. Weak localization manifests itself mainly in an increase in the probability of elastic backscattering in a narrow range of solid angles, of the order of λ/l, where λ is the length of an electron or light wave, and l is the mean free path of electrons and photons. Coherent phenomena associated with the scattering of external particles (such as electrons or neutrons) with fixed excitation energies of a disordered medium were studied in [10–13], where weak localization was observed for electron beams with energies from 10 to 1000 electron volts. Neutron beams were also the subject of this work [14]. According to these works, coherent phenomena can be observed in the enhancement of particle backscattering during elastic interaction with a disordered medium, despite the relatively high energies of the particle beams. The influence of inelastic processes on the conductivity under conditions of weak localization has been studied in many works [15–18]. In these cases, two basic assumptions are usually made. First, multiple scattering is represented as forward multiple scattering and single large-angle scattering [19]. Second, the scattering by each individual diffuser is considered to be isotropic. It was shown that the role of inelastic processes at weak localization is secondary and negative, since inelastic collisions violate phase relations, thereby reducing the probability of coherent processes. However, there are cases when inelastic processes do not lead to the loss of phase memory by the system. It was shown for the first time in [20]. In this case, in addition to the usual weak localization of the Anderson type, there is a weak localization of electrons of a new type or a new weak localization, which is essentially associated with inelastic processing. Moreover, the origin of inelasticity is not essential, for example, it can be a plasmon, photon, phonon, or exciton. Moreover, quantum coherence can exist even if the electron is exposed to an incoherent electromagnetic field. In these cases, the situation is considered when, in the case of an inelastic interaction, the particle loses a fixed energy and enters the inelastic channel, having an energy different from the initial value in the incident beam. In addition to inelastic collision, the particle can still participate in at least one elastic process, after which it leaves the medium and can be registered. This effect demonstrates itself in the scattering processes of high-energy electron beams, where electrons velocities are relativistic. The investigation of the process of the resonant spontaneous bremsstrahlung of ultrarelativistic electrons in the fields of a nucleus and a weak quasimonochromatic electromagnetic wave was done in [21,22]. There is a coherent scattering process of photons with inelastic interaction with ultrarelativistics electrons. In this case, a characteristic angular dependence with a frequency shift appears for photons. The process has been studied in a special kinematic region, where stimulated processes with correlated emission and absorption of photons of the first and second waves predominate (the effect of parametric interference). It is interesting that the described effect of a new type of weak localization for photons does not depend on the type of inelasticity, the creation of an electron-positron, or something else. The amount of energy loss is important. In this interference kinematics, correspondence is established between the emission angle and the energy of the final electron.

There are two ways to implement this process, so it can start or end with an inelastic collision. The interference of electron waves associated with these additional processes turns out to be constructive [23]. This manifests itself in an increase in the scattering of electrons at an angle other than π, and this difference can be significant. There are certain differences in the localization of electrons for different mechanisms of inelastic scattering, but it turns out that the general features of this phenomenon prevail. The most striking difference between localization of a new type in the case of inelastic scattering from ordinary weak localization is the difference in the characteristic scattering angles. The scattering probability here is maximal in the range of scattering angles close to $\pi/2$, and the effect manifests itself in a much wider range of angles than in the case of traditional localization, in which the beam enhancement is observed in forward or backward scattering at an angle of π. As already mentioned, the main difference between conventional and new weak localization is the typical electron scattering angle. The angular distribution of particles and radiation in the case of ordinary weak localization in a disordered medium is usually described using the maximum cross (or so-called "fan") diagrams, which are used to calculate the electron radiation cross-section. Regular weak localization, in particular, can be described by a simple graphical method [24,25], which gives an idea of this phenomenon and explains why the angle pi is specific for regular weak localization. This method takes into account that an electron with momentum k is scattered through two complementary series of intermediate scattering states $k \to k_1 \to k_2 \to \ldots \to k_{n-1} \to k_n = -k$ and $k \to k\prime_1 = -k_{n-1} \to k\prime_2 = -k_{n-2} \to \ldots \to k\prime_{n-1} = -k_1 \to k_n = -k$ to the state $-k$. Momentum changes: $q_1, q_2, \ldots q_{n-1}, q_n$ for the first scattering chain and $q_n, q_{n-1}, \ldots q_2, q_1$ for the second. The amplitudes in the final state $-k$ are the same, and add up, and the waves in the forward and reverse directions are superimposed on each other constructively, reinforcing each other. This is due to the fact that the complementary scattering processes have the same changes in momentum, both in a straight line and in the opposite sequence. An explanation of why the coherent enhancement of electron scattering in the inelastic scattering channel occurs at angles other than π is proposed in [25]. A simple kinematic model is used to determine the basic properties of weak localization of electrons in the inelastic scattering channel. It easily reproduces the range of scattering angles characteristic of weak localization of electrons with energy loss. The results are consistent with the results based on the dynamic theory associated with the calculation of crossed and ladder diagrams. It is possible to trace the transition from a new type of weak localization to the usual weak localization with a decrease in energy losses. The new type of weak localization is consistent with regular weak localization if the energy losses are approximately equal to zero.

The range of angles in the elastic channel is of the order of λ/l, and in the inelastic channel is of the order $\gamma/\omega = (\lambda/l)(E/\hbar\omega)$, where γ is the electron collision frequency, E is its energy, and $\hbar\omega$ is the energy transferred to the medium. It is assumed that the energy of an electron incident on the medium is high enough to excite plasmons or atoms. However, this energy should not be too large so that the de Broglie wavelength of electrons was less but remained comparable to the distance between the centers of scattering, so that constructive interference of waves of scattered electrons inside condensed media could exist. The corresponding energy values for electron beams are in the range from hundreds of eV to keV. Moreover, the above statements remain valid both for the case of a small number of elastic collisions, and for a sufficiently large number; the main thing in this case is only one inelastic collision [23]. The main reason for the difficulty of fixing the indicated effect in a solid is that the very phenomenon of the new weak localization and all possible measurable parameters have been considered and calculated for an infinite three-dimensional medium, while the role of surface effects at a boundary of a condensed medium is large.

Taking into account the principle of indistinguishability of electrons in a beam and a medium, as was shown for a neutral atomic medium [26], but also takes place in a solid, this shifts the scattering enhancement to the parameters of an ordinary weak localization. As it is shown in [26], the scattering intensity includes the electron exchange terms varying directly as $\cos\chi$, where χ is a scattering angle. The expression for intensity contains two main terms, the first one is specific for the new weak localization, and the exchange interaction here contributes as an increasing pre-factor. The second term is entirely produced by electron exchange and is proportional to $\cos\chi$. Despite the obtained interesting exchange effect in a new type of weak localization for hydrogen scatterers, there was no systematic method allowing the principle of indistinguishability and permutation symmetry to be taken into account when describing scattering processes in the general case. The exchange effect considered in [24] is a good example of the importance of developing a general formalism that makes it possible to take into account the permutation symmetry of a many-electron system in the scattering problem in general and for weak localization, in particular.

In the presented work, we discuss how exchange affects the scattering of electrons in a disordered medium in a general case. We will take into account the identity of the incident electrons and electrons belonging to multielectron scatterer atoms. An invariant exchange perturbation theory method [27,28] is applied for the development of the cross-section for the weak localization scattering process while taking into account exchange effects. It is shown that, in spite of isotropic scattering by each individual scatterer, the number of electrons and their spin state in the open shell of the atom radically affect the resulting angular dependence of the scattering cross-section of incident electrons. In this case, the electrons themselves of the open shell of the scattering atom can be in the s-state, that is, the shell of the atom remains spherically symmetric. An additional angular dependence of the scattering cross-section appears, which is proportional to $\cos\chi$ and does not depend on the number and state of the spin of electrons with an open or closed shell. The main difference, depending on the number of electrons and their spin states, is the coefficient in front of $\cos\chi$.

2. Probability of Plasmon Emission with Atom Excitation and Electron Exchange

We consider the interaction of electrons with a medium, accompanying the excitation of atoms when electrons are scattered by them, by the same way as in [20] by using the interaction operator in the form of the sum of two contributions:

$$V = V_{at} + V_{pl}, \tag{1}$$

where the operator V_{at} describes electron interaction with atoms of medium, numbered by indexes l:

$$V_{at} = \sum_{l} V(r - R_l), \tag{2}$$

and V_{pl} describes the interaction of the moving electron with the electric polarization field of the medium electrons, which arises under this action. Then, let the process of inelastic scattering of an electron be associated with the loss of energy on the plasmon:

$$V_{pl} = \int dr\hat{\rho}(\mathrm{r})\hat{\phi}(r), \tag{3}$$

where $\hat{\phi}(r)$ is the operator of the electric field potential due to plasma oscillation and $\hat{\rho}(r)$ is the charge density operator.

The initial state of the system corresponding to an incident electron and an atom of the medium (for example, a Zi-Ni alloy doped with Ca, Mg, or hydrogen or He) is described by the wave vector $|n,p\rangle$, antisymmetric taking into account the electronic permutations between the atom and the incident electron. Here, n is the set of quantum numbers of atoms in the initial state, and p is the momentum of the incident electron. Regarding the exchange perturbation theory method [27,28], the final state of the whole system can be described by the non-symmetric wave function $|m,p-Q) = |m\rangle \cdot |p-Q\rangle$, which is a simple product of the atomic function $|m\rangle$, antisymmetrized with internal electron permutations, and a free electron function $|p-Q\rangle$, where Q is the total momentum transferred to the medium. It should be underlined that the operator (2) describing electron interaction with atoms has a non-symmetric form regarding the interatomic electron permutations (the exchange of numbered atomic electrons with the free electron of incident):

$$\left[\hat{A}V_{at}\right] \neq 0, \tag{4}$$

where $\hat{A}$ is the antisymmetrization operator [27,28]. Acting on a non-symmetric wave function, operator $\hat{A}$ performs antisymmetrization:

$$|n,p\rangle = \hat{A}\left|n,p\right) = \frac{1}{f_{np}} \sum_{\nu} (-1)^{g_\nu} \left|n,p\right)_{\nu}, \tag{5}$$

where g_ν is the parity of the $\nu - th$ permutation, $(n,p|n,p\rangle = 1$, $f_{np} = \sum_\nu (-1)^{g_\nu} \left(n,p\,\middle|n,p\right)_\nu$ are the normalization condition and a normalization factor, $\left|n,p\right) = |n\rangle \cdot |p\rangle \equiv \left|n,p\right)_0$ is the wave vector of the zeroth permutation with the initial arrangement of the electrons, and the subindex shows that the number of the permutation is $\nu = 0$. Here, we deal with the antisymmetric non-orthogonal basis, while $0 < (m,p\prime|n,p\rangle < 1$, is the same as for $0 < \langle m,p\prime \left|n,p\right\rangle < 1$, but this set meet a completeness property:

$$\sum_{n,p} |n,p\rangle \frac{f_{np}}{N} (n,p| = \hat{1}, \tag{6}$$

where N is the total number of electron permutations. The proof of completeness property (6) is described in detail in Appendix A. Generally speaking, both the "zero" Hamiltonian $\hat{H}_0$ describing a multicentre many-electron system without interatomic interaction, and the perturbation operator $\hat{V}$ describing this interatomic interaction are not invariant to the operation of antisymmetrization, taking into account the rearrangement of electrons between the two subsystems: $[\hat{A}\hat{H}_0] \neq 0, [\hat{A}V] \neq 0$. At the same time, the complete Hamiltonian of the system $\hat{H} = \hat{H}_0 + \hat{V}$ retains its invariance: $\left[\hat{A}\hat{H}\right] = 0$. This discrepancy means a serious problem related to the fact that the zero-order wave function, antisymmetrized with respect to center-to-center electron permutations, is not an eigenfunction of the non-invariant Hamiltonian $\hat{H}_0$ [27], and the corrections are obtained by applying an asymmetric operator of interactions containing non-physical contributions. In [27,28], a special symmetric form of the perturbation operator and zeroth Hamiltonian were developed. It allows corrections of the wave functions, properly antisymmetrised, and the energy corrections to be obtained by using the correct antisymmetric basis of the wavefunctions:

$$\hat{H}_0 = \left(\sum_{\nu=0}^{\nu=N} H_\nu^0 \Lambda^\nu\right) = \sum_{\nu=0}^{N} H_\nu^0 \sum_n \frac{|n,p)_\nu (p,n|_\nu}{f_{np}},$$
$$\hat{V}_{at} = \left(\sum_{\nu=0}^{\nu=N} V_\nu \Lambda^\nu\right) = \sum_{\nu=0}^{N} V_\nu \sum_n \frac{|n,p)_\nu (p,n|_\nu}{f_{np}},$$

where Λ^ν denotes a projector onto an asymmetric state, corresponding to the $\nu - th$ permutation $\Lambda^\nu = \sum_n \frac{|n,p)_\nu (p,n|_\nu}{f_n^N}$, and $\Lambda^\nu |n,p\rangle = (-1)^{g_\nu} |n,p)$, and N, as mentioned before, is the total number of permutations. In this case, the Hamiltonian $\hat{H}_0$ describing a system without the interaction of two subsystems has an eigenvector, which has an antisymmetric form taking into account the electronic permutations between these subsystems:

$$\hat{H}_0 |n,p\rangle = \left(E_n + p^2/2m\right)|n,p\rangle. \tag{7}$$

In our case, the second term of the perturbation operator (1) V_{pl} has a symmetric form with respect to the permutations of electrons. It is important to emphasize that after obtaining all the corrections to the energy and the wave vector in all orders of the perturbation theory in the general formalism, all matrix elements containing the complex symmetric form (6) of the perturbation operator can be analytically reduced to a simple form. This is a symmetry-adapted form that includes the usual asymmetric perturbation operator corresponding to the initial zero permutation of electrons. Such a form is more convenient for practical applications. We transform all matrix elements in the following way:

$$\begin{gathered}(k,p\prime|\hat{V}|n,p\rangle = (k,p\prime|\sum_{\nu=0}^{N} V_\nu \Lambda^\nu |n,p\rangle = (k,p\prime|\sum_{\nu=0}^{N} \frac{(-1)^{g_\nu}}{f_{np}} V_\nu |n,p)_\nu = \\ \sum_{\nu=0}^{N} \frac{(-1)^{g_\nu}}{f_{kp}} (k,p\prime|_\nu V_{p=0}|n,p)_{\nu=0} = \frac{f_{np}}{f_{kp}} k,p\prime|V_{\nu=0}|n,p).\end{gathered} \tag{8}$$

The probability of a transition of the system in unit time from one state to another with transfer of momentum Q from the incident electron to the medium and with transition of the medium from state n to state m has the form obtained in [27,28], taking into account electron exchange

$$w_{mn,Q} = \frac{2\pi}{\hbar}\delta\left(E_n - E_m + E_p - E_{p-Q}\right)\left|\langle m,p-Q|\hat{T}|n,p)\right|^2, \tag{9}$$

where operator $\hat{T}$ is the operator of transition on the energy surface [27–29], the general operator equation for which with taking into account electron permutations between subsystems is:

$$\begin{gathered}\hat{T} = V^{\mathrm{N}} + V^{\mathrm{N}}(E_i - H_0 + i\eta)^{-1}\left(\frac{f}{N}\right)^{-1}\hat{T}, \\ V^{\mathrm{N}} = V\left(\frac{f}{N}\right), \_\_\_ f|n,p\rangle = f_{np}|n,p\rangle,\end{gathered} \tag{10}$$

where η is the relaxation frequency in the general case. We rewrite it as a series:

$$\begin{gathered}\hat{T} = V^{\mathrm{N}} + V^{\mathrm{N}}(E_i - H_0 + i\eta)^{-1}\left(\frac{f}{N}\right)^{-1} V^{\mathrm{N}} + \\ V^{\mathrm{N}}\left(E_i - H^0_{p=0} + i\eta\right)^{-1}\left(\frac{f}{N}\right)^{-1} V^{\mathrm{N}}\left(E_i - H^0_{p=0} + i\eta\right)^{-1}\left(\frac{f}{N}\right)^{-1} V^{\mathrm{N}} + \ldots\end{gathered} \tag{11}$$

In the matrix element $\langle m,p-Q|\hat{T}|n,p)$ in Equation (9), the *bra*-vector has an antisymmetric form with respect to electronic permutations, and the *ket*-vector has a simple non-symmetric form. Then, a transition amplitude will be:

$$\begin{aligned}&\langle m,p-Q|\hat{T}|n,p) = \langle m,p-Q|(V_{at}+V_{pl})^{\mathbb{N}}|n,p) + \langle m,p-Q|(V_{at}+\underline{V_{pl}})^{\mathbb{N}}\hat{G}_0^{\mathbb{N}}(\underline{V_{at}}+V_{pl})^{\mathbb{N}}|n,p)+\\ &\langle m,p-Q|(V_{at}+\underline{V_{pl}})^{\mathbb{N}}\hat{G}_0^{\mathbb{N}}(\underline{V_{at}}+V_{pl})^{\mathbb{N}}\hat{G}_0^{\mathbb{N}}(\underline{V_{at}}+V_{pl})^{\mathbb{N}}|n,p) + \ldots = \langle m,p-Q|(\underline{V_{pl}})^{\mathbb{N}}\hat{G}_a^{\mathbb{N}}(\underline{V_{at}})^{\mathbb{N}}|n,p)+\\ &\langle m,p-Q|(\underline{V_{at}})^{\mathbb{N}}\overline{\hat{G}_a^{\mathbb{N}}}(\underline{V_{pl}})^{\mathbb{N}}|n,p) + \langle m,p-Q|(V_{at})^{\mathbb{N}}\hat{G}_a^{\mathbb{N}}(\underline{V_{at}})^{\mathbb{N}}|n,p)\ \ldots,\end{aligned} \tag{12}$$

where we denote the normalized (indexed N) resolvent operator $\hat{G}_0^{\mathbb{N}}$ (or the Green's function for the coordinate representation) and the operator equation taking into account an antisymmetric nonorthogonal basis:

$$\begin{aligned}\hat{G}_0^{\mathbb{N}} &= (E_i - H_0 + i\eta)^{-1}\left(\frac{f}{N}\right)^{-1}\\ \hat{G}_a^{\mathbb{N}} &= \hat{G}_0^{\mathbb{N}} + \hat{G}_0^{\mathbb{N}}V_{at}^{\mathbb{N}}\hat{G}_a^{\mathbb{N}}.\end{aligned} \tag{13}$$

Since the plasmon excitation is small, we can consider this process in the Born approximation. We do not fix our attention on the features of plasmon excitation in different situations, as it was done in detail in [22,30], where the excitation and propagation of bulk and surface plasma waves by incident electrons were analyzed. The motion of electrons was considered both in vacuum when approaching the surface of the metal, and inside the metal, the boundary of which elastically and specularly reflected the internal nonequilibrium electrons. The effect of electron boundary scattering parameters on the structure of bulk and surface plasmon resonances was analyzed in [31]. The probability of transition radiation of bulk plasmon by an electron moving in vacuum was examined.

We leave only the first two terms in Equation (12), where the scattering process begins or ends with plasmon excitation, and omit all other terms where the inelastic process with a plasmon occurs between the processes of elastic scattering by atoms. This is because we are not interested in small-angle scattering by the medium as a whole. The mentioned two terms we shall write out in the form:

$$\begin{aligned}&\langle m,p-Q|\hat{T}|\Psi_p^+) = \Big\{\langle m,p-Q|(\underline{V_{pl}})|m,p-Q+q)\left[\langle m,p-Q+q|\delta_{Qq}|n,p) + \langle m,p-Q+q|\hat{G}_a(\underline{V_{at}})^{\mathbb{N}}|n,p)\right]+\\ &\left[\langle m,p-Q|(\underline{V_{at}})^{\mathbb{N}}\hat{G}_a|n,p-q) + \langle m,p-Q|\delta_{Qq}|n,p-q)\right]\langle n,p-q|\underline{V_{pl}}|n,p)\Big\} =\\ &\langle m,p-q|n,p-q)\langle n,p-q|V_{pl}|n,p)+\langle m,p-q|V_{pl}|m,p)\langle m,p|n,p)+\\ &\Big\{\langle m,p-Q|(\underline{V_{pl}})|m,p-Q+q)\,\langle m,p-Q+q|\hat{G}_a(\underline{V_{at}})^{\mathbb{N}}|n,p) + \langle m,p-Q|(\underline{V_{at}})^{\mathbb{N}}\hat{G}_a|n,p-q)\,\langle n,p-q|\underline{V_{pl}}|n,p)\Big\},\end{aligned} \tag{14}$$

where the first two matrix elements describe the process of plasmon excitation without atomic scattering. However, the exchange coefficients:

$$S_{mnp-q} = \langle m,p-q|n,p-q) \text{ and } S_{mnp} = \langle m,p|n,p), \tag{15}$$

mean the exchange density due to the entanglement of electronic states. The third term of Equation (14) corresponds to a process that begins with scattering by atoms and ends with excitation of plasmons. The fourth term corresponds to the process in which the events occur in the opposite order. Since we are not interested in small-angle scattering by the medium as a whole, as mentioned earlier, we can omit the first two terms.

Dividing the result of Equation (9) by the flux density of incident particles ($j_p = \hbar k_p/m_e$, where k_p is the wavevector of the relative movement of the incident electrons and m_e is their mass), we obtain an expression for the cross-section of scattering events and reactions. Then, we multiply this expression for the cross-section by the number of final states in the volume per unit energy interval for scattering along the unit vector n_f into a solid angle element $d\Omega$, $d\rho(E_f) = \frac{mk_f}{(2\pi)^3\hbar^2}d\Omega$, because the final state is within the continuous spectrum. Then, for the differential cross-section summed over the final states of the medium, it has the form:

$$d\sigma_{fi} = \frac{m_e^2 k_f}{(2\pi\hbar^2)^2 k_p}\left|\langle m, p-Q|\hat{T}|\Psi^+_{n,p})\right|^2 d\Omega d(\hbar\omega) \to \left(\frac{m_e}{2\hbar^2}\right)^{3/2}\frac{\left(p^2-2m_e\hbar\omega\right)^{1/2}}{2\pi^2 p}\Sigma_q\left|\left\langle m, p-Q|\hat{T}|\Psi^+_{n,p}\right)\right|^2 d\Omega_Q d\omega. \tag{16}$$

For electrons, we take into account the disorder of the medium, and for the plasmon electric field, we assume that the medium is completely homogeneous. Therefore, we can assume that the matrix elements in Equation (14) are approximately equal to each other:

$$\langle n, p-q|V_{pl}|n, p) \approx \langle m, p-q|V_{pl}|m, p) \approx \langle m, p-Q|\left(\underline{V_{pl}}\right)|m, p-Q+q)$$

and the exchange coefficients S_{mnp-q} and S_{mnp} have the same order and are proportional to $\frac{1}{[1+(p/\hbar)^2 a_B^2]^2} = \frac{1}{\left[1+\left(a_B/l_p\right)^2\right]^2}$, then we can rewrite Equation (16) in the form:

$$d\sigma_{fi} = \left(\frac{m_e}{2\hbar^2}\right)^{3/2}\frac{\left(p^2-2m_e\hbar\omega\right)^{1/2}}{2\pi^2 p}\times \Sigma_q\left|\langle m, p-Q|\left(\underline{V_{pl}}\right)|m, p-Q+q)\langle m, p-Q+q|\hat{G}_a(\underline{V_{at}})^{\mathrm{N}}|n, p) + \langle m, p-Q|(\underline{V_{at}})^{\mathrm{N}}\hat{G}_a|n, p-q)\langle n, p-q|\underline{V_{pl}}|n, p)\right|^2 d\Omega_Q d\omega, \tag{17}$$

which is illustrated in Figure 1 by the set of so-called fan diagrams determining the effect of new angle dependence of the effective cross-section.

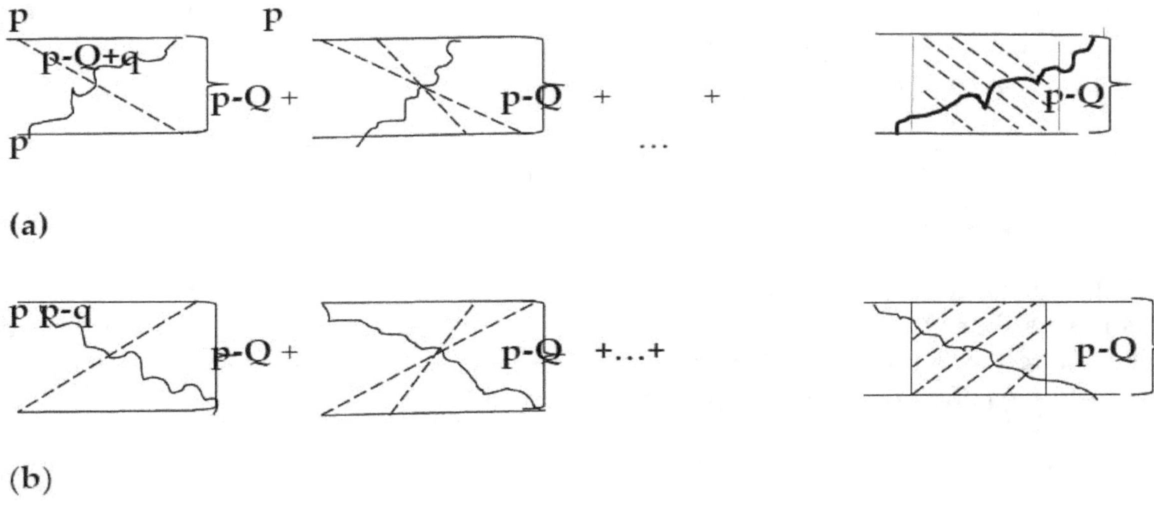

Figure 1. Crossed diagrams correspond to two, three, and so on, scattering events. The wavy line corresponds to the inelastic scattering process with plasmon excitation. Dotted lines link the same atom. (**a**) The processes, beginning from the elastic scattering events and ending by the inelastic, (**b**) processes ranging from inelastic scattering to a series of elastic scattering by atoms.

As usual, in the diagrams that match Equation (17), we will use the dotted line to connect two vertices belonging to the same atom. The lower parts of the diagrams correspond to the analytic expression, which is the complex conjugation of the expression equivalent to the upper parts. The wavy line corresponds to the plasmon. The set of ladder diagrams, correspondent to the terms contributed to the cross-section (17), are independent of the electron-scattering angle. The relation of the crossed diagram contributions and ladder diagrams determine a so-called coherency degree [25]:

$$\begin{aligned}
&\frac{d\sigma_{fi}}{d\Omega_Q} = \left(\frac{m_e}{2\hbar^2}\right)^{3/2} \frac{\left(p^2-2m_e\hbar\omega\right)^{1/2}}{2\pi^2 p} \times \\
&\sum_q w_{mnQ} \left| \langle m, p-Q+q | \hat{G}_a (\underline{V_{at}})^{\mathbb{N}} | n, p) + \langle m, p-Q | (\underline{V_{at}})^{\mathbb{N}} \hat{G}_a | n, p-q) \right|^2 d\omega \\
&\textit{where,} \\
&w_{mnQ} = \left| \langle m, p-q | n, p-q) \langle n, p-q | V_{pl} | n, p) \right|^2 \sim \\
&\left| \langle m, p-q | V_{pl} | m, p) \langle m, p | n, p) \right|^2 \sim \left| \langle m, p-Q \left| \left(\underline{V_{pl}} \right) \right| m, p-Q+q) \right|^2 .
\end{aligned} \tag{18}$$

Neglecting exchange effects, in the resulting expression (17), both *bra-* and *ket-* vectors become non symmetric, and we obtain the well-known result obtained in [25–27]. In contrast to the usual weak localization, one of the crossed lines in the new type of weak localization corresponds to inelastic scattering, while the others correspond to elastic interaction with randomly distributed power centers. According to formula (18), crossed diagrams together with the corresponding ladder diagrams contribute to the scattering probability factor $\Im(\omega, \chi)$ [20]:

$$\begin{aligned}
&\int d\Omega_Q \sum_q w_{mnQ}(q, \hbar\omega) \left| \langle m, p-Q+q | \hat{G}_a (\underline{V_{at}})^{\mathbb{N}} | n, p) + \langle m, p-Q | (\underline{V_{at}})^{\mathbb{N}} \hat{G}_a | n, p-q) \right|^2 d\omega \rightarrow \\
&\sum_q w_{mnQ}(q, \hbar\omega) d\omega \left| m, p-Q+q V_{at} | n, p) \right|^2 \times \\
&\int d\Omega_Q \left| p-Q+q \hat{G}_a p-Q+q) + \right| m, p-Q \hat{G}_a m, p-Q) \Big|^2 d\omega, \\
&\Im(\omega, \chi) = \frac{\hbar^2}{(2\pi\hbar)^3} \int_0^\infty q dq w_{mn}(q, \hbar\omega) \int \left| G\left(p-Q+q, E_p\right) + G\left(p-q, E_p-\hbar\omega\right) \right|^2 d\Omega_Q .
\end{aligned} \tag{19}$$

It was shown in [24–26] that the angular dependence characteristic of a new weak localization arises due to the term in the integrand, which describes the contribution of the interference of two electron waves propagating along the same path in opposite directions:

$$\begin{aligned}
&G^*\left(p-Q+q, E_p\right) G\left(p-q, E_p-\hbar\omega\right) + G\left(p-Q+q, E_p\right) G^*\left(p-q, E_p-\hbar\omega\right) = \\
&\frac{1}{E_p - E_{p-Q+q} + i\gamma} \cdot \frac{1}{E_p - E_{p-q} - \hbar\omega - i\gamma} + \frac{1}{E_p - E_{p-Q+q} - i\gamma} \cdot \frac{1}{E_p - E_{p-q} - \hbar\omega + i\gamma},
\end{aligned} \tag{20}$$

where γ is the electron collision frequency.

This means that weak localization occurs due to such collisions of electrons in which each subsequent scattering begins earlier than the end of the previous one. Thus, a new weak localization is realized when two conditions are met simultaneously:

$$\left\{ \begin{aligned} &\left| E_p - E_{p-Q+q} = 0, \right. \\ &\left| E_p - E_{p-q} - \hbar\omega = 0 \right. \end{aligned} \right. . \tag{21}$$

If we denote $p\prime = p - Q$, then a simple kinematic approach explains the range of the scattering angle $\chi = cos^{-1} \frac{p \cdot p\prime}{p \cdot p\prime}$ typical for the new type of weak localization:

$$cos\chi = -\frac{p^2 + p\prime^2}{2pp\prime} + \frac{2(m\hbar\omega)^2}{pp\prime q^2}, . \tag{22}$$

For more details, [24,25], where it is explained why the angles typical of the new type of weak localization differ from π, and by using the simple kinematic approach, these angles can be estimated very accurately.

Considering Equation (18) taking into account the exchange contributions, we can write the following:

$$\begin{aligned}
&\sigma_{fi}(\chi,\omega) = \left(\tfrac{m_e}{2\hbar^2}\right)^{3/2} \tfrac{\left(p^2-2m_e\hbar\omega\right)^{1/2}}{2\pi^2 p} \times \\
&\int d\Omega_Q \sum_q w_{mnQ}(q,\hbar\omega) \left| \langle m,p-Q+q|\hat{G}_a(\underline{V_{at}})^{\mathbb{N}}|n,p) + \langle m,p-Q|(\underline{V_{at}})^{\mathbb{N}}\hat{G}_a|n,p-q) \right|^2 = \\
&\left(\tfrac{m_e}{2\hbar^2}\right)^{3/2} \tfrac{\left(p^2-2m_e\hbar\omega\right)^{1/2}}{2\pi^2 p} \sum_q w_{mnQ}(q,\hbar\omega) \left| \langle m,p-Q+q|(\underline{V_{at}})^{\mathbb{N}}|n,p) \right|^2 \times \\
&\int d\Omega_Q \left| \langle m,p-Q+q|\hat{G}_a|m,p-Q+q) + \langle n,p-q|\hat{G}_a|n,p-q) \right|^2.
\end{aligned} \tag{23}$$

In the case of an antisymmetric basis, the calculation of matrix elements $\langle n,p-q|\hat{G}_a|n,p-q)$ requires additional comments. Instead of the resolvent operator $\hat{G}_a$, we will use a one-electron Green function $G_a^{ex}(r,r\prime)$ taking into account that exchange contributions influence the one-particle system:

$$G_a^{ex}(r,r\prime,R) = \langle n,p-q|\hat{G}_a|n,p-q)_{rr\prime} = \frac{\sum_{lp} f_{nl}(r,R) f*_{ln}(r\prime,R)}{E_p - E_{p-q} - \hbar\omega + i\gamma}, \tag{24}$$

where:

$$\begin{aligned}
f_{nl}(r,R) = \int dr_1 dr_2 \ldots r_n &\left(\sum_\nu^N (-1)^{g_\nu} \Psi_n{}^*(r_1-R, r_2-R, \ldots (r-R)_\nu \ldots r_k - R) \psi*_{p-q}(r_\nu) \right) \times \\
&\sqrt{\tfrac{f_{lp}}{N}} \Psi_l(r_1-R, r_2-R, \ldots r_\nu - R \ldots r_k - R) \psi_{p-q}(r),
\end{aligned} \tag{25}$$

$$\begin{aligned}
f_{ln}(r\prime,R) = \int dr_1 dr_2 \ldots r_n &\left(\sum_\nu^N (-1)^{g_\nu} \Psi_l{}^*(r_1-R, r_2-R, \ldots (r\prime-R)_\nu \ldots r_k - R) \psi*_{p-q}(r_\nu) \right) \times \\
&\sqrt{\tfrac{f_{lp}}{N}} \Psi_n(r_1-R, r_2-R, \ldots r_\nu - R \ldots r_k - R) \psi_{p-q}(r\prime).
\end{aligned} \tag{26}$$

Integration over the conductor volume V means averaging over randomly distributed scatterers. Here, we used the completeness property (6). In the same way, we rewrite the matrix element of another resolvent operator:

$$\begin{aligned}
G^{ex}_{a(p-Q+q)}(r,r\prime) = \langle m,p-Q+q|\hat{G}_a|m,p-Q+q)_{r,r\prime} &= \tfrac{\sum_{lp} \overline{f_{ml}(r,R) f_{lm}(r\prime,R)}_R}{E_p - E_{p-Q+q} + i\gamma}, \\
\overline{f_{ml}(r,R) f_{lm}(r\prime,R)}_R = \tfrac{1}{V} \int_V f_{ml}(r,R) f_{lm}(r\prime,R) dR.&
\end{aligned} \tag{27}$$

The spectral density $\sigma_{fi}(\chi,\omega)$ of the cross-section (23) of the process of inelastic scattering of electrons on disordered media consisting of atomic defects in a metal alloy (metal alloy doped with di- or trivalent atoms), taking into account exchange effects, has the form:

$$\begin{aligned}
\sigma_{fi}(\chi,\omega) = \left(\tfrac{m_e}{2\hbar^2}\right)^{3/2} \tfrac{\left(p^2-2m_e\hbar\omega\right)^{1/2}}{2\pi^2 p} &\sum_q w_{mnQ}(q,\hbar\omega) \left| \langle m,p-Q+q|(\underline{V_{at}})^{\mathbb{N}}|n,p) \right|^2 \times \\
\int d\Omega_Q &\left| \tfrac{\sum_{lp} \overline{f_{ml}(r,R) f_{lm}(r\prime,R)}_R}{E_p - E_{p-Q+q} + i\gamma} + \tfrac{\sum_{lp} \overline{f_{nl}(r,R) f_{ln}(r\prime,R)}_R}{E_p - E_{p-q} - \hbar\omega + i\gamma} \right|^2.
\end{aligned} \tag{28}$$

Using the obtained general expression for the cross-section (18) for the case considered in [24], the alloy doped by hydrogen atoms as disordered centers, we obtain the same expression for the Green functions and for the scattering probability. It proves a limit transition of the obtained general expression (18) for the cross-section to the known case described in [24].

3. The Electron Identity Principle in Quantum Interference for Multielectron Scatterers

In our consideration, we take into account the identity of the beam electrons and the electrons belonging to the atoms of the scatterer. Firstly, we use a scatterer as a helium-like atom in the ground state ($1s^2$, ^1S) and in the excited state ($1s^1 2s^1$, ^3S).

3.1. Helium-Like Scatterer ^1S

The wave functions of the ground state of a helium-like atom and a free (incident) electron are:

$$\begin{gathered}\Phi_{He}(R-r_{11},R-r_{22};\xi_{11},\xi_{22}) = (\varphi_{1s}(R-r_1)\varphi_{1s}(R-r_2))\cdot\tfrac{1}{\sqrt{2}}(\alpha 1\beta 2-\beta 1\alpha 2),\\ \psi_p(r)=e^{i\frac{p\cdot r}{\hbar}},\end{gathered} \tag{29}$$

where $\varphi_{1s}(R-r_i)=\left(\alpha^3/\pi\right)^{1/2}exp(-\alpha|R-r_i|)$, $\alpha=27/16$.

The vector of the initial state, corresponding to the permutation $\nu = 0$, but antisymmetrized only accounting for the internal permutations in the helium atom, see Equation (29), has the form:

$$\left|\Phi_i^{\nu=0}\right) = \Phi_{He}(r_1,r_2)X_{He}(\xi_1\xi_2)\psi_e(r)\chi_e(\xi) = \left|n=0,p\right). \tag{30}$$

We obtain the antisymmetric wave function by applying the normalized Young operator [29,32] to the wave function of the free electron–helium system as follows: Antisymmetrization of the atomic wavefunction given by Equation (30) over interatomic electrons permutations performed using the four independent Young's operators: $\omega_{11}^{[21]};\omega_{12}^{[21]};\omega_{21}^{[21]};\omega_{22}^{[21]}$ (see Appendix A), where:

$$\Psi=\frac{1}{f_0}(\Psi_{11}(r_1,r_2,r)X_{22}(\xi_1,\xi_2,\xi)+\Psi_{12}(r_1,r_2,r)X_{21}(\xi_1,\xi_2,\xi)),$$

$$\begin{aligned}&\Psi_{11}(r_1,r_2,r)=\omega_{11}^{[21]}\Phi_{He}(r_1,r_2)\psi_p(r)=\tfrac{1}{\sqrt{3}}\left(2\Phi_{He}(r_1,r_2)\psi_p(r)-\Phi_{He}(r_1,r)\psi_p(r_2)-\Phi_{He}(r,r_2)\psi_p(r_1)\right)\\&\Psi_{12}(r_1,r_2,r)=\omega_{12}^{[21]}\Phi_{He}(r_1,r_2)\psi_p(r)=\left(\Phi_{He}(r_1,r)\psi_p(r_2)-\Phi_{He}(r,r_2)\psi_p(r_1)\right)\\&X_{21}(\xi_1,\xi_2,\xi)=\omega_{21}^{[21]}X_{He}(\xi_1\xi_2)\chi_e(\xi)=\omega_{21}^{[21]}\tfrac{1}{\sqrt{2}}(\alpha_1\beta_2-\beta_1\alpha_2)\chi_e(\xi)=\\&\tfrac{1}{\sqrt{2}}[(\alpha_1\beta-\beta_1\alpha)\chi_2(\xi)-(\alpha\beta_2-\beta\alpha_2)\chi_1(\xi)]\\&X_{22}(\xi_1,\xi_2,\xi)=\omega_{22}^{[21]}X_{He}(\xi_1\xi_2)\chi_e(\xi)=\\&\tfrac{1}{\sqrt{3}}\tfrac{1}{\sqrt{2}}[2(\alpha_1\beta_2-\beta_1\alpha_2)\chi_e(\xi)+(\alpha_1\beta-\beta_1\alpha)\chi_2(\xi)+(\alpha\beta_2-\beta\alpha_2)\chi_1(\xi)],\end{aligned} \tag{31}$$

where f_0 is the normalized factor, determined by:

$$\begin{gathered}\overline{\langle(\Psi_{11}(r_1,r_2,r)X_{22}(\xi_1,\xi_2,\xi)+\Psi_{12}(r_1,r_2,r)X_{21}(\xi_1,\xi_2,\xi))\left|\Phi_{He}(R-r_1,R-r_2,\xi_1,\xi_2)\psi_p(r)\chi(\xi)\right)_R}=\\ f_0=\tfrac{4\pi n}{3}\left(\tfrac{2\pi\hbar}{p}\right)^3\left(1-\tfrac{(2p/\hbar\alpha)}{1+(p/\hbar\alpha)}\right).\end{gathered} \tag{32}$$

For this case, Equations (25) and (26) will have the forms:

$$\begin{aligned}&f_{nl}(r,R)\to f_{00}(r,R)=\\&\int dr_1dr_2\tfrac{1}{\sqrt{3}}\{2\varphi_{1s}(r_1-R)\varphi_{1s}(r_2-R)\psi*_{p-q}(r)-\\&\varphi_{1s}(r_1-R)\varphi_{1s}(r-R)\psi*_{p-q}(r_2)-\varphi_{1s}(r_2-R)\varphi_{1s}(r-R)\psi*_{p-q}(r_1)\}\times\\&\sqrt{\tfrac{f_0}{6}}(\varphi_{1s}(R-r_1)\varphi_{1s}(R-r_2))\psi_{p-q}(r'')\tfrac{2}{\sqrt{3}}=\\&\tfrac{4}{3}\sqrt{\tfrac{f_0}{6}}\left\{e^{i\frac{(p-q)(r''-r)}{\hbar}}-\frac{8}{\left(1+\frac{|p-q|^2}{\alpha^2\hbar}^2\right)^2}exp\left(i\frac{(p-q)r''}{\hbar}-\alpha|r-R|\right)\right\}.\end{aligned} \tag{33}$$

That is, the Green's function (28), averaged over randomly distributed centers, can be written as:

$$
\begin{aligned}
&G^{ex}_{(p-Q+q)}(r,r\prime)=\tfrac{1}{V}\int_V dRG^{ex}_{a(p-Q+q)}(r,r\prime,R)\rightarrow\tfrac{1}{V}\int_V dR\frac{\sum_p f_{00}(r)f*_{00}(r\prime)}{E_p-E_{p-Q+q}+i\gamma}=\\
&\left(\tfrac{4}{3}\right)^2\tfrac{f_0}{6}\tfrac{1}{V}\int_V dR\sum_p\left\{\frac{exp\left(-i\frac{p(r-r\prime)}{\hbar}\right)}{E_p-E_{p-Q+q}+i\gamma}\left[1+\frac{8^2}{\left(1+\frac{|p|^2}{\alpha^2\hbar}{}^2\right)^4}\right]-\right.\\
&\frac{8}{\left(1+\frac{|p|^2}{\alpha^2\hbar}{}^2\right)^2}\frac{exp\left(i\frac{p(R+r\prime-r)}{\hbar}-\alpha|r-R|\right)+exp\left(-i\frac{p(R-r\prime+r)}{\hbar}-\alpha|r\prime-R|\right)}{E_p-E_{p-Q+q}+i\gamma}+\\
&\left.\frac{8^2}{\left(1+\frac{|p|^2}{\alpha^2\hbar}{}^2\right)^4}\frac{exp(-\alpha|r-R|-\alpha|r\prime-R|)}{E_p-E_{p-Q+q}+i\gamma}\right\}=\\
&G^{ex}_{0(p-Q+q)}(r,r\prime)-G^{ex}_{1(p-Q+q)}(r,r\prime)+G^{ex}_{a(p-Q+q)}(r,r\prime),
\end{aligned}
\tag{34}
$$

which has the same construction as in [26] but with a well-defined sign in front of the second term.

Here, $G^{ex}_0(r,r\prime)$ is the ordinary Green's function, previously considered in [22,24,26], making it the main contribution to the new weak localization, corrected only by adding a normalization factor due to exchange effects. The specific exchange contribution has the form:

$$
G^{ex}_{1p-Q+q)}(r,r\prime)=\left(\frac{4}{3}\right)^2\frac{f_0}{6V\alpha^3}2\pi\sum_p\frac{8^2}{\left(1+\frac{p^2}{\alpha^2\hbar}{}^2\right)^4}\frac{\left(1+\frac{ip(r-r\prime)}{2\hbar}\right)e^{-i\frac{p(r-r\prime)}{\hbar}}}{E_p-E_{p-Q+q}+i\gamma},. \tag{35}
$$

Since we are interested in the correlation length $|\boldsymbol{r}-\boldsymbol{r}\prime|\leq\lambda_p$, we can neglect the higher-order terms of the parameter $\frac{p(r-r\prime)}{\hbar}$. Then, this term in the p-representation after the Fourier transform and an ordinary Green function in the p-representation have the following form:

$$
\begin{aligned}
G^{ex}_1(p-Q+q)=G^{ex}_1(p\prime+q)&=\frac{2^{10}}{3^4}\left(1-\frac{\left(\frac{2p}{\hbar\alpha}\right)^3}{\left(1+\left(\frac{p}{\hbar\alpha}\right)^2\right)^4}\right)\frac{\left(\frac{\hbar}{\alpha p}\right)^3}{\left(1+\frac{p^2}{\alpha^2\hbar}{}^2\right)^4}\frac{(2\pi)^2n^2}{E_p-E_{p-Q+q}+i\gamma},\\
G_0(p\prime+q)&=\frac{1}{E_p-E_{p-Q+q}+i\gamma}.
\end{aligned}
\tag{36}
$$

The third term $G^{ex}_a(r,r\prime)$ describes a process that begins and ends at the same atom; therefore, it does not contribute to the quantum transfer of electrons.

Now, consider the main pre-factor Ξ in the cross-section (28):

$$
\begin{aligned}
\Xi=\int d\Omega_Q&\left|\frac{\sum_{lp}\overline{f_{ml}(r,R)f_{lm}(r\prime,R)}_R}{E_p-E_{p-Q+q}+i\gamma}+\frac{\sum_{lp}\overline{f_{nl}(r)f_{ln}(r\prime)}_R}{E_p-E_{p-q}-\hbar\omega+i\gamma}\right|^2=\\
&\int d\Omega_Q\left|G^{ex}(p-Q+q)+G^{ex}(p-q)\right|^2,
\end{aligned}
\tag{37}
$$

where we denoted the Green's function (34) in the p-representation as $G^{ex}(p-Q+q)=G^{ex}_0(p\prime+q)-G^{ex}_1(p\prime+q)$, taking into account $p-Q=p\prime$, and in exactly the same way the Green's function of the second term in the same representation as $G^{ex}(p-q)=G^{ex}_0(p-q)-G^{ex}_1(p-q)$. Then, the factor Ξ consists of three contributions, background, "the simple new weak localization", and the exchange term:

$$
\Xi=\Xi_{bg}+\Xi_0-\Xi_{exc}. \tag{38}
$$

The background term has the following form:

$$\Xi_{bg} = \int d\Omega_Q \Big\{ \big| G_0^{ex}(p\prime + q) - G_1^{ex}(p\prime + q) \big|^2 + \big| G_0^{ex}(p - q) - G_1^{ex}(p - q) \big|^2 \Big\} = \int d\Omega_Q \Big\{ \big| G_0^{ex}(p\prime + q) \big|^2 + \big| G_1^{ex}(p\prime + q) \big|^2 + \big| G_0^{ex}(p - q) \big|^2 + \big| G_1^{ex}(p - q) \big|^2 - \Big[G_0^{ex*}(p\prime + q) G_1^{ex}(p\prime + q) + G_1^{ex*}(p\prime + q) G_0^{ex}(p\prime + q) + G_0^{ex*}(p - q) G_1^{ex}(p - q) + G_1^{ex*}(p - q) G_0^{ex}(p - q) \Big\} \Big]. \quad (39)$$

Substituting Expression (36), we obtain:

$$\Xi_{bg} = \left(1 - \frac{\pi}{\left(1 + \frac{p^2}{(\hbar\alpha)^2}\right)^6} \frac{2^{10}(\pi n)^2}{3\alpha^6} \right) \times \left[5 \left(\frac{1}{(v\prime q)^2} ln \left(\frac{(v\prime q - \hbar\omega)^2 + (\hbar\gamma)^2}{(v\prime q + \hbar\omega)^2 + (\hbar\gamma)^2} \right) - \frac{1}{(vq)^2} ln \left(\frac{(vq - \hbar\omega)^2 + (\hbar\gamma)^2}{(vq + \hbar\omega)^2 + (\hbar\gamma)^2} \right) \right) - \left(4 - \frac{25}{\left(1 + \frac{p^2}{(\hbar\alpha)^2}\right)^2} \right) \frac{\Gamma^2(5)(vq)^2}{\left(\frac{\hbar^2\alpha^2}{2m}\right)^4 \left(1 + \frac{p^2}{(\hbar\alpha)^2}\right)^4} \right]. \quad (40)$$

"The ordinary new weak localization" contribution has the form:

$$\Xi_0 = \int d\Omega_Q \Big\{ G_0^{ex*}(p\prime + q) G_0^{ex}(p - q) + G_0^{ex*}(p - q) G_0^{ex}(p\prime + q)^2 \Big\} = \frac{4\pi}{(qv)\hbar\gamma} \cdot \left[arctg\left(\frac{qv + \hbar\omega_p}{\hbar\gamma} \right) + arctg\left(\frac{qv - \hbar\omega_p}{\hbar\gamma} \right) - \frac{\gamma}{\sqrt{2\omega_p^2(1 - cos\chi) - \frac{q^2v^2}{\hbar^2} sin^2\chi}} \right] \times ln\left(\frac{\hbar^2\omega_p^2 - q^2v^2 cos\chi + qv\sqrt{2\hbar^2\omega_p^2(1 - cos\chi) - q^2v^2 sin^2\chi}}{\hbar^2\omega_p^2 - q^2v^2 cos\chi - qv\sqrt{2\hbar^2\omega_p^2(1 - cos\chi) - q^2v^2 sin^2\chi}} \right). \quad (41)$$

This expression was first obtained in [20] and used in [26]. The Langmuir frequency is denoted as ω_p, and γ is the collision frequency. The specific exchange contribution in the cross-section is:

$$\Xi_{exc} = \int d\Omega_Q \Big\{ G_1^{ex*}(p + q) G_1^{ex}(p - q) + G_1^{ex*}(p\prime + q) G_0^{ex}(p - q) + G_0^{ex*}(p - q) G_1^{ex}(p\prime + q) + G_1^{ex*}(p - q) G_0^{ex}(p\prime + q) - \big(G_1^{ex*}(p\prime + q) G_1^{ex}(p - q) + G_1^{ex*}(p - q) G_1^{ex}(p\prime + q) \big) \Big\} = \frac{2^8 \frac{n\pi}{\alpha^3}}{\left(1 + \frac{p^2}{(\hbar\alpha)^2}\right)^3} \left\{ \Xi_0 \left(1 - \frac{2^5 \frac{n\pi}{\alpha^3}}{\left(1 + \frac{p^2}{(\hbar\alpha)^2}\right)^5} \right) - cos\chi \left(4 + \frac{\hbar\omega_L}{vq} ln\left(\frac{(vq - \hbar\omega_p)^2 + (\hbar\gamma)^2}{(vq + \hbar\omega_p)^2 + (\hbar\gamma)^2} \right) \right) \frac{(2m)^2 \cdot 4\pi}{(\hbar\alpha)^4 \left(1 + \frac{p^2}{(\hbar\alpha)^2}\right)^2} \left(1 - \frac{2^5 \frac{n\pi}{\alpha^3} \Gamma^2(5)}{\left(1 + \frac{p^2}{(\hbar\alpha)^2}\right)^5} \right) \right\}, \quad (42)$$

where $\Gamma(5) = 4!$ is a Γ-function. Then, taking into account Equations (39)–(42), the main factor determining the angular dependence of the scattering cross-section is:

$$\Xi(\chi) = \Xi_{bg} + \Xi_0 \left(1 - \frac{2^8 \frac{n\pi}{\alpha^3}}{\left(1 + \frac{p^2}{(\hbar\alpha)^2}\right)^3} + \frac{2^{13} \left(\frac{n\pi}{\alpha^3} \right)^2}{\left(1 + \frac{p^2}{(\hbar\alpha)^2}\right)^8} \right) + \frac{2^8 \frac{n\pi}{\alpha^3}}{\left(1 + \frac{p^2}{(\hbar\alpha)^2}\right)^3} cos\chi \left(4 + \frac{\hbar\omega_L}{vq} ln\left(\frac{(vq - \hbar\omega_p)^2 + (\hbar\gamma)^2}{(vq + \hbar\omega_p)^2 + (\hbar\gamma)^2} \right) \right) \frac{(2m)^2 \cdot 4\pi}{(\hbar\alpha)^4 \left(1 + \frac{p^2}{(\hbar\alpha)^2}\right)^2} \left(1 - \frac{2^5 \frac{n\pi}{\alpha^3} \Gamma^2(5)}{\left(1 + \frac{p^2}{(\hbar\alpha)^2}\right)^5} \right) = \Xi_{bg} + \Xi_{0exc} + \Xi_{1exc}. \quad (43)$$

Thus, the cross-section $\sigma_{fi}(\chi, \omega) \propto \Xi(\chi)$. is proportional to the factor $\Xi(\chi)$ determined by Equation (43), which contains the total dependence on the scattering angle.

3.2. Helium-Like Scatterer ^3S

The wave functions of the excited state helium-like atom and a free (incident) electron are:

$$\Phi_{He}(R-r_{11},R-r_{22};\xi_1,\xi_2)=\frac{1}{\sqrt{2}}(\varphi_{1s}(R-r_1)\varphi_{2s}(R-r_2)-\varphi_{1s}(R-r_2)\varphi_{2s}(R-r_1))\cdot X(\alpha 1\alpha 2), \quad (44)$$

where:

$$\begin{aligned}&\varphi_{2s}(R-r_i)=(\alpha^3/2^3\pi)^{1/2}\Big(1-\tfrac{\alpha|R-r_i|}{2}\Big)exp\Big(-\tfrac{\alpha|R-r_i|}{2}\Big),\\&\alpha=27/16.\end{aligned} \quad (45)$$

The vector of the initial state, corresponding to the permutation $\nu=0$, but antisymmetrized only accounting for the internal permutations in the helium atom, see Equation (44), has the form: $\left|\Phi_i^{\nu=0}\right)=\Phi_{He}(r_1,r_2)X_{He}(\xi_1\xi_2)\psi_e(r)\chi_e(\xi)=\left|n=1*,p\right)$.
The antisymmetrized vector of the final state is:

$$\begin{aligned}&\Psi(r_1,r_2,r_3\alpha_1\alpha_2\alpha_3)=\frac{(\alpha 1\alpha 2\alpha 3)}{f_1\sqrt{3!}}\begin{vmatrix}\phi_{1s}(R-r_1) & \phi_{2s}(R-r_1) & \psi_p(r_1)\\ \phi_{1s}(R-r_2) & \phi_{2s}(R-r_2) & \psi_p(r_2)\\ \phi_{1s\alpha}(R-r_3) & \phi_{2s}(R-r_3) & \psi_p(r_3)\end{vmatrix}=\\&\frac{(\alpha 1\alpha 2\alpha 3)}{f_1\sqrt{3}}\Big(\tfrac{1}{\sqrt{2}}[\phi_{1s}(R-r_1)\phi_{2s}(R-r_2)-\phi_{1s}(R-r_2)\phi_{2s}(R-r_1)]\cdot\psi_p(r_3)-\\&[\phi_{1s}(R-r_1)\phi_{2s}(R-r_3)-\phi_{2s}(R-r_1)\phi_{1s}(R-r_3)]\cdot\tfrac{1}{\sqrt{2}}\psi_p(r_2)+\\&[\phi_{1s}(R-r_2)\phi_{2s}(R-r_3)-\phi_{1s}(R-r_3)\phi_{2s}(R-r_2)]\cdot\tfrac{1}{\sqrt{2}}\psi_p(r_1)\Big).\end{aligned} \quad (46)$$

Then, an exchange normalization factor:

$$\begin{aligned}&f_1=\iiint dr_1dr_2dr_3\tfrac{1}{\sqrt{2}}\overline{(\varphi*_{1s}(R-r_1)\varphi*_{2s}(R-r_2)-\varphi*_{1s}(R-r_2)\varphi*_{2s}(R-r_1))\psi_p(r_3)}\times\\&\tfrac{1}{\sqrt{3}}\Big(\tfrac{1}{\sqrt{2}}[\phi_{1s}(R-r_1)\phi_{2s}(R-r_2)-\phi_{1s}(R-r_2)\phi_{2s}(R-r_1)]\cdot\psi_p(r_3)-[\phi_{1s}(R-r_1)\phi_{2s}(R-r_3)-\phi_{2s}(R-r_1)\phi_{1s}(R-r_3)]\cdot\tfrac{1}{\sqrt{2}}\psi_p(r_2)+\\&[\phi_{1s}(R-r_2)\phi_{2s}(R-r_3)-\phi_{1s}(R-r_3)\phi_{2s}(R-r_2)]\cdot\tfrac{1}{\sqrt{2}}\psi_p(r_1)\Big)_R=\\&\tfrac{1}{\sqrt{3}}\Big\{1-\Big[\big|\langle\phi_{2s}|\psi_p\rangle\big|^2+\big|\langle\phi_{1s}|\psi_p\rangle\big|^2\Big]\Big\}_R=\\&\tfrac{4\pi n}{3}\Big(\tfrac{2\pi\hbar}{p}\Big)^{3^3}\tfrac{1}{\sqrt{3}}\left\{1-2^6\left[\left(\frac{(\frac{2p}{\hbar}\alpha)^4+7\cdot 2(\frac{2p}{\hbar}\alpha)^2-11}{\big(1+(\frac{2p}{\hbar}\alpha)^2\big)^4}\right)^2+\frac{4}{\big(4+(\frac{2p}{\alpha\hbar})^2\big)^4}\right]\right\}.\end{aligned}$$

A one-electron amplitude with respect to the electron permutations has the form:

$$\begin{aligned}&f_{11}(r,R)=\frac{(\alpha 1\alpha 2\alpha 3)}{f_1\sqrt{3}}\int dr_1dr_2\Big(\tfrac{1}{\sqrt{2}}[\varphi_{1s}(R-r_1)\varphi_{2s}(R-r_2)-\varphi_{1s}(R-r_2)\varphi_{2s}(R-r_1)]\cdot\psi_p(r)-\\&[\varphi_{1s}(R-r_1)\varphi_{2s}(R-r)-\varphi_{2s}(R-r_1)\varphi_{1s}(R-r)]\cdot\tfrac{1}{\sqrt{2}}\psi_p(r_2)+\\&[\varphi_{1s}(R-r_2)\varphi_{2s}(R-r)-\varphi_{1s}(R-r)\varphi_{2s}(R-r_2)]\cdot\tfrac{1}{\sqrt{2}}\psi_p(r_1)\Big)\times\\&\sqrt{\tfrac{f_1}{6}}\tfrac{1}{\sqrt{2}}(\varphi*_{1s}(R-r_1)\varphi*_{2s}(R-r_2)-\varphi*_{1s}(R-r_2)\varphi*_{2s}(R-r_1))\psi*_p(r'')=\\&\frac{1}{\sqrt{6f_1}\sqrt{3}}\Big(e^{ip(r-r'')/\hbar}-(\alpha^3/\pi)^{1/2}exp\Big(-\tfrac{\alpha|R-r|}{2}-\tfrac{ipr''}{\hbar}\Big)\times\\&\Big[-2^{-3/2}\Big(1-\tfrac{\alpha|R-r|}{2}\Big)\frac{4\pi\Gamma(5)(\alpha/2)^3\big((\alpha/2)^2-(p/\hbar)^2\big)}{\big((\alpha/2)^2+(p/\hbar)^2\big)^4}+\\&\Big(2^{-3/2}\Big(1-\tfrac{\alpha|R-r|}{2}\Big)+exp\Big(-\tfrac{\alpha|R-r|}{2}\Big)\frac{8\pi(\alpha/2)}{\big((\alpha/2)^2+(p/\hbar)^2\big)^2}\Big]\Big).\end{aligned} \quad (47)$$

These amplitudes determine a one-electron Green function while accounting for the exchange contributions:

$$\begin{gathered}
G(\boldsymbol{r},\boldsymbol{r\prime}) = \frac{1}{18f_1}\sum_p \frac{1}{E_p - E_{\boldsymbol{p}-\boldsymbol{Q}+\boldsymbol{q}} + i\gamma}\left(e^{i\boldsymbol{p}(\boldsymbol{r}-\boldsymbol{r\prime})/\hbar} - (\alpha^3/\pi)^{1/2} e^{i\boldsymbol{p}(\boldsymbol{r}-\boldsymbol{r\prime})/\hbar}\left(e^{-\frac{i\boldsymbol{p}\boldsymbol{r}}{\hbar}} + e^{\frac{i\boldsymbol{p}\boldsymbol{r\prime}}{\hbar}}\right)\frac{8\pi}{(\alpha/2)^3}\right)\times \\
\left[2^{-3/2}\left(\frac{\Gamma(5)(\alpha/2)^3\left((\alpha/2)^2-(p/\hbar)^2\right)}{\left((\alpha/2)^2+(p/\hbar)^2\right)^4} - 2\right) + \frac{(\alpha/2)}{8\left((\alpha/2)^2+(p/\hbar)^2\right)^2}\right]\Bigg) + G_{aa}(\boldsymbol{r},\boldsymbol{r\prime}) = \\
\frac{1}{18f_1}\sum_p \frac{e^{ip(r-r\prime)/\hbar}}{E_p-E_{p-Q+q}+i\gamma} - \frac{1}{18f_1}\left(\sum_p \frac{e^{ip(r-r\prime)/\hbar}}{E_p-E_{p-Q+q}+i\gamma}\left(2-\frac{ip(r-r\prime)}{\hbar}\right)8^2 n(\alpha/2)^{-3}\right)\times \\
\left[\left(\frac{\Gamma(5)\left(1-(2p/\hbar\alpha)^2\right)}{\left(1+(2p/\hbar\alpha)^2\right)^4} - 2\right) + \frac{1}{8\left(1+(2p/\hbar\alpha)^2\right)^2}\right]\Bigg) + G_{aa}(r,r\prime) = \\
\frac{1}{18f_1}\left(G_0(r,r\prime) - G_1^{exc}(r,r\prime) + G_{aa}\right). \\
G_1^{exc}(r,r\prime) = \sum_p \frac{e^{ip(r-r\prime)/\hbar}}{E_p-E_{p-Q+q}+i\gamma}\left(2-\frac{ip(r-r\prime)}{\hbar}\right)8^2 n(\alpha/2)^{-3}\left[\left(\frac{\Gamma(5)\left(1-(2p/\hbar\alpha)^2\right)}{\left(1+(2p/\hbar\alpha)^2\right)^4} - 2\right) + \frac{1}{8\left(1+(2p/\hbar\alpha)^2\right)^2}\right].
\end{gathered} \tag{48}$$

The mentioned Green function in p-representation is:

$$\begin{gathered}
G(p-Q+q) = G(p\prime+q) = \frac{n\alpha^{-3}}{18f_1}G_0(p\prime+q)\left(1+2^{11}\pi\right) - \frac{2^{10}\pi n\alpha^{-3}}{18f_1}G_1^{exc}(p\prime+q) = \\
\frac{n\alpha^{-3}}{18f_1}G_0((p\prime+q)\left(1+2^{11}\pi\right) - \\
\frac{2^{10}\pi n\alpha^{-3}}{18f_1}\frac{1}{E_p-E_{p-Q+q}+i\gamma}\cdot\left[\frac{1}{8\left(1+(2|p\prime+q|/\hbar\alpha)^2\right)^2} - \frac{24}{\left(1+(2|p\prime+q|/\hbar\alpha)^2\right)^3} + \frac{24}{\left(1+(2|p\prime+q|/\hbar\alpha)^2\right)^4}\right]\Bigg).
\end{gathered} \tag{49}$$

where the general $G_0(p\prime+q)$ and the exchange $G_1^{exc}(p\prime+q)$ contributions are:

$$\begin{gathered}
G_0(p\prime+q) = \frac{1}{E_p-E_{p-Q+q}+i\gamma} \\
G_1^{exc}(p\prime+q) = \\
\frac{1}{E_p-E_{p-Q+q}+i\gamma}\cdot\left[\frac{1}{8\left(1+(2|p\prime+q|/\hbar\alpha)^2\right)^2} - \frac{24}{\left(1+(2|p\prime+q|/\hbar\alpha)^2\right)^3} + \frac{24}{\left(1+(2|p\prime+q|/\hbar\alpha)^2\right)^4}\right].
\end{gathered} \tag{50}$$

The contribution Ξ in the cross-section (28) in the form (38) for our case of the excited helium atom has the same structure as Equation (40), where the background contribution is:

$$\begin{gathered}
\Xi_{bg} = \left(\frac{2^{10}\pi n\alpha^{-3}}{18f_1}\right)^2 \int d\Omega_Q\{4\left(\left|G_0^{ex}(p\prime+q)\right|^2 + \left|G_0^{ex}(p-q)\right|^2\right) + \left(\left|G_1^{ex}(p\prime+q)\right|^2 + \left|G_1^{ex}(p-q)\right|^2\right) - \\
2\left[G_0^{ex*}(p\prime+q)G_1^{ex}(p\prime+q) + G_1^{ex*}(p\prime+q)G_0^{ex}(p\prime+q) + G_0^{ex*}(p-q)G_1^{ex}(p-q) + G_1^{ex*}(p-q)G_0^{ex}(p-q)\right\}\right].
\end{gathered} \tag{51}$$

The exchange contribution (42) in this case has the form:

$$\begin{gathered}
\Xi_{exc} = \int d\Omega_Q\left\{G_0^{ex*}(p\prime+q)G_1^{ex}(p-q) + G_1^{ex*}(p\prime+q)G_0^{ex}(p-q) + G_0^{ex*}(p-q)G_1^{ex}(p\prime+q) + \right. \\
\left. G_1^{ex*}(p-q)G_0^{ex}(p\prime+q) - \left(G_1^{ex*}(p\prime+q)G_1^{ex}(p-q) + G_1^{ex*}(p-q)G_1^{ex}(p\prime+q)\right)\right\}.
\end{gathered} \tag{52}$$

The explicit form is presented in Appendix A (see Equations (A3) and (A4)). In this expression, we neglected by the second order small terms proportional to q^2, while the inelastic momentum loss is $q \ll p$. Thus, the exchange contribution to the cross-section has the form:

$$\begin{gathered}
\Xi_{exc} = -\left(\frac{(\hbar\alpha)^2}{8m}\right)^2 \left\{\frac{1}{8}\frac{\partial}{\partial\mathbb{R}} + 12\left(\frac{(\hbar\alpha)^2}{8m}\right)\frac{\partial^2}{\partial\mathbb{R}^2} + 4\left(\frac{(\hbar\alpha)^2}{8m}\right)^2 \frac{\partial^3}{\partial\mathbb{R}^3}\right\}\left[\frac{1}{(\mathbb{R}-\hbar\omega-i\gamma)} + c.c.\right](J_0 + J_2) - \\
\left(\frac{(\hbar\alpha)^2}{8m}\right)^2 \times \left\{\frac{1}{8}\frac{\partial}{\partial\mathbb{R}'} + 12\frac{(\hbar\alpha)^2}{8m}\frac{\partial^2}{\partial\mathbb{R}'^2} + 4\left(\frac{(\hbar\alpha)^2}{8m}\right)^2 \frac{\partial^3}{\partial\mathbb{R}'^3}\right\}\left[\frac{1}{\mathbb{R}'+\hbar\omega+i\gamma} + c.c.\right](J_0 + J_1) - \\
\left(\frac{(\hbar\alpha)^2}{8m}\right)^4 \left[\frac{1}{8}\frac{\partial}{\partial\mathbb{R}'} + 12\left(\frac{(\hbar\alpha)^2}{8m}\right)\frac{\partial^2}{\partial\mathbb{R}'^2} + 4\left(\frac{(\hbar\alpha)^2}{8m}\right)^2 \frac{\partial^3}{\partial\mathbb{R}'^3}\right] \times \\
\left[\frac{1}{8}\frac{\partial}{\partial\mathbb{R}} + 12\left(\frac{(\hbar\alpha)^2}{8m}\right)\frac{\partial^2}{\partial\mathbb{R}^2} + 4\left(\frac{(\hbar\alpha)^2}{8m}\right)^2 \frac{\partial^3}{\partial\mathbb{R}^3}\right] \times \\
\frac{1}{\mathbb{R}-i\gamma}\frac{1}{\mathbb{R}'-i\gamma}\{J_0 + J_1 + J_2 + J_3\}.,
\end{gathered} \tag{53}$$

where the contribution $J_0 = \Xi_0$ is determined by Equation (41), c.c. is denoted by a complex conjugate expression. Here:

$$\begin{gathered}
\mathbb{R} = \left(\frac{(\hbar\alpha)^2}{8m} + E_p\right) \\
\mathbb{R}' = \mathbb{R} - \hbar\omega = \left(\frac{(\hbar\alpha)^2}{8m} + E_p - \hbar\omega\right).
\end{gathered} \tag{54}$$

The total exchange contribution consists of four terms:

$$\Xi_{exc} = \Xi_{exc0} + \Xi_{exc1} + \Xi_{exc2} + \Xi_{exc3}, \tag{55}$$

where each term is determined in detail in Appendix A, and the results of the calculations for each term are the following:

(1) The usual term "new weak localization" Ξ_{exc0}, which retains the angular dependence of the kinematic model with a factor depending on the exchange contributions:

$$\begin{gathered}
\Xi_{exc_0} = \left\{\frac{1}{2^6}\left(\frac{(\hbar\alpha)^2}{2m}\right)^2 \frac{1}{(\mathbb{R})^2}\left[\frac{(\mathbb{R})^2}{(\mathbb{R}-\hbar\omega)^2} + 1\right] - 12\left(\frac{(\hbar\alpha)^2}{2m}\right)^3 \frac{1}{(\mathbb{R})^3}\left[\frac{(\mathbb{R})^3}{(\mathbb{R}-\hbar\omega)^3} + 1\right] + \right. \\
3\left(\frac{(\hbar\alpha)^2}{2m}\right)^4 \frac{1}{(\mathbb{R})^4}\left[\frac{(\mathbb{R})^4}{(\mathbb{R}-\hbar\omega)^4} + 1\right] - \frac{1}{2^{14}}\left(\frac{(\hbar\alpha)^2}{2m}\right)^4 \frac{1}{(\mathbb{R}-\hbar\omega)^2(\mathbb{R})^2} + \\
\frac{3}{2^{10}}\left(\frac{(\hbar\alpha)^2}{2m}\right)^5 \frac{1}{(\mathbb{R})^5}\left(\frac{(\mathbb{R})^2}{(\mathbb{R}-\hbar\omega)^2} + \frac{(\mathbb{R})^3}{(\mathbb{R}-\hbar\omega)^3}\right) - \frac{9}{2^6}\left(\frac{(\hbar\alpha)^2}{2m}\right)^6 \frac{1}{(\mathbb{R})^6}\frac{(\mathbb{R})^3}{(\mathbb{R}-\hbar\omega)^3} - \\
\frac{3}{2^{13}}\left(\frac{(\hbar\alpha)^2}{2m}\right)^6 \frac{1}{(\mathbb{R})^6}\left(\frac{(\mathbb{R})^2}{(\mathbb{R}-\hbar\omega)^2} + \frac{(\mathbb{R})^4}{(\mathbb{R}-\hbar\omega)^4}\right) + \\
\left.\frac{9}{2^9}\left(\frac{(\hbar\alpha)^2}{2m}\right)^7 \frac{1}{(\mathbb{R})^7}\left(\frac{(\mathbb{R})^4}{(\mathbb{R}-\hbar\omega)^4} + \frac{(\mathbb{R})^3}{(\mathbb{R}-\hbar\omega)^3}\right) - \frac{9}{2^{12}}\left(\frac{(\hbar\alpha)^2}{2m}\right)^8 \frac{1}{(\mathbb{R}-\hbar\omega)^4(\mathbb{R})^4}\right\} J_0.
\end{gathered} \tag{56}$$

(2) There are three typical exchange contributions, which differ from each other, each of which contains, in addition to the background exchange terms, a term proportional to $\cos\chi$

$$\begin{gathered}
\Xi_{exc1} = \frac{1}{8}\frac{\pi}{\mathbb{R}^4}\left(\frac{(\hbar\alpha)^2}{2m}\right)^2 \left\{1 + \frac{1}{2^5}\left[\frac{1}{8}\frac{\left(\frac{(\hbar\alpha)^2}{2m}\right)^2}{(\mathbb{R}-\hbar\omega)^2} - 6\frac{\left(\frac{(\hbar\alpha)^2}{2m}\right)^3}{(\mathbb{R}-\hbar\omega)^3} + \frac{3}{4}\frac{\left(\frac{(\hbar\alpha)^2}{2m}\right)^4}{(\mathbb{R}-\hbar\omega)^4}\right]\right\} \times \\
\left\{\frac{\mathbb{R}}{vq} ln\left(\frac{(vq-\hbar\omega_p)^2+(\hbar\gamma)^2}{(vq+\hbar\omega_p)^2+(\hbar\gamma)^2}\right)\left[\frac{1}{8}\left(2 + \frac{3\hbar\omega}{\mathbb{R}}\right) - 3\frac{(\hbar\alpha)^2}{2m}\left(\frac{6}{\mathbb{R}} + \frac{12\hbar\omega}{\mathbb{R}^2}\right) + \left(\frac{(\hbar\alpha)^2}{2m}\right)^2\left(\frac{6}{\mathbb{R}^2} + \frac{15\hbar\omega}{\mathbb{R}^3}\right)\right] + \right. \\
\left. 3\frac{v'}{v}\cos\chi\left(2 + \frac{\hbar\omega}{2v\cdot q} ln\left(\frac{(vq-\hbar\omega_p)^2+(\hbar\gamma)^2}{(vq+\hbar\omega_p)^2+(\hbar\gamma)^2}\right)\right)\left[\frac{1}{4} - 24\frac{\left(\frac{(\hbar\alpha)^2}{2m}\right)}{\mathbb{R}} + 10\frac{\left(\frac{(\hbar\alpha)^2}{2m}\right)^2}{\mathbb{R}^2}\right]\right\}.
\end{gathered} \tag{57}$$

$$\begin{gathered}\Xi_{exc2} = \frac{1}{8}\left(\frac{\pi}{\mathbb{R}^4}\right)\left(\frac{(\hbar\alpha)^2}{2m}\right)^2\left\{1 + \frac{1}{2^5}\frac{\left(\frac{(\hbar\alpha)^2}{2m}\right)^2}{(\mathbb{R}-\hbar\omega)^2}\left[\frac{1}{8} - 6\frac{\left(\frac{(\hbar\alpha)^2}{2m}\right)}{(\mathbb{R}-\hbar\omega)} + \frac{3}{4}\frac{\left(\frac{(\hbar\alpha)^2}{2m}\right)^2}{(\mathbb{R}-\hbar\omega)^2}\right]\right\}\times \\ \left\{\frac{2\mathbb{R}}{v\prime q}ln\left(\frac{(v\prime q-\hbar\omega_p)^2+(\hbar\gamma)^2}{(v\prime q+\hbar\omega_p)^2+(\hbar\gamma)^2}\right)\left[\frac{1}{8} - 9\frac{\frac{(\hbar\alpha)^2}{2m}}{\mathbb{R}} + 3\frac{\left(\frac{(\hbar\alpha)^2}{2m}\right)^2}{\mathbb{R}^2}\right] + \right. \\ \left. 3\frac{v}{v\prime}cos\chi\left(2 + \frac{\hbar\omega+i\gamma}{2v\cdot q}ln\left(\frac{(v\prime q-\hbar\omega_p)^2+(\hbar\gamma)^2}{(v\prime q+\hbar\omega_p)^2+(\hbar\gamma)^2}\right)\right)\left[\frac{1}{4} - \left(\frac{(\hbar\alpha)^2}{2m}\right)\frac{24}{\mathbb{R}} + \left(\frac{(\hbar\alpha)^2}{2m}\right)^2\frac{10}{\mathbb{R}^2}\right]\right\}.\end{gathered} \tag{58}$$

$$\begin{gathered}\Xi_{exc3} = -\pi\cdot cos\chi\frac{1}{(\mathbb{R}-\hbar\omega)^2}\frac{\left(\frac{(\hbar\alpha)^2}{2m}\right)^4}{(4\mathbb{R})^4}\left[-\frac{1}{2} + 24\frac{\left(\frac{(\hbar\alpha)^2}{2m}\right)}{(\mathbb{R}-\hbar\omega)} + 3\frac{\left(\frac{(\hbar\alpha)^2}{2m}\right)^2}{(\mathbb{R}-\hbar\omega)^2}\right]\times \\ \left\{\left[-\frac{1}{8}\left(3 + \frac{4\hbar\omega}{\mathbb{R}}\right) + \frac{5}{3}\frac{(vq)(v\prime q)}{\mathbb{R}^2}\right] + 6\frac{\left(\frac{(\hbar\alpha)^2}{2m}\right)}{\mathbb{R}}\left[\left(6 + \frac{10\hbar\omega}{\mathbb{R}}\right) - 5\frac{(vq)(v\prime q)}{\mathbb{R}^2}\right] + \right. \\ \left. \frac{5}{2}\frac{\left(\frac{(\hbar\alpha)^2}{2m}\right)^2}{\mathbb{R}^2}\left[-6\left(1 + \frac{2\hbar\omega}{\mathbb{R}}\right) + 7\frac{(vq)(v\prime q)}{\mathbb{R}^2}\right]\right\}.\end{gathered} \tag{59}$$

These three exchange terms (57)–(59) depend on the initial atomic states of the scatterers, the number of electrons, and their multiplicity, but this dependence is expressed in the value of the background contribution and the value of the factor in front of cosχ. Variance in the number of electrons for different scatterers and their states does not lead to a new angular dependence in the scattering cross-section under conditions of a new weak localization.

In our work, we consider the identity of the incident electrons and electrons on helium-like scatterers. Helium-doped Zi-Ni alloys were used as a disordered medium. Doping the samples with helium increases disorder and enhances quantum interference. For our estimates, we take $\omega_p \sim 10^{15}$, then the energy losses due to inelastic scattering will be ~1 eV, then the contribution from the usual weak localization of a new type, determined by Equation (42), to the factor in the cross-section is shown in Figure 2a. It is easy to see the fluctuations of the factor with increasing energy of the incident electron (the value determined by Equation (52)) and the singularity at an angle of 2.9 rad, which corresponds to 166 angular degrees. This angle is in a good agreement with the results of the kinematic model Equation (22):

$$cos\chi = -\frac{E_p + \left(E_p - \hbar\omega\right)}{2\sqrt{E_p\left(E_p - \hbar\omega\right)}} + 2\frac{(\hbar\omega)^2}{vv\prime q^2} \approx 0.98, \Rightarrow \chi \approx 2,94, \tag{60}$$

where it was taken $\frac{\hbar\omega}{vq} \approx 0.1$, and $\frac{\hbar\omega}{E_p} \approx 0.05$. Taking into account the identity of incident electrons with atomic electrons increases the contribution Ξ_{exc0} in Equation (57) by 3–15 times dependently from the incident electron energy and does not change the singularity angle (see Figure 2b). True, it should be noted that a special contribution of exchange effects occurs at low energies, of the order of E_p ~2 (a.u.) atomic energy units, which is natural, since the de Broglie wavelength of an incident electron becomes comparable to the wavelength of an atomic electron. Under such conditions, exchange effects become dominant. Both exchange eigenvalues Ξ_{exc1}, Ξ_{exc2}, proportional to cosχ, give the same contribution as shown in Figure 2c. Naturally, it makes the main contribution, only at low energies of incident electrons, less than 3 eV. The third exchange term also proportional to cosχ makes the main exchange contribution in this effect. It is shown in Figure 2d. Then, the total angle factor in the scattering cross-section has the form shown in Figure 2e. Figure 2f,h show the shape of the peak for the incident electron energy ~2 a.u. for the intrinsic contribution J_0 (41), and its increasing exchange effects and for the total factor in the cross-section taking into account the exchange terms. It can be argued that the exchange effects do not change the shape and angular position of this peak. The peak size depends on the energy of the incident electrons due to the exchange effects associated with the ratio of the de Broglie wavelength to the size of the atom, as mentioned above. Figure 3 shows all contributions to the angle factor of the cross-section for the singlet state of the helium atom. Here, the intrinsic contribution J_0 (41) is the same, but the exchange coefficient differs from that shown in Figure 2b and is determined by the exchange effects from the overlap of the wave function of freely falling electrons with the singlet state of the helium atom. The exchange factor of the singlet state is shown in Figure 3a, and it is 1.5 times less then the triplet's exchange factor. The eigent exchange contribution Ξ_{exc1}, proportional to $\cos\chi$, presented by Equation (44), is shown in Figure 3b, and it has the same shape as the Ξ_{exc3} of the triplet state but is five times less compared with the last. The total angle factor to the cross-section, including the exchange effects from the singlet state, is shown in Figure 3c.

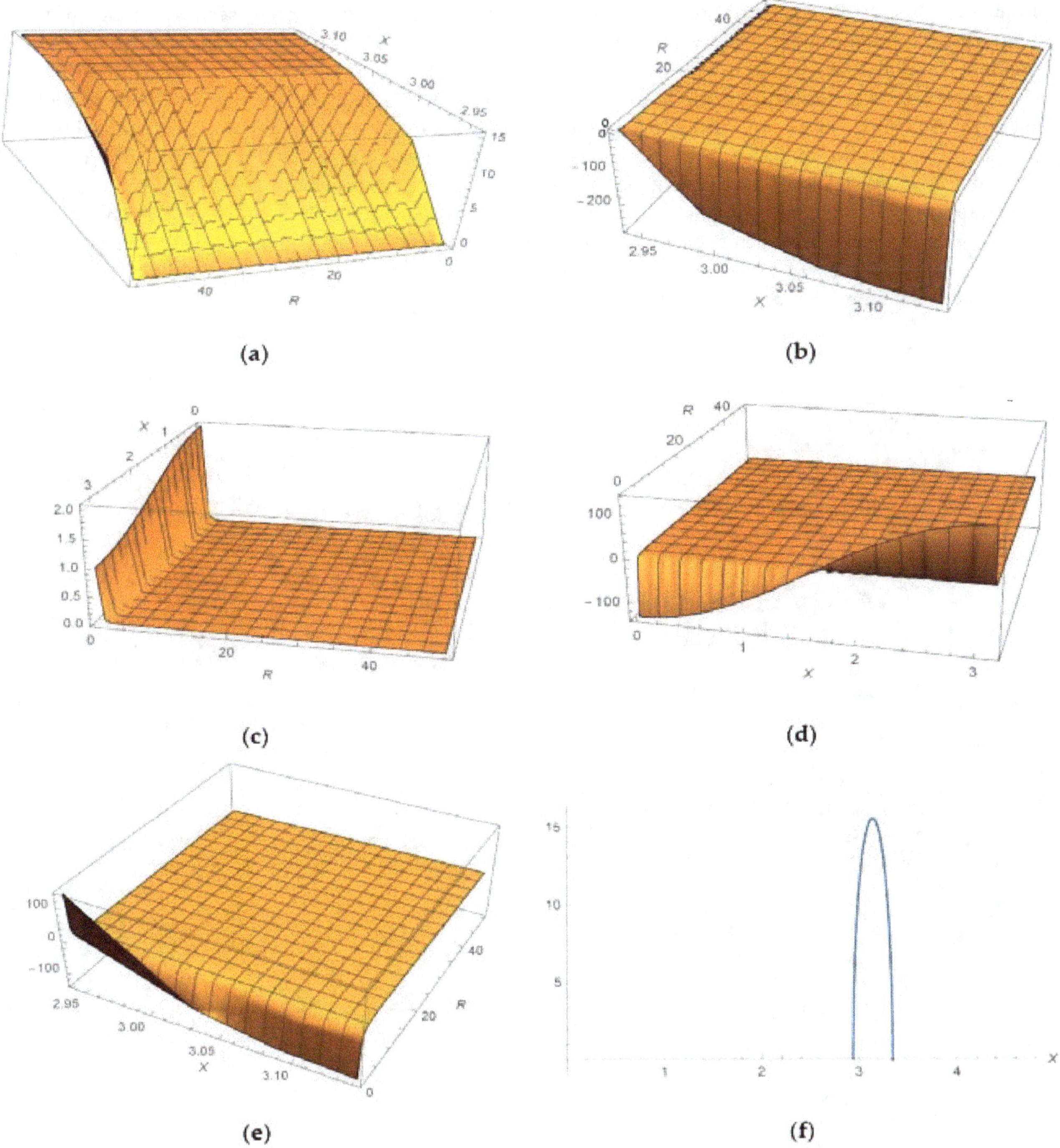

Figure 2. *Cont.*

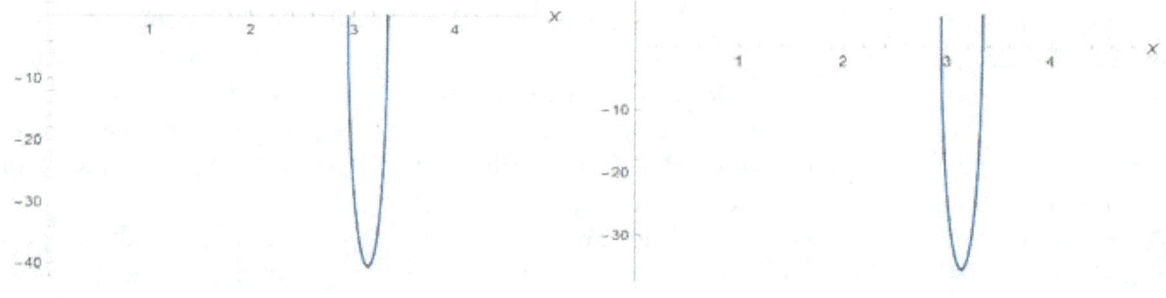

(g) (h)

Figure 2. This is a figure of the contributions to the angular factor in the cross-section for electron scattering by a helium atom in the triplet state. Here, one axis corresponds to the energy of the incident electron, R, measured from the energy of the bound state in the atomic units (a.u.). The second axis corresponds to scattering angle X, measured in radian (rad.). The third axis corresponds to the measureless factor. (**a**) A contribution J_0 from the usual weak localization of a new type, determined by Equation (42), to the factor in the cross-section; (**b**) corresponds to the term Ξ_{exc0}, presented by Equation (58); (**c**) Both exchange eigenvalues Ξ_{exc1}, Ξ_{exc2}, proportional to $\cos\chi$, give the same contribution; (**d**) The third "exchange-exchange" term Ξ_{exc3} is also proportional to $\cos\chi$. (**e**) The total angle factor in the scattering cross-section; (**f**) The peak for the incident electron energy ~2 a.u. for the intrinsic contribution J_0 (42); (**g**) The increase of the peak by exchange effects; (**h**) The total factor in the cross-section taking into account the exchange terms.

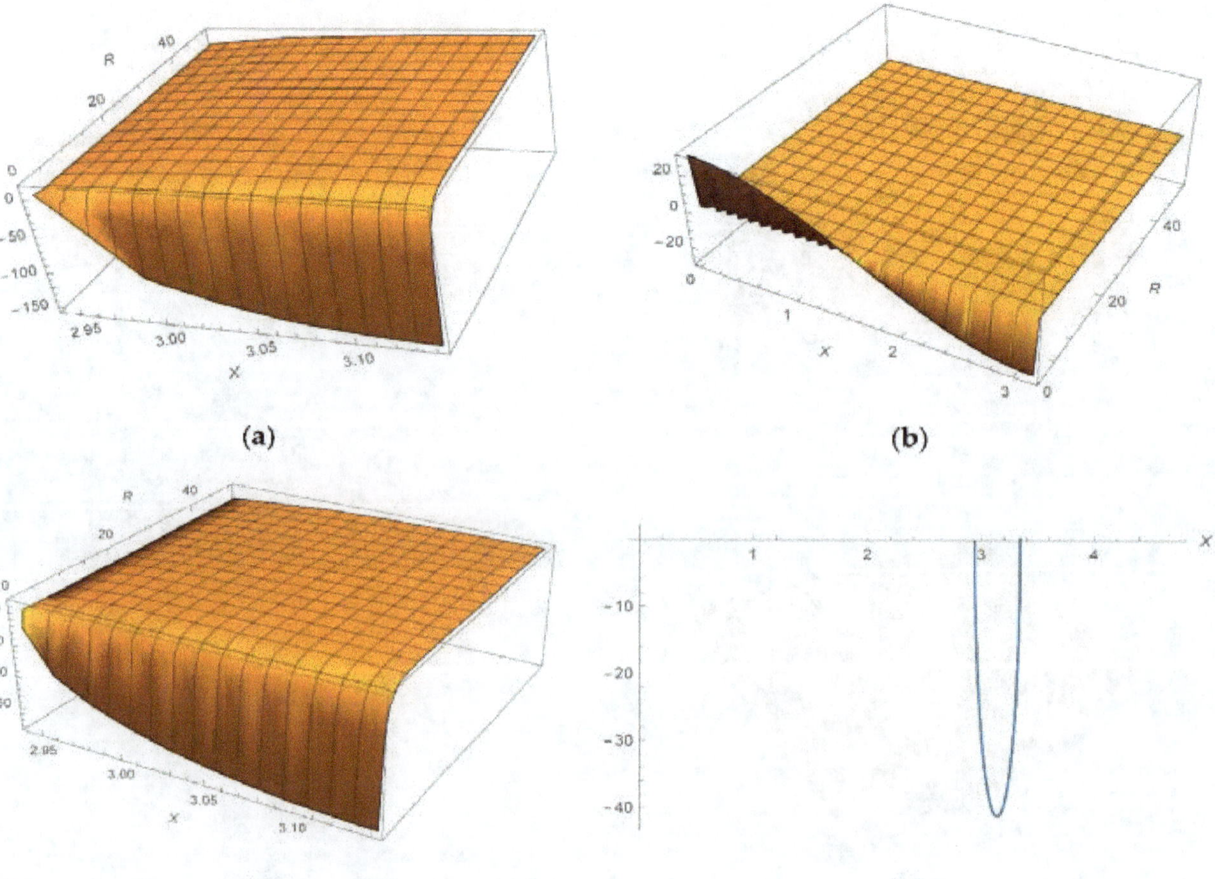

(a) (b)

(c) (d)

Figure 3. *Cont.*

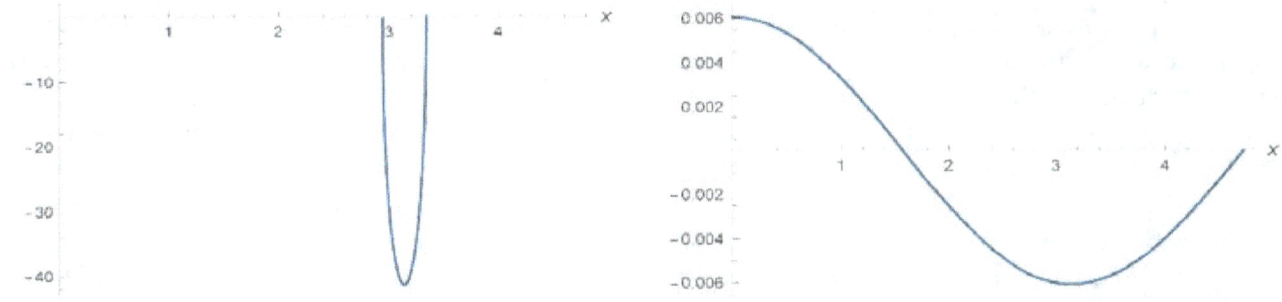

(e) (f)

Figure 3. This is a figure of the contributions to the angular factor in the cross-section for electron scattering by a helium atom in the singlet state. Here, one axis corresponds to the energy of the incident electron, R, measured from the energy of the bound state (in the atomic units (a.u.)) and the second axis corresponds to scattering angle X, measured in radians (rad.). The third axis corresponds to the measureless factor. (**a**) A contribution Ξ_{0exc} from the usual weak localization of a new type, determined by Equation (44), to the factor in the cross-section; (**b**) corresponds to the eigent exchange contribution Ξ_{exc1}, proportional to $\cos\chi$ and presented by the Equation (44); (**c**) The total angle factor in the scattering cross-section; (**d**). The increase of the peak by exchange effects peak for the incident electron energy ~2 a.u. (42); (**e**) The total factor in the cross-section taking into account the exchange terms; (**f**) The exchange contribution Ξ_{exc1}, proportional to $\cos\chi$.

4. Discussion

Investigation of the influence of exchange effects on the process of coherent backscattering of electrons by disordered media is a very important task both for weak localization of the Anderson type and for a new type of weak localization with an inelastic process. For a new type of weak localization, this problem stood and was solved theoretically only for the case of hydrogen-like scatterers in metal alloys [26], where an additional angular dependence was mentioned, proportional to $\cos\chi$ in the cross-section. It is now clear that this new dependence is not clearly visible, in contrast to the angular dependence of the new weak localization, explained by the kinematic consideration. The latter has a sharper peak and remains invariable even in the range of beam energies, where exchange effects are strong. In this work, we investigated the exchange effects from the first principle consideration for general case. We used the formulas for the S-matrix and scattering cross-section developed generally [27] while taking into account the exchange contributions. We developed the Green function (13), taking into account the wave function overlapping due to permutation symmetry. After analyzing all exchange contributions in the example of helium-like scatterers in metallic alloys, we come to the conclusion that exchange effects do not change the angle dependence in the new weak localization but could change the size of the peak very dramatically. A detailed comparison of the influence of exchange effects on the new weak localization for the singlet and triplet states of the helium atom shows exactly that the role of exchange effects in the case of a singlet is negligible. While, for the triplet state, it is decisive, especially for those values of the energy of incident electrons when de Broglie's waves are commensurate with the atomic.

5. Conclusions

In our work, we focused on two aspects of the theoretical description of non-Anderson weak localization of electrons in disordered media. First, we made an attempt to create a general method for taking into account the exchange of electrons between the incident beam and the belonging of electrons to atomic scatterers, using the formalism of the invariant exchange perturbation theory. We applied the general expressions for the scattering amplitude and scattering probability with the allowance for permutation symmetry obtained in [27,28] to the problem of the process of electron scattering by disordered media under new weak localization. The result of this development of the theory is expressed by Equations (27) and (28). Second, we applied the developed general equations to a specific system: a scattering electron beam on disordered helium-like scatterers in Zi-Ni alloys doped with helium. We performed a detailed analytical calculation of the angle factor in the scattering cross-section for this system. We showed that the exchange contribution in the angle dependence of the electrons' cross-section under the "new weak localization" condition appears in two apostasies, in the additional angle dependence, proportional to cosχ, and in the exchange factor to the "kinematic" angle dependence. In our opinion, such consideration would also be useful for the so-called anti-Anderson localization considered experimentally and theoretically in the works [33–36] for the different disordered systems.

Author Contributions: Conceptualization, E.V.O.; methodology, E.V.O.; software, E.V.O. and F.E.O.; validation, E.V.O. and F.E.O.; formal analysis, E.V.O. and F.E.O.; investigation, E.V.O. and F.E.O.; resources, F.E.O.; data curation, F.E.O.; writing—original draft preparation, E.V.O.; writing—review and editing, E.V.O. and F.E.O.; visualization, F.E.O. All authors have read and agreed to the published version of the manuscript.

Acknowledgments: The authors are grateful to Alexey I. Borovkov, Vice-Rector for Innovative Projects at Peter the Great St. Petersburg Polytechnic University, for financial support of the presented publication.

Appendix A. Mathematical Additions

Appendix A.1. The Proof of Completeness Property of Antisymmetric Basis

It is easy to show this property by the acting operator (6) on to any antisymmetric vector $|m,p\prime\rangle$ and by using Equation (5):

$$\begin{gathered}\left(\sum_{n,p}|n,p\rangle\tfrac{f_{np}}{N}(n,p|\right)|m,p\prime\rangle=\sum_{n,p}|n,p\rangle\tfrac{f_{np}}{N}(n,p|m,p\prime\rangle=\sum_{\nu}\sum_{n,p}(-1)^{g_\nu}\frac{1}{\boxed{f_{np}}}\frac{\boxed{f_{np}}}{N}|n,p)_\nu(n,p|m,p\prime\rangle=\\\sum_{\nu}\sum_{n,p}\boxed{(-1)^{g_\nu}}\frac{1}{\boxed{f_{np}}}\frac{\boxed{f_{np}}}{N}|n,p)_\nu(n,p|_\nu m,p\prime\rangle\boxed{(-1)^{g_\nu}}=\sum_{\nu}\tfrac{1}{N}\sum_{n,p}|n,p)_\nu(n,p|_\nu m,p\prime\rangle=\tfrac{1}{N}\sum_{\nu}|m,p\prime\rangle=|m,p\prime\rangle\end{gathered}\tag{A1}$$

Appendix A.2. Young Operators

$$\begin{aligned}\omega_{11}^{[21]}&=\tfrac{1}{\sqrt{12}}(2+2P_{12}-P_{23}-P_{13}-P_{123}-P_{132}),\\\omega_{12}^{[21]}&=\tfrac{1}{2}(P_{23}-P_{13}+P_{123}-P_{132}),\\\omega_{21}^{[21]}&=\tfrac{1}{2}(P_{23}-P_{13}-P_{123}+P_{132})\\\omega_{22}^{[21]}&=\tfrac{1}{\sqrt{12}}(2-2P_{12}+P_{23}+P_{13}-P_{123}-P_{132}).\end{aligned}\tag{A2}$$

Appendix A.3. The Explicit Form of the Cross-Section Pre-Factor $\Xi(\chi)$ for Expression (52)

Here, we took into account the obvious relations:

$$\begin{aligned}&E_p-E_{\boldsymbol{p}-\boldsymbol{Q}+\boldsymbol{q}}=\tfrac{1}{2m}\left(p^2-p\prime^2-2p\prime q-q^2\right)=\hbar\omega-v\prime q,\\&E_p-E_{\boldsymbol{p}-\boldsymbol{q}}=\tfrac{1}{2m}\left(p^2-p^2+2p\cdot q-q^2-\hbar\omega\right)=v\cdot q-\hbar\omega,\\&\left(2|p-q|/\hbar\alpha\right)^2=\left(\tfrac{2}{\hbar\alpha}\right)^2\tfrac{2m}{2m}\left(p^2-2p\cdot q+q^2\right)=4\tfrac{2m}{(\hbar\alpha)^2}\left(E_p-v\cdot q+\boxed{q^2}\right),\\&\left(2|p\prime+q|/\hbar\alpha\right)^2=\left(\tfrac{2}{\hbar\alpha}\right)^2\tfrac{2m}{2m}\left(p\prime^2+2p\prime\cdot q+q^2\right)=4\tfrac{2m}{(\hbar\alpha)^2}\left(E_p-\hbar\omega+v\prime\cdot q+\boxed{q^2}\right),\end{aligned}\tag{A3}$$

$$\begin{aligned}
&\Xi_{exc} = \int d\Omega_Q \Big\{ \frac{1}{\hbar\omega - v'q + i\gamma} \cdot \frac{1}{v\cdot q - \hbar\omega - i\gamma} \left[\frac{1}{8\left(1+(2|p-q|/\hbar\alpha)^2\right)^2} - \frac{24}{\left(1+(2|p-q|/\hbar\alpha)^2\right)^3} + \frac{24}{\left(1+(2|p-q|/\hbar\alpha)^2\right)^4} \right] + \\
&\frac{1}{v\cdot q - \hbar\omega - i\gamma} \cdot \frac{1}{\hbar\omega - v'q + i\gamma} \left[\frac{1}{8\left(1+(2|p'+q|/\hbar\alpha)^2\right)^2} - \frac{24}{\left(1+(2|p'+q|/\hbar\alpha)^2\right)^3} + \frac{24}{\left(1+(2|p'+q|/\hbar\alpha)^2\right)^4} \right] + \\
&\frac{1}{v\cdot q - \hbar\omega + i\gamma} \cdot \frac{1}{\hbar\omega - v'q - i\gamma} \left[\frac{1}{8\left(1+4\frac{2m}{(\hbar\alpha)^2}(E_p - \hbar\omega + v'\cdot q)\right)^2} - \frac{24}{\left(1+4\frac{2m}{(\hbar\alpha)^2}(E_p - \hbar\omega + v'\cdot q)\right)^3} + \frac{24}{\left(1+4\frac{2m}{(\hbar\alpha)^2}(E_p - \hbar\omega + v'\cdot q)\right)^4} \right] + \\
&\frac{1}{\hbar\omega - v'q - i\gamma}\, \frac{1}{v\cdot q - \hbar\omega + i\gamma} \left[\frac{1}{8\left(1+4\frac{2m}{(\hbar\alpha)^2}\left(E_p - v\cdot q + \boxed{q^2}\right)\right)^2} - \frac{24}{\left(1+4\frac{2m}{(\hbar\alpha)^2}(E_p - v\cdot q)\right)^3} + \frac{24}{\left(1+4\frac{2m}{(\hbar\alpha)^2}(E_p - v\cdot q)\right)^4} \right] - \\
&\frac{1}{\hbar\omega - v'q + i\gamma} \left[\frac{1}{8\left(1+(2|p'+q|/\hbar\alpha)^2\right)^2} - \frac{24}{\left(1+(2|p'+q|/\hbar\alpha)^2\right)^3} + \frac{24}{\left(1+(2|p'+q|/\hbar\alpha)^2\right)^4} \right] \times \\
&\frac{1}{v\cdot q - \hbar\omega - i\gamma} \left[\frac{1}{8\left(1+(2|p-q|/\hbar\alpha)^2\right)^2} - \frac{24}{\left(1+(2|p-q|/\hbar\alpha)^2\right)^3} + \frac{24}{\left(1+(2|p-q|/\hbar\alpha)^2\right)^4} \right] - \\
&\frac{1}{v\cdot q - \hbar\omega + i\gamma} \left[\frac{1}{8\left(1+4\frac{2m}{(\hbar\alpha)^2}\left(E_p - v\cdot q + \boxed{q^2}\right)\right)^2} - \frac{24}{\left(1+4\frac{2m}{(\hbar\alpha)^2}(E_p - v\cdot q)\right)^3} + \frac{24}{\left(1+4\frac{2m}{(\hbar\alpha)^2}(E_p - v\cdot q)\right)^4} \right] \times \\
&\frac{1}{\hbar\omega - v'q - i\gamma} \left[\frac{1}{8\left(1+4\frac{2m}{(\hbar\alpha)^2}(E_p - \hbar\omega + v'\cdot q)\right)^2} - \frac{24}{\left(1+4\frac{2m}{(\hbar\alpha)^2}(E_p - \hbar\omega + v'\cdot q)\right)^3} + \frac{24}{\left(1+4\frac{2m}{(\hbar\alpha)^2}(E_p - \hbar\omega + v'\cdot q)\right)^4} \right] \Big\},
\end{aligned} \tag{A4}$$

In this expression, we neglected by the second order small terms proportional to q^2, while inelastic momentum loss is $q << p$.

Appendix A.4. The Integrals Used to Derive the Expression for the Factor $\Xi(\chi)$

Here are the integrals used to derive the expression for the factor $\Xi(\chi)$ ((39) and (57)), which contains the angular dependence introduced into Expression (28) for the cross-section:

$$\begin{aligned}
&J_0 = \Xi_0 = \int d\Omega_Q \{ G_0^{ex*}(p'+q) G_0^{ex}(p-q) + G_0^{ex*}(p-q) G_0^{ex}(p'+q)^2 \} = \\
&\frac{4\pi\hbar\omega^2 p}{qv\gamma} \cdot \left[arctg\left(\frac{qv+\hbar\omega_p}{\hbar\gamma}\right) + arctg\left(\frac{qv-\hbar\omega_p}{\hbar\gamma}\right) - \frac{\gamma}{\sqrt{2\omega_p^2(1-cos\chi) - \frac{q^2v^2}{\hbar^2} sin^2\chi}} \right] \times \\
&ln\left(\frac{\hbar^2\omega_p^2 - q^2v^2 cos\chi + qv\sqrt{2\hbar^2\omega_p^2(1-cos\chi) - q^2v^2 sin^2\chi}}{\hbar^2\omega_p^2 - q^2v^2 cos\chi - qv\sqrt{2\hbar^2\omega_p^2(1-cos\chi) - q^2v^2 sin^2\chi}} \right).
\end{aligned} \tag{A5}$$

$$\begin{aligned}
&J_1 = \int d\Omega_q \frac{1}{v\cdot q - \hbar\omega + i\gamma} \frac{1}{\left(\left(\frac{(\hbar\alpha)^2}{8m} + E_p - \hbar\omega\right) + v'\cdot q\right)} = \\
&\frac{1}{\left(\frac{(\hbar\alpha)^2}{8m} + E_p\right)} \left\{ \left(1 + \frac{\hbar\omega}{\left(\frac{(\hbar\alpha)^2}{8m} + E_p\right)} \right) \frac{2\pi}{vq} ln\left| \frac{v\cdot q - \hbar\omega - i\gamma}{v\cdot q + \hbar\omega + i\gamma} \right| - \frac{2\pi}{vq} \frac{v'q}{\left(\frac{(\hbar\alpha)^2}{8m} + E_p\right)} cos\chi \left(2 + z ln\left| \frac{v\cdot q - \hbar\omega - i\gamma}{v\cdot q + \hbar\omega + i\gamma} \right| \right) \right\}
\end{aligned} \tag{A6}$$

$$\begin{aligned}
&J_2 = \int d\Omega_{Q'} \frac{1}{\hbar\omega - v'q + i\gamma} \frac{1}{\left(\left(\frac{(\hbar\alpha)^2}{8m} + E_p\right) - v\cdot q\right)} = \frac{-1}{\left(\frac{(\hbar\alpha)^2}{8m} + E_p\right)} \left\{ \frac{2\pi}{v'q} ln\left| \frac{v'\cdot q - \hbar\omega - i\gamma}{v'\cdot q + \hbar\omega + i\gamma} \right| + \right. \\
&\left. \frac{vq}{v'q} cos\chi \frac{2\pi}{\left(\frac{(\hbar\alpha)^2}{8m} + E_p\right)} \left(2 + z' ln\left| \frac{v'\cdot q - \hbar\omega - i\gamma}{v'\cdot q + \hbar\omega + i\gamma} \right| \right) \right\}, \\
&z' = \frac{\hbar\omega + i\gamma}{v'\cdot q}.
\end{aligned} \tag{A7}$$

$$J_3 = \int d\Omega_{Q'} \frac{1}{\left(\left(\frac{(\hbar\alpha)^2}{8m}+E_p\right)-v\cdot q\right)} \frac{1}{\left(\left(\frac{(\hbar\alpha)^2}{8m}+E_p-\hbar\omega\right)+v'\cdot q\right)} = \frac{4\pi\cdot\cos\chi}{\left(\frac{(\hbar\alpha)^2}{8m}+E_p\right)^2}\left[\left(1+\frac{\hbar\omega}{\left(\frac{(\hbar\alpha)^2}{8m}+E_p\right)}\right)-\frac{1}{3}\frac{(vq)(v'q)}{\left(\frac{(\hbar\alpha)^2}{8m}+E_p\right)^2}\right] \tag{A8}$$

Here, the c.c. complex conjugate expression is denoted:

$$\begin{aligned}\Xi_{exc\_0} = &\left\{\frac{1}{2^6}\left(\frac{(\hbar\alpha)^2}{2m}\right)^2\left[\frac{1}{(\mathbb{R}-\hbar\omega)^2}+\frac{1}{(\mathbb{R})^2}\right]-12\left(\frac{(\hbar\alpha)^2}{2m}\right)^3\left[\frac{1}{(\mathbb{R}-\hbar\omega)^3}+\frac{1}{(\mathbb{R})^3}\right]+\right.\\ &3\left(\frac{(\hbar\alpha)^2}{2m}\right)^4\left[\frac{1}{(\mathbb{R}-\hbar\omega)^4}+\frac{1}{(\mathbb{R})^4}\right]-\frac{1}{2^{14}}\left(\frac{(\hbar\alpha)^2}{2m}\right)^4\frac{1}{(\mathbb{R}-\hbar\omega)^2(\mathbb{R})^2}+\\ &\frac{3}{2^{10}}\left(\frac{(\hbar\alpha)^2}{2m}\right)^5\left(\frac{1}{(\mathbb{R}-\hbar\omega)^2(\mathbb{R})^3}+\frac{1}{(\mathbb{R}-\hbar\omega)^3(\mathbb{R})^2}\right)-\frac{9\cdot}{2^6}\left(\frac{(\hbar\alpha)^2}{2m}\right)^6\frac{1}{(\mathbb{R}-\hbar\omega)^3(\mathbb{R})^3}-\\ &\frac{3}{2^{13}}\left(\frac{(\hbar\alpha)^2}{2m}\right)^6\left(\frac{1}{(\mathbb{R}-\hbar\omega)^2(\mathbb{R})^4}+\frac{1}{(\mathbb{R}-\hbar\omega)^4(\mathbb{R})^2}\right)+\\ &\left.\frac{9}{2^9}\left(\frac{(\hbar\alpha)^2}{2m}\right)^7\left(\frac{1}{(\mathbb{R}-\hbar\omega)^4(\mathbb{R})^3}+\frac{1}{(\mathbb{R}-\hbar\omega)^3(\mathbb{R})^4}\right)-\frac{9}{2^{12}}\left(\frac{(\hbar\alpha)^2}{2m}\right)^8\frac{1}{(\mathbb{R}-\hbar\omega)^4(\mathbb{R})^4}\right\}J_0.\end{aligned} \tag{A9}$$

$$\begin{aligned}\Xi_{exc1} = &-\left(\frac{(\hbar\alpha)^2}{8m}\right)^2\times\left\{\frac{1}{8}\frac{\partial}{\partial\mathbb{R}'}+12\frac{(\hbar\alpha)^2}{8m}\frac{\partial^2}{\partial\mathbb{R}'^2}+4\left(\frac{(\hbar\alpha)^2}{8m}\right)^2\frac{\partial^3}{\partial\mathbb{R}'^3}\right\}\left[\frac{1}{\mathbb{R}'+\hbar\omega+i\gamma}+c.c.\right]J_1-\\ &\left(\frac{(\hbar\alpha)^2}{8m}\right)^4\left[\frac{1}{8}\frac{\partial}{\partial\mathbb{R}'}+12\left(\frac{(\hbar\alpha)^2}{8m}\right)\frac{\partial^2}{\partial\mathbb{R}'^2}+4\left(\frac{(\hbar\alpha)^2}{8m}\right)^2\frac{\partial^3}{\partial\mathbb{R}'^3}\right]\times\\ &\left[\frac{1}{8}\frac{\partial}{\partial\mathbb{R}}+12\left(\frac{(\hbar\alpha)^2}{8m}\right)\frac{\partial^2}{\partial\mathbb{R}^2}+4\left(\frac{(\hbar\alpha)^2}{8m}\right)^2\frac{\partial^3}{\partial\mathbb{R}^3}\right]\frac{1}{\mathbb{R}-i\gamma}\frac{1}{\mathbb{R}'-i\gamma}J_1 = \\ &\frac{1}{8}\left(\frac{(\hbar\alpha)^2}{2m}\right)^2\left\{1+\frac{1}{2^5}\left[\frac{1}{8}\frac{\left(\frac{(\hbar\alpha)^2}{2m}\right)^2}{(\mathbb{R}-\hbar\omega)^2}-6\frac{\left(\frac{(\hbar\alpha)^2}{2m}\right)^3}{(\mathbb{R}-\hbar\omega)^3}+\frac{3}{4}\frac{\left(\frac{(\hbar\alpha)^2}{2m}\right)^4}{(\mathbb{R}-\hbar\omega)^4}\right]\right\}\times\\ &\left\{\frac{\pi}{vq}\frac{1}{\mathbb{R}^3}ln\left(\frac{(vq-\hbar\omega_p)^2+(\hbar\gamma)^2}{(vq+\hbar\omega_p)^2+(\hbar\gamma)^2}\right)\left[\frac{1}{8}\left(2+\frac{3\hbar\omega}{\mathbb{R}}\right)-3\frac{(\hbar\alpha)^2}{2m}\left(\frac{6}{\mathbb{R}}+\frac{12\hbar\omega}{\mathbb{R}^2}\right)+\left(\frac{(\hbar\alpha)^2}{2m}\right)^2\left(\frac{6}{\mathbb{R}^2}+\frac{15\hbar\omega}{\mathbb{R}^3}\right)\right]+\right.\\ &\left.\frac{3\pi}{\mathbb{R}^4}\frac{v'}{v}cos\chi\left(2+\frac{\hbar\omega}{2v\cdot q}ln\left(\frac{(vq-\hbar\omega_p)^2+(\hbar\gamma)^2}{(vq+\hbar\omega_p)^2+(\hbar\gamma)^2}\right)\right)\left[\frac{1}{4}-24\frac{\left(\frac{(\hbar\alpha)^2}{2m}\right)}{\mathbb{R}}+10\frac{\left(\frac{(\hbar\alpha)^2}{2m}\right)^2}{\mathbb{R}^2}\right]\right\}.\end{aligned} \tag{A10}$$

$$\begin{aligned}\Xi_{exc2} = &\frac{1}{8}\left(\frac{(\hbar\alpha)^2}{2m}\right)^2\left\{1+\frac{1}{2^5}\left(\frac{(\hbar\alpha)^2}{2m}\right)^2\left[\frac{1}{8}\frac{1}{(\mathbb{R}-\hbar\omega)^2}-6\left(\frac{(\hbar\alpha)^2}{2m}\right)\frac{1}{(\mathbb{R}-\hbar\omega)^3}+\frac{3}{4}\left(\frac{(\hbar\alpha)^2}{2m}\right)^2\frac{1}{(\mathbb{R}-\hbar\omega)^4}\right]\right\}\times\\ &\left\{\frac{\pi}{v'q}ln\left(\frac{(v'q-\hbar\omega_p)^2+(\hbar\gamma)^2}{(v'q+\hbar\omega_p)^2+(\hbar\gamma)^2}\right)\left[\frac{1}{8}\left(\frac{2}{\mathbb{R}^3}\right)-3\frac{(\hbar\alpha)^2}{2m}\left(\frac{6}{\mathbb{R}^4}\right)+\left(\frac{(\hbar\alpha)^2}{2m}\right)^2\left(\frac{6}{\mathbb{R}^5}\right)\right]+\right.\\ &\left\{\frac{\pi}{v'q}ln\left(\frac{(v'q-\hbar\omega_p)^2+(\hbar\gamma)^2}{(v'q+\hbar\omega_p)^2+(\hbar\gamma)^2}\right)\left[\frac{1}{8}\left(\frac{2}{\mathbb{R}^3}\right)-3\frac{(\hbar\alpha)^2}{2m}\left(\frac{6}{\mathbb{R}^4}\right)+\left(\frac{(\hbar\alpha)^2}{2m}\right)^2\left(\frac{6}{\mathbb{R}^5}\right)\right]+\right.\\ &\left.\frac{v}{v'}cos\chi\left(2+\frac{\hbar\omega+i\gamma}{2v\cdot q}ln\left(\frac{(v'q-\hbar\omega_p)^2+(\hbar\gamma)^2}{(v'q+\hbar\omega_p)^2+(\hbar\gamma)^2}\right)\right)\left[+\frac{3\pi}{4\mathbb{R}^4}-\left(\frac{(\hbar\alpha)^2}{2m}\right)\frac{72\pi}{\mathbb{R}^5}+\left(\frac{(\hbar\alpha)^2}{2m}\right)^2\frac{30\pi}{\mathbb{R}^6}\right]\right\}\end{aligned} \tag{A11}$$

$$
\begin{aligned}
&\Xi_{exc3} = -\left(\frac{(\hbar\alpha)^2}{8m}\right)^4\left[\frac{1}{8}\frac{\partial}{\partial\mathbb{R}'} + 12\left(\frac{(\hbar\alpha)^2}{8m}\right)\frac{\partial^2}{\partial\mathbb{R}'^2} + 4\left(\frac{(\hbar\alpha)^2}{8m}\right)^2\frac{\partial^3}{\partial\mathbb{R}'^3}\right]\times \\
&\left[\frac{1}{8}\frac{\partial}{\partial\mathbb{R}} + 12\left(\frac{(\hbar\alpha)^2}{8m}\right)\frac{\partial^2}{\partial\mathbb{R}^2} + 4\left(\frac{(\hbar\alpha)^2}{8m}\right)^2\frac{\partial^3}{\partial\mathbb{R}^3}\right]\frac{1}{\mathbb{R}-i\gamma}\,\frac{1}{\mathbb{R}'-i\gamma}J_3 = \\
&-\left(\frac{(\hbar\alpha)^2}{8m}\right)^4\left[\frac{1}{8}\frac{\partial}{\partial\mathbb{R}'} + 12\left(\frac{(\hbar\alpha)^2}{8m}\right)\frac{\partial^2}{\partial\mathbb{R}'^2} + 4\left(\frac{(\hbar\alpha)^2}{8m}\right)^2\frac{\partial^3}{\partial\mathbb{R}'^3}\right]\times \\
&\left[\frac{1}{8}\frac{\partial}{\partial\mathbb{R}} + 12\left(\frac{(\hbar\alpha)^2}{8m}\right)\frac{\partial^2}{\partial\mathbb{R}^2} + 4\left(\frac{(\hbar\alpha)^2}{8m}\right)^2\frac{\partial^3}{\partial\mathbb{R}^3}\right]\frac{1}{\mathbb{R}-i\gamma}\,\frac{1}{\mathbb{R}'-i\gamma}\,\frac{4\pi\cdot cos\chi}{\mathbb{R}^2}\left[\left(1+\frac{\hbar\omega}{\mathbb{R}}\right)-\frac{1}{3}\frac{(vq)(v'q)}{\mathbb{R}^2}\right] = \\
&-\pi\cdot cos\chi\frac{1}{(\mathbb{R}-\hbar\omega)^2}\frac{\left(\frac{(\hbar\alpha)^2}{2m}\right)^4}{(4\mathbb{R})^4}\left[-\frac{1}{2}+24\frac{\left(\frac{(\hbar\alpha)^2}{2m}\right)}{(\mathbb{R}-\hbar\omega)}+3\frac{\left(\frac{(\hbar\alpha)^2}{2m}\right)^2}{(\mathbb{R}-\hbar\omega)^2}\right]\times \\
&\left\{\left[-\frac{1}{8}\left(3+\frac{4\hbar\omega}{\mathbb{R}}\right)+\frac{5}{3}\frac{(vq)(v'q)}{\mathbb{R}^2}\right]+6\frac{\left(\frac{(\hbar\alpha)^2}{2m}\right)}{\mathbb{R}}\left[\left(6+\frac{10\hbar\omega}{\mathbb{R}}\right)-5\frac{(vq)(v'q)}{\mathbb{R}^2}\right]+\right. \\
&\left.\frac{5}{2}\frac{\left(\frac{(\hbar\alpha)^2}{2m}\right)^2}{\mathbb{R}^2}\left[-6\left(1+\frac{2\hbar\omega}{\mathbb{R}}\right)+7\frac{(vq)(v'q)}{\mathbb{R}^2}\right]\right\}
\end{aligned}
\tag{A12}
$$

References

1. Pollak, M.; Ortuño, M.; Frydman, A. Disordered electronic systems. In *The Electron Glass*; Cambridge University Press: Cambridge, UK, 2013; pp. 7–39. [CrossRef]
2. Bergmann, G. Weak localization in thin films. A time-of-light experiment with conduction electrons. *Phys.Rep.* **1984**, *107*, 1–58.
3. Rammer, J. Quantum transport theory of electrons in solids: A single-particle approach. *Rev. Mod. Phys.* **1991**, *63*, 781–817. [CrossRef]
4. Kramer, B.; MacKinnon, A. Localization: Theory and experiment. *Rep. Prog. Phys.* **1993**, *56*, 1469.
5. Aronov, A. Fourteen years of quantum interference in disordered metals. *Phys. Scr.* **1993**, *1993*, 28–33. [CrossRef]
6. Van Der Mark, M.B.; Van Albada, M.P.; Lagendijk, A. Light scattering in strongly scattering media: Multiple scattering and weak localization. *Phys. Rev. B* **1988**, *37*, 3575–3592. [CrossRef]
7. Kravzov, Y.A. Propagation of electromagnetic waves through a turbulent atmosphere. *Rep. Prog. Phys.* **1992**, *55*, 39–51.
8. Kaveh, M. Electron and optical wave phenomena: Some new statistical properties. *Physica B* **1991**, *17*, 1–5.
9. Sheng, P. *Scattering and Localization of Classical Waves in Random Media*; World Scientific Pub. Co. Pte. Lt.: Singapore, 1990; p. 648.
10. Bergmann, G. Physical interpretation of weak localization: A time-of-flight experiment with conduction electrons. *Phys. Rev. B* **1983**, *28*, 2914–2920. [CrossRef]
11. Berkovits, R.; Kaveh, M. Back scattering of high-energy electrons from disordered media: Antienhancement due to spin-orbit interaction. *Phys. Rev. B* **1988**, *37*, 584–587.
12. Berkovits, R.; Eliyahu, D.; Kaveh, M. Enhanced backscattering of electrons in a magnetic field. *Phys. Rev. B* **1990**, *41*, 407–412. [CrossRef]
13. Gorodnichev, E.E.; Dudarev, S.L.; Rogozkin, D.B. Polarization effects in coherent backscattering of particles from random media. *JETP* **1990**, *70*, 853–863.
14. Igarashi, J.-I. Coherent backscattering of neutrons. *Phys. Rev. B* **1987**, *35*, 8894–8897. [CrossRef]
15. Washburn, S.; Webb, R.A. Aharonov-Bohm effect in normal metal. Quatnum coherence and transport. *Adv. Phys.* **1986**, *35*, 375–422.
16. Kouwenhoven, L.P.; Harmans, C.J.P.M.; Timmering, C.E.; Hekking, F.W.J.; Van Wees, B.J.; Foxon, C.T. Transport through a finite one-dimensional crystal. *Phys. Rev. Lett.* **1990**, *65*, 361–364. [CrossRef] [PubMed]
17. Maschke, K.; Schreiber, M. Unified description of coherent and dissipative electron transport. *Phys. Rev. B* **1991**, *44*, 3835–3841. [CrossRef]
18. Kanzieper, E.A.; Freilikher, V. Coherent phenomena in inelastic backscattering of electrons from disordered media. *Phys. Rev. B* **1995**, *51*, 2759–2769.

19. Tsang, L.; Ishimaru, A. Backscattering enhancement of random discrete scatterers. *J. Opt. Soc. Am. A* **1984**, *1*, 836–839.
20. Libenson, B.N.; Platonov, K.Y.; Rumyantsev, V.V. New type of weak localization of electrons in disordered media. *Sov. Phys. JETP* **1992**, *74*, 326–333.
21. Lebed', A.A.; Padusenko, E.A.; Roshchupkin, S.P.; Dubov, V.V. Resonant parametric interference effect in spontaneous bremsstrahlung of an electron in the field of a nucleus and two pulsed laser waves. *Phys. Rev. A* **2018**, *97*, 043404. [CrossRef]
22. Dubov, A.; Dubov, V.V.; Roshchupkin, S.P. Resonant high-energy bremsstrahlung of ultrarelativistic electrons in the field of a nucleus and a weak electromagnetic wave. *Laser Phys. Lett.* **2020**, *17*, 045301. [CrossRef]
23. Rumyantsev, V.V.; Doubov, V.V. Quantum transport of electrons scattered inelastically from disordered media. *Phys. Rev. B* **1994**, *49*, 8643–8654. [CrossRef] [PubMed]
24. Rumyantsev, V.; Orlenko, E.; Libenson, B. Nature of coherent enhancement of inelastic electron scattering from disordered media. *Eur. Phys. J. B* **1997**, *103*, 53–60. [CrossRef]
25. Rumyantsev, V.V.; Orlenko, E.V.; Libenson, B.N. On the theory of quantum interference between inelastic and elastic electron scattering events. *J. Exp. Theor. Phys.* **1997**, *84*, 552–559. [CrossRef]
26. Orlenko, E.V.; Rumyantsev, V.V. The effect of particle identity on a new type of weak localization. *J. Phys. Condens. Matter* **1995**, *7*, 3557–3564. [CrossRef]
27. Orlenko, E.V. Exchange Perturbation theory. In *Perturbation Theory: Advances in Research and Applications, Nova*; Science Publishers, Inc.: New York, NY, USA, 2018; pp. 1–59.
28. Orlenko, E.; Latychevskaia, T.; Evstafev, A.V.; Orlenko, F.E. Invariant time-dependent exchange perturbation theory and its application to the particles collision problem. *Theor. Chem. Acc.* **2015**, *134*. [CrossRef]
29. Davydov, A.S. *Quantum Mechanics*, 2nd ed.; Pergamon Press: Oxford, UK; New York, NY, USA, 1976; p. 680.
30. Libenson, B.N. Trinal nature of the excitation of bulk plasmons by a fast charged particle at grazing incidence on the interface between vacuum and a metal with strong spatial dispersion. *J. Exp. Theor. Phys.* **2018**, *127*, 37–47. [CrossRef]
31. Libenson, B.N. Effect of metallic surface microroughness on the probability and spectrum of plasmon excitation by a fast charged particle moving parallel to the surface. *J. Exp. Theor. Phys.* **2012**, *114*, 194–204. [CrossRef]
32. Dixon, R.N. Symmetry of many electron systems. *Phys. Bull.* **1975**, *26*, 546. [CrossRef]
33. Wang, H.; Liu, H.; Chang, C.-Z.; Zuo, H.; Zhao, Y.; Sun, Y.; Xia, Z.; He, K.; Ma, X.; Xie, X.C.; et al. Crossover between weak antilocalization and weak localization of bulk states in ultrathin Bi2Se3 films. *Sci. Rep.* **2014**, *4*, 5817. [CrossRef]
34. Lu, H.Z.; Shen, S.Q. Finite-Temperature Conductivity and Magnetoconductivity of Topological Insulators. *Phys. Rev. Lett.* **2014**, *112*, 146601. [CrossRef]
35. Hasan, M.Z.; Kane, C.L. Colloquium: Topological insulators. *Rev. Mod. Phys.* **2010**, *82*, 3045–3067. [CrossRef]
36. Gornyi, I.V.; Kachorovskii, V.Y.; Ostrovsky, P.M. Interference-induced magnetoresistance in HgTe quantum wells. *Phys. Rev. B* **2014**, *90*, 085401. [CrossRef]

3

Perturbation Theory near Degenerate Exceptional Points

Miloslav Znojil [1,2]

[1] The Czech Academy of Sciences, Nuclear Physics Institute, Hlavní 130, 25068 Řež, Czech Republic; znojil@ujf.cas.cz

[2] Department of Physics, Faculty of Science, University of Hradec Králové, Rokitanského 62, 50003 Hradec Králové, Czech Republic

Abstract: In an overall framework of quantum mechanics of unitary systems a rather sophisticated new version of perturbation theory is developed and described. The motivation of such an extension of the list of the currently available perturbation-approximation recipes was four-fold: (1) its need results from the quick growth of interest in quantum systems exhibiting parity-time symmetry ($\mathcal{PT}$-symmetry) and its generalizations; (2) in the context of physics, the necessity of a thorough update of perturbation theory became clear immediately after the identification of a class of quantum phase transitions with the non-Hermitian spectral degeneracies at the Kato's exceptional points (EP); (3) in the dedicated literature, the EPs are only being studied in the special scenarios characterized by the spectral geometric multiplicity L equal to one; (4) apparently, one of the decisive reasons may be seen in the complicated nature of mathematics behind the $L \geq 2$ constructions. In our present paper we show how to overcome the latter, purely technical obstacle. The temporarily forgotten class of the $L > 1$ models is shown accessible to a feasible perturbation-approximation analysis. In particular, an emergence of a counterintuitive connection between the value of L, the structure of the matrix elements of perturbations, and the possible loss of the stability and unitarity of the processes of the unfolding of the singularities is given a detailed explanation.

Keywords: non-Hermitian quantum dynamics; unitary vicinity of exceptional points; degenerate perturbation theory; Hilbert-space geometry near EPs

---

1. Introduction

The Bender's and Boettcher's [1] idea of replacement of Hermiticity $H = H^\dagger$ by parity-time symmetry ($\mathcal{PT}$-symmetry) $H\mathcal{PT} = \mathcal{PT}H$ of a Hamiltonian responsible for unitary evolution opened, after an appropriate mathematical completion [2–5] of the theory, a way towards the building of quantum models exhibiting non-Hermitian degeneracies [6] *alias* exceptional points (EPs, [7]). For a fairly realistic illustrative example of possible applications opening multiple new horizons in phenomenology we could recall, e.g., the well-known phenomenon of Bose–Einstein condensation is a schematic simplification described by the non-Hermitian but $\mathcal{PT}$-symmetric three-parametric Hamiltonian

$$H^{(GGKN)}(\gamma, v, c) = -\mathrm{i}\gamma\left(a_1^\dagger a_1 - a_2^\dagger a_2\right) + v\left(a_1^\dagger a_2 + a_2^\dagger a_1\right) + \frac{c}{2}\left(a_1^\dagger a_1 - a_2^\dagger a_2\right)^2 . \quad (1)$$

This Hamiltonian represents an interesting analytic-continuation modification of the conventional Hermitian Bose–Hubbard Hamiltonian [8–10]. In this form the model was recently paid detailed attention in Ref. [11]. A consequent application of multiple, often fairly sophisticated forms of perturbation theory has been shown there to lead to surprising results. In particular, the behavior of the bound and resonant states of the system was found to lead to the new and unexpected phenomena in the dynamical regime characterized by the small coupling constant c. The authors of Ref. [11] emphasized that new physics may be expected to emerge precisely in the vicinity of the EP-related dynamical singularities.

These phenomena (simulating not necessarily just the Bose–Einstein condensation of course) were analyzed, in [11], using several ad hoc, not entirely standard perturbation techniques. The role of an unperturbed Hamiltonian H_0 was assigned, typically, to the extreme EP limits of H. Unfortunately, only too often the perturbed energies appeared to be complex as a consequence. In other words, the systems exhibiting $\mathcal{PT}$-symmetry seemed to favor the spontaneous breakdown of this symmetry near EPs.

In the light of similar results one immediately must ask the question whether such a "wild behavior" of the EP-related quantum systems is generic. Indeed, an affirmative answer is often encountered in the studies by mathematicians (see, e.g., [12]). A hidden reason is that they usually tacitly keep in mind just the "effective theory" and/or the so-called "open quantum system" dynamical scenario [13].

In the more restrictive context of the unitary quantum mechanics the situation is different: the "wild behavior" of systems is usually not generic there (cf. also the recent explanatory commentary on the sources of possible misunderstandings in [14]). Several non-numerical illustrative models may be found in [15] where typically, the $\mathcal{PT}$-symmetric Bose–Hubbard model of Equation (1) (which behaves as unstable near its EP singularities [11])) has been replaced by its "softly perturbed" alternative in which in an arbitrarily small vicinity of its EP singularity, the system remains stable and unitary under admissible perturbations.

The physics of stability covered by paper [15] can be perceived as one of the main sources of inspiration of our present study. We intend to replace here the very specific model of Equation (1) (in which the geometric multiplicity L of all of its EP-related degeneracies was always equal to one) by a broader class of quantum systems. In a way motivated by the idea of a highly desirable extension of the currently available menu of the tractable and eligible dynamical scenarios beyond their $L = 1$ subclass, we will turn attention here to the EP-related degeneracies of the larger, nontrivial geometric multiplicities $L \geq 2$. We will reveal that such a study opens new horizons not only in phenomenology (where the influence of perturbations becomes strongly dependent on the detailed structure of the non-Hermitian degeneracy) but also in mathematics (where a rich menu of physical consequences will be shown reflected by an unexpected adaptability of the geometry of the Hilbert space to the detailed structure of the perturbation).

The presentation of our results will be organized as follows. First, in Section 2 we will recall a typical quantum system (viz., a version of the non-Hermitian Bose–Hubbard multi-bosonic model) in which the EP degeneracies play a decisive phenomenological role. We will explain that although the model itself only exhibits the maximal-order $L = 1$ EP degeneracies, such an option represents, from the purely formal point of view, just one of the eligible dynamical scenarios. In Appendix A a full classification of the EPs is presented therefore, showing, i.a., that the number of the "anomalous" EPs of our present interest with $L \geq 2$ exhibits an almost exponential growth at the larger matrix dimension N.

The goals of our considerations are subsequently explained in section 3. For the sake of brevity, we just pick up the first nontrivial case with $L = 2$, and we emphasize that even in such a case the basic features of an appropriate adaptation of perturbation theory may be explained, exhibiting also, not quite expectedly, a survival of the fairly user-friendly mathematical structure.

In order to make our message self-contained, the known form of the EP-related perturbation formalism restricted to $L = 1$ is reviewed in Appendix B. On this background, in a way based on a not too dissimilar constructive strategy, our present main $L \geq 2$ results are then presented and described in Section 4. We emphasize there the existence of the phenomenological as well as mathematical subtleties of the large-L models. We show that in our generalized, degenerate-perturbation-theory formalism a key role is played by an interplay between its formal mathematical background (viz, the non-Hermiticity of the Hamiltonians) and its phenomenological aspects (typically, the knowledge of L must be complemented by an explicit knowledge of the partitioning of Schrödinger equation).

In Section 5, all these aspects of the $L > 1$ perturbation theory are summarized and illustrated by a detailed description of the characteristic, not always expected features of the leading-order approximations. Several related applicability aspects of our present degenerate-perturbation-theory formalism are finally discussed in Section 6 and in two Appendices. We point out there that some of the features of the $L > 1$ theory (e.g., a qualitative, fairly counterintuitive clarification of the concept of the smallness of perturbations) may be treated as not too different from their $L = 1$ predecessors. At the same time, a wealth of new formal challenges is emphasized to emerge, in particular, in the analyses of the role of perturbations in the specific quantum systems which are required unitary.

2. Exceptional Points

2.1. *Bose–Hubbard Model and Exceptional Points of Geometric Multiplicity One, $L = 1$*

Illustrative $\mathcal{PT}$-symmetric Bose–Hubbard Hamiltonian operator (1) commutes with the number operator

$$\widehat{N} = a_1^\dagger a_1 + a_2^\dagger a_2 \,. \tag{2}$$

This means that the number of bosons N is conserved so that after its choice the Hamiltonian may be represented by a finite-dimensional non-Hermitian K by K matrix $H^{(GGKN)}(K,\gamma,v,c)$ with $K = N+1$ (see its explicit construction in [11]). Once we fix the units (such that $v = 1$) and once we set $c = 0$ (preserving, for the sake of simplicity, just the first two components of the Hamiltonian), the resulting one-parametric family of Hamiltonian matrices $H^{(GGKN)}(K,\gamma)$ can be assigned the closed-form energy spectra

$$E_n^{(GGKN)}(K,\gamma) = (1-\gamma^2)^{1/2}\,(1-K+2n)\,, \qquad n = 0,1,\ldots,K-1\,. \tag{3}$$

These energies remain real and non-degenerate (i.e., observable) if and only if $\gamma^2 < 1$.

In *loc. cit.* it has also been proved that the two interval-boundary values of $\gamma = \pm 1$ are, in the terminology of the Kato's mathematical monograph [7], exceptional points (EPs, $\gamma_+^{(EP)} = 1$ and $\gamma_-^{(EP)} = -1$). More precisely, one should speak about the very special EPs of maximal order (i.e., of order K, abbreviated as EPK). The latter observation may be given a more general, model-independent linear-algebraic background via relation

$$H^{(K)}(\gamma^{(EPK)})\,Q^{(EPK)} = Q^{(EPK)}\,J^{(K)}(\eta) \tag{4}$$

where

$$\eta = \lim_{\gamma\to\gamma^{(EPK)}} E_n(K,\gamma)\,, \qquad n = 0,1,\ldots,K-1 \tag{5}$$

is the limiting degenerate energy. Relation (4) contains the so-called transition matrix $Q^{(EPK)}$ and the non-diagonal, EPK-related canonical-representation Jordan-block matrix

$$J^{(K)}(\eta) = \left(\begin{array}{ccccc} \eta & 1 & 0 & \ldots & 0 \\ 0 & \eta & 1 & \ddots & \vdots \\ 0 & 0 & \eta & \ddots & 0 \\ \vdots & \ddots & \ddots & \ddots & 1 \\ 0 & \ldots & 0 & 0 & \eta \end{array} \right) . \tag{6}$$

As a certain limiting analogue of the conventional set of eigenvectors the transition matrix is obtainable via the solution of the EPK-related analogue (4) of conventional Schrödinger equation. For our illustrative example $H^{(GGKN)}(K,\gamma)$, in particular, all of the K- and $\gamma^{(EPK)}$-dependent explicit, closed forms of solutions $Q^{(EPK)}$ remain non-numerical and may be found constructed in dedicated paper [15].

2.2. *Generic Non-Hermitian Degeneracies with Geometric Multiplicities Larger than One, $L > 1$*

In the common model-building scenarios the γ-dependence of Hamiltonians $H^{(K)}(\gamma)$ is analytic. Under this assumption the exceptional points of maximal order as discussed in preceding subsection represent just one of several possible realizations of a non-Hermitian degeneracy (NHD) with its characteristic EP-related confluence of eigenvalues (5). Besides the maximal, EPK-related complete confluence of eigenvectors as described in preceding subsection we may encounter, in general, multiple other, incomplete confluences of eigenvectors

$$\lim_{\gamma\to\gamma^{(NHD)}} |\psi_{m_{k[j]}}(\gamma)\rangle = |\chi_j\rangle\,, \qquad k[j]=1,2,\ldots,M_j\,, \quad M_j \geq 2\,, \quad j=1,2,\ldots,L\,. \tag{7}$$

Here we have $M_1 + M_2 + \ldots + M_L = K$ where L is called the geometric multiplicity of the EP degeneracy [7] (see also Appendix A for more details).

Every EP instant $\gamma^{(NHD)}$ may be characterized not only by the overall Hilbert-space dimension K and by the number $L \geq 1$ of linearly independent η-related states $|\chi_j\rangle$ of Equation (7) but also by a suitably ordered L-plet of the related subspace dimensions M_j. Thus, in the present $L \geq 2$ extension of preceding subsection the fully non-diagonal Jordan block of Equation (6) must be replaced by the more general block-diagonal canonical representation of the Hamiltonian,

$$\mathcal{J}^{(K)}(\eta) = \left(\begin{array}{cccc} J^{(M_1)}(\eta) & 0 & \ldots & 0 \\ 0 & J^{(M_2)}(\eta) & \ddots & \vdots \\ \vdots & \ddots & \ddots & 0 \\ 0 & \ldots & 0 & J^{(M_L)}(\eta) \end{array} \right) = \bigoplus_{j=1}^{L} J^{(M_j)}(\eta)\,. \tag{8}$$

In parallel, EPK relation (4) must be replaced by its generalization

$$H^{(K)}(\gamma^{(NHD)})\, Q^{(NHD)} = Q^{(NHD)}\, \mathcal{J}^{(K)}(\eta)\,. \tag{9}$$

Naturally, the direct-sum structure of $\mathcal{J}^{(K)}(\eta)$ becomes reflected by a partitioned-matrix structure of transition matrices $Q^{(NHD)}$ which are, in general, not block-diagonal of course.

3. Unitary Processes of Collapse at L = 2

A priori one may expect that the existence of anisotropy of the Hilbert space as realized, in Equation (A18) at $L = 1$, by an elementary rescaling $B(\lambda)$ of the basis will also exist at any larger geometric multiplicity $L > 1$, i.e., in the unitary quantum systems with the more complicated structure of the NHD limiting *alias* quantum phase transition.

3.1. *Quantum Physics behind "Degenerate Degeneracies" with $L = 2$*

Hypothetically, the unitary evolution of any $\mathcal{PT}$-symmetric quantum system moving towards a hiddenly Hermitian EP degeneracy with geometric multiplicity two can be perceived as generated by a suitable diagonalizable Hamiltonian $H^{(K)}(\gamma)$ with real spectrum [3]. What is only necessary is that its (perhaps, properly renumbered) eigenvectors $|\Phi_j(\gamma)\rangle$ obey the $L = 2$ EP-degeneracy rule

$$\lim_{\gamma\to\gamma^{(EP)}} |\Phi_m(\gamma)\rangle = |\chi_a\rangle\,, \qquad m = 0,1,\ldots,M-1\,, \tag{10}$$

$$\lim_{\gamma \to \gamma^{(EP)}} |\Phi_{M+n}(\gamma)\rangle = |\chi_b\rangle\,, \qquad n = 0, 1, \ldots, N-1\,. \tag{11}$$

The two limiting eigenvectors $|\chi_a\rangle$ and $|\chi_b\rangle$ are, by our assumption, linearly independent so that partitioned matrix (8), i.e., at $L = 2$ and $\eta = 0$, matrix

$$\mathcal{J}^{(M \oplus N)}(0) = \left[\begin{array}{cc} J^{(M)}(0) & 0 \\ 0 & J^{(N)}(0) \end{array} \right] \tag{12}$$

will represent the canonical form of our Hamiltonian in the EP limit. The corresponding transition matrix $Q^{(M \oplus N)}$ can be then obtained by the solution of the limiting version

$$H^{(K)}(\gamma^{(EP)})\, Q^{(M \oplus N)} = Q^{(M \oplus N)}\, \mathcal{J}^{(M \oplus N)}(0) \tag{13}$$

of the initial Schrödinger equation. A few exactly solvable samples of the latter $L = 2$ degeneracy process $\gamma \to \gamma^{(EP)}$ can be found, e.g., in our recent paper [16].

In all the similar dynamical scenarios one can always find parallels with their simpler $L = 1$ predecessors. In particular, a return to the situation before the collapse as sampled, at $L = 1$, by Equation (A4) below, can be also given the following analogous form

$$\left[Q^{((M \oplus N)}\right]^{-1} H^{(K)}(\gamma)\, Q^{(M \oplus N)} = J^{(M \oplus N)}(0) + \lambda\, V^{(K)}(\gamma)\,, \quad \gamma \neq \gamma^{(EP)} \tag{14}$$

of a "unitarity-compatible" perturbation-theoretic reinterpretation in which the perturbation $\lambda\, V^{(K)}(\gamma)$ is fully determined by the input matrix Hamiltonian $H^{(K)}(\gamma)$.

3.2. *Unfoldings of Degeneracies under Random Perturbations at $L = 2$*

In a close parallel to the $L = 1$ Schrödinger's bound-state problem (A6) let us now start the study of its $L > 1$ generalizations by considering the first nontrivial choice of degeneracy with the geometric multiplicity $L = 2$. In our present notation the corresponding Schrödinger equation then reads

$$\left[\mathcal{J}^{(M \oplus N)}(0) + \lambda\, V^{(K)}\right] |\Psi\rangle = \varepsilon\, |\Psi\rangle\,. \tag{15}$$

Without any loss of generality, we set again $\eta = 0$. After such a choice all the eigenvalues $\varepsilon = \varepsilon(\lambda)$ will remain small, and they will vanish in the formal unperturbed-system limit $\lambda \to 0$.

In a way paralleling Equation (12), any given matrix of perturbations has to be partitioned as well,

$$V^{(K)} = \left[\begin{array}{cc} V^{(M,M)} & V^{(M,N)} \\ V^{(N,M)} & V^{(N,N)} \end{array} \right]. \tag{16}$$

Since the four submatrices have dimensions indicated by the superscripts, we shall assume, at the beginning at least, that all the individual matrix elements of the perturbation matrix $V^{(K)}$ remain bounded at small λ. This means that the Hamiltonian is dominated by its unperturbed part (12) so that also the whole perturbed $L = 2$ Schrödinger Equation (15) has to be partitioned. In order to simplify the notation we shall write

$$|\Psi\rangle = \left(\begin{array}{c} |\psi^{(a)}\succ \\ |\psi^{(b)}\succ \end{array} \right), \quad |\psi^{(a)}\succ = \left(\begin{array}{c} \psi_1^{(a)} \\ \psi_2^{(a)} \\ \vdots \\ \psi_M^{(a)} \end{array} \right), \quad |\psi^{(b)}\succ = \left(\begin{array}{c} \psi_1^{(b)} \\ \psi_2^{(b)} \\ \vdots \\ \psi_N^{(b)} \end{array} \right) \tag{17}$$

using the curly-ket symbols for subvectors. This will enable us to proceed in a partial parallel with the widely studied non-degenerate-EP cases where one has $L = 1$ (see Refs. [17,18] or Appendix B below for a compact review).

4. Perturbation Theory at $L = 2$

4.1. The Recent Change of the Unitary-Evolution Paradigm

From the historical point of view the use of EPs in physics has not been immediate. Only during the last circa 20 years one notices a perceivable increase of the relevance of the concept in various branches of theoretical as well as experimental physics. Various innovative EP applications emerged ranging from the analyses of resonances in classical mechanics [19] and of the so called non-Hermitian degeneracies in classical optics [6] up to the studies of the wealth of phenomena in quantum physics of open quantum systems [13,20] or, finally, even of the closed, stable and unitarily evolving quantum systems [1,2].

All these developments contributed to the motivation of our present study. For the sake of definiteness, we restricted our attention to the framework of quantum physics, unitary or non-unitary. In this setting the traditional role of the EPs $\gamma^{(EP)}$ has always been two-fold. First, in the context of mathematics, the conventional analyticity assumptions about Hamiltonians

$$H(\gamma) = H(\gamma_0) + (\gamma - \gamma_0)\, H^{(1)} + (\gamma - \gamma_0)^2\, H^{(2)} + \dots\,, \tag{18}$$

and the conventional power-series ansatz for energies

$$E_n(\gamma) = E_n(\gamma_0) + (\gamma - \gamma_0)\, E_n^{(1)} + (\gamma - \gamma_0)^2\, E_n^{(2)} + \dots \tag{19}$$

(etc.) gave birth to the so-called Rayleigh–Schrödinger perturbation-expansion constructions of the Schrödinger-equation solutions. It has been revealed that the radius of convergence R of these perturbation-series solutions is determined by the position of the nearest EP in the complex plane of the parameter, $R = \min |\gamma_0 - \gamma^{(EP)}|$ [7].

In the other, direct applications of EPs, the localization of singularities $\gamma^{(EP)}$ only played an important traditional role in non-unitary, open quantum systems [13]. An explanation is easy: for any self-adjoint Hamiltonian $H(\gamma)$ characterizing a closed quantum system, the necessary reality of the parameter ($\mathrm{Im}\,\gamma = 0$) cannot be made compatible with the fact that *all* of the values of the eligible (i.e., not accumulation-point) EP parameters are complex, $\mathrm{Im}\,\gamma^{(EP)} \neq 0$.

The traditional paradigm has only been changed recently, after Bender with Boettcher [1] managed to turn attention of physicists' community to the existence of a broad class of Hamiltonians $H(\gamma)$ which happen to be non-Hermitian but parity-time symmetric ($\mathcal{PT}$-symmetric) in $\mathcal{K}$. One of the characteristic mathematical features of these Hamiltonians is that despite their non-Hermiticity, their *whole* spectrum $\{E_n\}$ may remain strictly real in a suitable *real* interval $\mathcal{D}$ of the unitarity-compatible parameters γ (see, e.g., monograph [4] for more details).

4.2. Rearrangement of Schrödinger Equation

With the two subscripts $j_{(a)}$ and $j_{(b)}$ running, in the curly-ket subvectors in (17), from 1 to $N_{(a)} = M$ and $N_{(b)} = N$, respectively, we will now only partially fix the norm by setting $\psi_1^{(a)} = \omega_{(a)}$ and $\psi_1^{(b)} = \omega_{(b)}$ or, in a self-explanatory shorthand, $\psi_1^{(a,b)} = \omega_{(a,b)}$. Next, a parallel to the $L = 1$ redefinition (A8) of wave functions will be found in its $L = 2$ extension

$$|\vec{y}^{(a,b)}\succ = \begin{pmatrix} y_1^{(a,b)} \\ y_2^{(a,b)} \\ \vdots \\ y_{N_{(a,b)}-1}^{(a,b)} \\ y_{N_{(a,b)}}^{(a,b)} \end{pmatrix} = \begin{pmatrix} \psi_2^{(a,b)} \\ \psi_3^{(a,b)} \\ \vdots \\ \psi_{N_{(a,b)}}^{(a,b)} \\ \Omega_{(a,b)} \end{pmatrix} \tag{20}$$

where the two new, temporarily variable elements $\Omega_{(a)}$ and $\Omega_{(b)}$ will have to be determined later. In terms of the four auxiliary symbols

$$|e^{(a,b)}\succ = \begin{pmatrix} 1 \\ 0 \\ \vdots \\ 0 \end{pmatrix}, \quad \Pi^{(a,b)} = \begin{bmatrix} 0 & 0 & 0 & \dots & 0 \\ 1 & 0 & 0 & \ddots & \vdots \\ 0 & \ddots & \ddots & \ddots & 0 \\ \vdots & \ddots & 1 & 0 & 0 \\ 0 & \dots & 0 & 1 & 0 \end{bmatrix} \tag{21}$$

of dimensions $N_{(a)} = M$ and $N_{(b)} = N$ we will further decompose

$$|\psi^{(a,b)}\succ = |e^{(a,b)}\succ \, \omega_{(a,b)} + \Pi^{(a,b)} \, |\vec{y}^{(a,b)}\succ \ .$$

In the next step we introduce the $L = 2$ analogue of the unpartitioned $L = 1$ vector (A7),

$$|r\rangle = \begin{pmatrix} |r^{(a)}\succ \\ |r^{(b)}\succ \end{pmatrix}, \quad |r^{(a,b)}\succ = \begin{pmatrix} r_1^{(a,b)} \\ r_2^{(a,b)} \\ \vdots \\ r_{N_{(a,b)}}^{(a,b)} \end{pmatrix} \tag{22}$$

with components

$$r_i^{(a,b)} = \varepsilon\, \omega_{(a,b)}\, \delta_{i,1} - \lambda\, V_{i,1}^{(N_{(a,b)},M)}\, \omega_{(a)} - \lambda\, V_{i,1}^{(N_{(a,b)},N)}\, \omega_{(b)} = r_i^{(a,b)}(\varepsilon, \vec{\omega})\,. \tag{23}$$

Treating, temporarily, the two not yet specified quantities $\Omega_{(a)}$ and $\Omega_{(b)}$ as adjustable matrix-regularization parameters, and replacing the $L = 1$ auxiliary matrix (A9) by its partitioned $L = 2$ counterpart

$$\mathcal{A}(\varepsilon) = \begin{bmatrix} A(M,\varepsilon) & 0 \\ 0 & A(N,\varepsilon) \end{bmatrix}$$

we are just left with the problem of finding a suitable $L = 2$ analogue of relation (A11).

A key to the resolution of the puzzle is found in Equation (21) and in its partitioned direct-sum extension

$$\Pi^{(M\oplus N)} = \begin{bmatrix} \Pi^{(a)} & 0 \\ 0 & \Pi^{(b)} \end{bmatrix}.$$

Using this symbol, we can now rewrite our homogeneous Schrödinger Equation (15) in the inhomogeneous matrix-inversion representation

$$\left(\mathcal{A}^{-1}(\varepsilon) + \lambda\, V^{(K)}\, \Pi^{(M\oplus N)}\right) \begin{pmatrix} |\vec{y}^{(a)}\succ \\ |\vec{y}^{(b)}\succ \end{pmatrix} = \begin{pmatrix} |r^{(a)}\succ \\ |r^{(b)}\succ \end{pmatrix} \tag{24}$$

or, equivalently,

$$\left(I+\lambda\,\mathcal{A}(\varepsilon)\,V^{(K)}\,\Pi^{(M\oplus N)}\right)\,|\vec{y}\rangle = \mathcal{A}\,(\varepsilon)|r\rangle\,. \tag{25}$$

Once we drop the redundant superscripts, and once we add the relevant parameter-dependences in (25) we obtain relation

$$(I+\lambda\,\mathcal{A}(\varepsilon)\,V\,\Pi)\;|\vec{y}\rangle = \mathcal{A}(\varepsilon)\,|r(\lambda,\varepsilon,\vec{\omega})\rangle\,. \tag{26}$$

This is our ultimate, iteration-friendly exact form of our perturbed Schrödinger equation.

4.3. Solutions

Equation (26) yields the ket-vector part of the solution in closed form,

$$|\vec{y}\rangle = (I+\lambda\,\mathcal{A}(\varepsilon)\,V\,\Pi)^{-1}\;\mathcal{A}(\varepsilon)\,|r(\lambda,\varepsilon,\vec{\omega})\rangle = |\vec{y}^{(solution)}(\lambda,\varepsilon,\vec{\omega})\rangle\,. \tag{27}$$

In the small-perturbation regime the latter formula may be given the conventional Taylor-series form with

$$|\vec{y}^{(solution)}(\lambda,\varepsilon,\vec{\omega})\rangle = \mathcal{A}(\varepsilon)\,|r(\lambda,\varepsilon,\vec{\omega})\rangle - \lambda\,\mathcal{A}(\varepsilon)\,V\,\Pi\,\mathcal{A}(\varepsilon)\,|r(\lambda,\varepsilon,\vec{\omega})\rangle +$$
$$+\lambda^2\,\mathcal{A}(\varepsilon)\,V\,\Pi\,\mathcal{A}(\varepsilon)\,V\,\Pi\,\mathcal{A}(\varepsilon)\,|r(\lambda,\varepsilon,\vec{\omega})\rangle - \ldots\,. \tag{28}$$

Naturally, the construction is not yet finished because what is missing is the guarantee of equivalence between the eigenvalue problem (15) and its matrix-inversion reformulation (24) containing two redundant parameters. This is the last obstacle, easily circumvented by our setting, in solution (28), both of the redundant upper-case constants $\Omega_{(a)} = y_M^{(a)}$ and $\Omega_{(b)} = y_N^{(b)}$ equal to zero. In the light of explicit formula (28), this requirement is equivalent to the pair of relations

$$y_M^{(solution)}(\lambda,\varepsilon,\omega_{(a)},\omega_{(b)}) = 0\,, \quad y_{M+N}^{(solution)}(\lambda,\varepsilon,\omega_{(a)},\omega_{(b)}) = 0\,. \tag{29}$$

Both left-hand-side functions of the three unknown quantities ε, $\omega_{(a)}$ and $\omega_{(b)}$ are available, due to formula (28), in closed form. Although both can vary with the three independent unknowns (i.e., with $\varepsilon, \omega_{(a)}$ and $\omega_{(b)}$), one of these variables merely plays the role of an optional normalization constant so that we may set, say, $\omega_{(a)}^2 + \omega_{(b)}^2 = 1$. Thus, the implicit version of the perturbation-expansion construction of bound states is completed.

5. Schrödinger Equation in Leading-Order Approximation

Two coupled equations (29) determine the bound state. In the spirit of perturbation theory one may expect that the perturbations happen to be, in some sense, small. At the same time, even the analysis of the comparatively elementary $L = 1$ secular Equation (A16) determining the single free variable (viz., the energy) led to the necessity of a strongly counterintuitive scaling (A18) reflecting, near the extreme EP boundary of unitarity, the strong anisotropy of the geometry of the physical Hilbert space. Naturally, at least comparable complications must be expected to be encountered during the analysis of the more complicated set of two coupled equations (29) representing the secular equation in its exact $L = 2$ version, constructed as particularly suitable for systematic approximations.

5.1. Generic Case: Perturbations without Vanishing Elements

Even the most drastic truncation of the formal power series (28) yields already a nontrivial ket vector

$$|\vec{y}^{(solution)}(\lambda,\varepsilon,\vec{\omega})\rangle = \mathcal{A}(\varepsilon)\,|r(\lambda,\varepsilon,\vec{\omega})\rangle\,. \tag{30}$$

Needless to add that what must vanish are the auxiliary variables $\Omega_{(a,b)}$ *alias* two functions which are available in closed form. Thus, we must solve the following two simplified equations

$$\sum_{k=1}^{N_{(\varrho)}} A^{(\varrho)}_{N_{(\varrho)},k}(\varepsilon)\, r^{(\varrho)}_k(\lambda,\varepsilon,\omega_{(a)},\omega_{(b)}) = 0\,, \qquad \varrho = a,b\,. \tag{31}$$

After the insertion of the respective matrix elements $A^{(a,b)}_{N_{(a,b)},k}(\varepsilon)$ [cf. Equation (A9)] we obtain the pair of relations

$$\left[\sum_{m=0}^{M-1} \epsilon^m\, \lambda\, V^{(M,M)}_{M-m,1}\right] \omega_{(a)} + \left[\sum_{m=0}^{M-1} \epsilon^m\, \lambda\, V^{(M,N)}_{M-m,1}\right] \omega_{(b)} = \epsilon^M\, \omega_{(a)}\,, \tag{32}$$

$$\left[\sum_{n=0}^{N-1} \epsilon^n\, \lambda\, V^{(N,M)}_{N-n,1}\right] \omega_{(a)} + \left[\sum_{n=0}^{N-1} \epsilon^n\, \lambda\, V^{(N,N)}_{N-n,1}\right] \omega_{(b)} = \epsilon^N\, \omega_{(b)}\,. \tag{33}$$

Obviously, this set can be read as a generalized eigenvalue problem which determines generalized eigenvectors $\vec{\omega}$ at a K-plet of eigenenergies ϵ which are all defined as roots of the corresponding generalized secular determinant.

5.2. Hierarchy of Relevance and Reduced Approximations

In the generic case one must assume that the matrix elements of the perturbation do not vanish and that λ is small, i.e., that also the eigenvalues ε remain small. This enables one to omit all the asymptotically subdominant corrections and to consider just the linear algebraic system

$$\left[\lambda\, V^{(M,M)}_{M,1} - \epsilon^M\right] \omega_{(a)} + \lambda\, V^{(M,N)}_{M,1}\, \omega_{(b)} = 0\,, \tag{34}$$

$$\lambda\, V^{(N,M)}_{N,1} \omega_{(a)} + \left[\lambda\, V^{(N,N)}_{N,1} - \epsilon^N\right] \omega_{(b)} = 0\,. \tag{35}$$

The solution of these two simplified coupled linear relations exists if and only if the determinant of the system vanishes,

$$\det \left[\begin{array}{cc} \lambda\, V^{(M,M)}_{M,1} - \epsilon^M & \lambda\, V^{(M,N)}_{M,1} \\ \lambda\, V^{(N,M)}_{N,1} & \lambda\, V^{(NN)}_{N,1} - \epsilon^N \end{array} \right] = 0\,. \tag{36}$$

Thus, an ordinary eigenvalue problem is encountered when $M = N$.

Lemma 1. *In the generic equipartitioned cases with $M = N \geq 3$ the spectrum ceases to be all real under bounded perturbations. The loss of unitarity is encountered.*

Proof. Both roots $\epsilon^N = \lambda\, x$ of the exactly solvable quadratic algebraic secular Equation (36) may be guaranteed to be real in a certain domain of parameters. Still, some of the energies themselves are necessarily complex since $\epsilon = \sqrt[N]{\lambda\, x}$ is an N-valued function with values lying on a complex circle. □

In the other, non-equipartitioned dynamical scenarios with, say, $M > N$, the behavior of the system in an immediate vicinity of its EP extreme is still determined by the asymptotically dominant part of the secular equation. After an appropriate modification, the above proof still applies.

Lemma 2. *In the generic case with $M > N \geq 3$ we get, from the dominant part of the generalized eigenvalue problem (36), a subset (N-plet) of asymptotically dominant eigenvalues $\varepsilon = \mathcal{O}(\lambda^{1/N})$ which cannot be all real.*

5.3. Unitary Case: Re-Scaled Perturbations

The loss of unitarity occurring in the $L = 1$ models was associated with the use of the too broad a class of norm-bounded perturbations. From our preliminary results described in preceding subsection one can conclude that a similar loss of unitarity may be also expected to occur, in the generic case, at $L = 2$. Indeed, the anisotropy of the physical Hilbert space which reflects the influence of an

EP-related singularity of the Hamiltonian may be expected to lead again to a selective enhancement of the weight of certain specific matrix elements of perturbations $\lambda\, V^{(K)}$.

Once we recall the $L = 1$ scenario of Appendix B.3 we immediately imagine that the main source of the apparent universality of the instability under norm-bounded perturbations should be sought, paradoxically, in the routine but, in our case, entirely inadequate norm-boundedness assumption itself. Indeed, in a way documented by Lemmas 1 and 2, the loss of the reality of spectra may directly be attributed to the conventional and comfortable but entirely random, unfounded and formal assumption of the uniform boundedness of the matrix elements of the perturbations.

Most easily the latter result may be illustrated using the drastically simplified version (36) of the leading-order secular equation. Dominant role is played there by the quadruplet of matrix elements $V_{(P,1)}^{(P,Q)}$ with superscripts P and Q equal to M or N. Once they are assumed λ-independent and non-vanishing, the leading-order energies read $\epsilon = \sqrt[N]{\lambda\, x}$. Thus, at any $N \geq 3$ their N-plet forms an equilateral N-angle in the complex plane of λ.

The latter observation inspires a remedy. In a way eliminating the $N \geq 3$ complex-circle obstruction one simply has to re-scale the energies as well as all the relevant matrix elements of the perturbation. In this manner the ansatz

$$\epsilon = \epsilon(E) = \sqrt{\lambda}\, E \tag{37}$$

opens the possibility of the spectrum being real. Another multiplet of postulates

$$V_{M-m,1}^{(M,M)} = \lambda^{(M-m)/2}\, W_{M-m,1}^{(M,M)}\,, \quad V_{M-m,1}^{(M,N)} = \lambda^{(M-m)/2}\, W_{M-m,1}^{(M,N)}\,, \quad m = 0, 1, \ldots, M-1\,, \tag{38}$$

and

$$V_{N-n,1}^{(N,M)} = \lambda^{(N-n)/2}\, W_{N-n,1}^{(N,M)}\,, \quad V_{N-n,1}^{(N,N)} = \lambda^{(N-n)/2}\, W_{N-n,1}^{(N,N)}\,, \quad n = 0, 1, \ldots, N-1 \tag{39}$$

now contains a new partitioned matrix $W = W^{(K)}$ which is assumed uniformly bounded.

Lemma 3. *There always exists a non-empty $(2M + 2N)$-dimensional domain $\mathcal{D}$ of the "physical" matrix elements of W for which the leading-order spectrum is all real and non-degenerate, i.e., in the language of physics, tractable as stable bound-state energies.*

Proof. The main consequence of the amended, necessary-condition perturbation-smallness requirements (38) and (39) is that in our initial, unreduced leading-order generalized eigenvalue problem (32) + (33), all terms in the sums become of the same order of magnitude. By the scaling we managed to eliminate the explicit presence of the measure of smallness λ. In other words the input information about dynamics is formed now by the re-scaled perturbation matrix W which offers an $(2M + 2N)$-plet of free $\mathcal{O}(1)$ parameters. At the same time, the spectral-definition output is given by the $M + N$ roots of the corresponding secular equation, i.e., by the roots of an $(M + N)$-th-degree polynomial in E, with all its separate coefficients bounded and, in general, non-vanishing. Under these conditions the assertion of the lemma is obvious. □

6. Discussion

6.1. Schrödinger Picture and Quasi-Hermitian Hamiltonians

In the context of the new, Bender- and Boettcher-inspired paradigm the non-Hermiticity of $H(\gamma)$ in $\mathcal{K}$ may be considered compatible with unitarity whenever the spectrum itself is found real. The explanation of the apparent paradox is easy: Under certain reasonable mathematical assumptions (cf. [21]) one can find a non-unitary invertible mapping $\Omega = \Omega(\gamma)$ with property $\Omega^\dagger\Omega = \Theta \neq I$ which makes our Hamiltonian self-adjoint,

$$H(\gamma) \to \mathfrak{h}(\gamma) = \Omega(\gamma)\, H(\gamma)\, \Omega(\gamma) = \mathfrak{h}^\dagger(\gamma)\,. \tag{40}$$

In practice the idea enables one to avoid the use of the "conventional" Hamiltonian $\mathfrak{h}(\gamma)$ (self-adjoint, by construction, in another Hilbert space $\mathcal{L}$) whenever it happens to be "prohibitively complicated". In the literature such a type of simplification of calculations is usually attributed to Dyson [22]. Expectedly, the strategy (also known as preconditioning) proved efficient as a tool of construction of bound states in nuclear physics [21]. In spite of a certain initial doubts [23,24], the approach also proved applicable in the quantum physics of scattering [25,26].

From a more abstract theoretical point of view the reference to the "Dyson's" isospectral mapping (40) becomes redundant when we reclassify the space $\mathcal{K}$ as "unphysical", and when we redefine its inner product yielding another, "physical" Hilbert space $\mathcal{H}$,

$$\langle \psi_1|\psi_2\rangle_{\mathcal{H}} = \langle \psi_1|\Theta|\psi_2\rangle_{\mathcal{K}}\,, \quad |\psi_{1,2}\rangle \in \mathcal{K}\,. \tag{41}$$

A new, equivalent, "two-Hilbert-space" version of the Schrödinger picture is obtained in which the physics described in the correct Hilbert space $\mathcal{H}$ is "translated" to its mathematically easier representation in $\mathcal{K}$. In this sense, the self-adjointness of hypothetical $\mathfrak{h}$ in hypothetical $\mathcal{L}$ is found equivalent to the self-adjointness of H in $\mathcal{H}$, represented by the relation

$$H^\dagger(\gamma)\, \Theta(\gamma) = \Theta(\gamma)\, H(\gamma)\,. \tag{42}$$

Dieudonné [27] and Scholtz et al [21] suggested to call relation (42) the quasi-Hermiticity of H in $\mathcal{K}$. In the context of quantum phenomenology a decisive amendment of the formalism may be seen in the split of the description of dynamics with information carried by the two operators $H(\gamma)$ and $\Theta(\gamma)$ defined in $\mathcal{K}$ in place of one (viz., of $\mathfrak{h}(\gamma)$ living in $\mathcal{L}$).

6.2. Non-Hermitian Degeneracies with $L > 2$

After the present decisive clarification of the possibility of the replacement of the $L = 1$ formalism by its $L = 2$ generalization, the next move to the further, $L > 2$ dynamical scenarios is now an almost elementary exercise. Indeed, in Sections 3–5 it would be sufficient to move from the $L = 2$ partitioning of $K = N_{(a)} + N_{(b)}$ to its arbitrary $L > 2$ analogues (A1) and to the related partitioned vector sets [sampled, e.g., by Equation (7)] and matrices [sampled, e.g., by Equation (8)]. The $L = 2$ doublets of the eligible superscripts $^{(a)}$ and $^{(b)}$ marking the curly kets [cf. (17) or (31), etc.] may be very easily extended to the L-plets with $L > 2$, etc.

Along these lines the form of our basic power-series expansion of the perturbed bound-state kets (28) remains unchanged. The related $L = 2$ compatibility constraint (29) must only be replaced by an L-plet of equations

$$y^{(solution)}_{N_{(a_j)}}(\lambda, \varepsilon, \vec{\omega}) = 0\,, \quad j = 1, 2, \ldots, L \tag{43}$$

where the unknown vector $\vec{\omega}$ has L components.

6.3. The Next-to-Leading-Order Approximation

In our present paper we did not pay too much attention to the next-to-leading-order (NLO) approximation where one would have to set

$$|\vec{y}^{(solution)}(\lambda, \varepsilon, \vec{\omega})\rangle = \mathcal{A}(\varepsilon)\, |r(\lambda, \varepsilon, \vec{\omega})\rangle - \lambda\, \mathcal{A}(\varepsilon)\, V\, \Pi\, \mathcal{A}(\varepsilon)\, |r(\lambda, \varepsilon, \vec{\omega})\rangle\,. \tag{44}$$

Our main reason was that such a generalization would make the associated compatibility conditions (43) perceivably more complicated. Indeed, in contrast to the leading-order case, an almost inessential improvement of the insight in the qualitative features of the quantum system in question would be accompanied, in the NLO formulae, by the emergence of multiple new matrix elements of perturbations $\lambda\, V^{(K)}$ including even the terms which would be quadratic functions of these matrix elements.

This being said, it is necessary to add that even on the pragmatic and qualitative level, the omission of the NLO corrections only seems completely harmful in the open-system setting using bounded matrices $\lambda\, V^{(K)}$ where one does not insist on having the strictly real spectrum. The point is that whenever one must guarantee the unitarity of the system, the class of the admissible perturbations must be further restricted. In this sense, unfortunately, an analysis using NLO might prove necessary. Indeed, the use and precision of the leading-order approximation need not be sufficiently reliable in general. Even a quick glimpse at the underlying assumptions (38) and (39) reveals that the leading-order approximation does not incorporate the influence of a large subset of the matrix elements of $\lambda\, V^{(K)}$. At the same time, one must keep in mind that in our present approach the dimension of the matrices K has been assumed finite. For this reason, in the case of doubts, a turn to the more universal and brute-force numerical methods might prove to be, in practical calculations, a reasonable alternative to the rather lengthy and complicated NLO calculations.

Acknowledgments: The author acknowledges the financial support from the Excellence project PřF UHK 2020 Nr. 2212.

Appendix A. An Exhaustive Classification of the Degeneracies of Exceptional Points

Up to an arbitrary permutation, every partitioning

$$K = M_1 + M_2 + \ldots + M_L \tag{A1}$$

of the full matrix dimension characterizes a different system and its NHD limit. Thus, we must postulate, say,

$$M_1 \geq M_2 \geq \ldots \geq M_L \geq 2$$

and introduce the schematic multi-indices $\boxed{M_1 + M_2 + \ldots + M_L}$ in a way illustrated in Table A1.

Table A1. Eligible EP-related partitioning.

K	List
2	$\boxed{2}$
3	$\boxed{3}$
4	$\boxed{4}$ $\boxed{2+2}$
5	$\boxed{5}$ $\boxed{3+2}$
6	$\boxed{6}$ $\boxed{4+2}$ $\boxed{3+3}$ $\boxed{2+2+2}$
7	$\boxed{7}$ $\boxed{5+2}$ $\boxed{4+3}$ $\boxed{3+2+2}$
8	$\boxed{8}$ $\boxed{6+2}$ $\boxed{5+3}$ $\boxed{4+4}$ $\boxed{4+2+2}$ $\boxed{3+3+2}$ $\boxed{2+2+2+2}$
9	$\boxed{9}$ $\boxed{7+2}$ $\boxed{6+3}$ $\boxed{5+4}$ $\boxed{5+2+2}$ $\boxed{4+3+2}$ $\boxed{3+3+3}$ $\boxed{3+2+2+2}$
10	$\boxed{10}$ $\boxed{8+2}$ $\boxed{7+3}$ $\boxed{6+4}$ $\boxed{6+2+2}$ $\boxed{5+5}$ $\boxed{5+3+2}$
	$\boxed{4+4+2}$ $\boxed{4+3+3}$ $\boxed{4+2+2+2}$ $\boxed{3+3+2+2}$ $\boxed{2+2+2+2+2}$
$\vdots$	...

The Table indicates that the nontrivial $L > 1$ partitioning not containing trivial items $M_j = 1$ only exists at $K \geq 4$. The value of the count $\mathcal{N}(K)$ of the separate nontrivial, $L \geq 2$ items is seen to exceed one only at $K = 6$. Nevertheless, this count starts growing quickly at the larger Ks (see Table A2). Thus, the current practice of studying just the EPK models with $L = 1$ misses in fact the huge majority of the alternative NHD scenarios. Marginally, it is worth adding that the counts $\mathcal{N}(K)$ form a well-known sequence. In the open-access on-line encyclopedia of integer sequences [28] it is assigned the code number A083751.

Table A2. The sequence of counts $\mathcal{N}(K)$ with $K = 2, 3, \ldots$.

0, 0, 1, 1, 3, 3, 6, 7, 11, 13, 20, 23, 33, 40, 54, 65, 87, 104, 136, 164,
209, 252, 319, 382, 477, 573, 707, 846, 1038, 1237, 1506, 1793, … .

For our present purposes, the asymptotic growth of sequence $\mathcal{N}(K)$ (which is slightly slower than exponential) as well as its precise mathematical definition is less relevant. At the realistic, not too large dimensions K, for example, we might also use some alternative definitions. One of them leads to values $\mathcal{N}(K)$ equal to the first differences (diminished by one) of the special partitions (of the code number $A000041$, cf. [29]), or to the first differences of the numbers of trees of diameter four (see the integer sequence with code number $A000094$ in [30]). Nevertheless, irrespectively of the choice of definition let us point out that in the NHD vicinity, the partitioning multi-indices will classify the phenomenologically non-equivalent physical systems in general. As long as the superscripts $^{(K)}$ are in fact redundant, we will omit or replace them by the more relevant information about the partitioning, therefore. In particular, symbol

$$\mathcal{J}^{(M_1 \oplus M_2 \oplus \ldots \oplus M_L)}(\eta) \tag{A2}$$

will represent the general block-diagonal canonical representation $\mathcal{J}^{(K)}(\eta)$ of the Hamiltonian defined in Equation (8).

Appendix B. Perturbation Theory Near Non-Degenerate Exceptional Points

Appendix B.1. The Choice of Basis at $L = 1$

Relation (4) can be read as a definition of transition matrix responsible for the canonical K by K Jordan-block representation

$$J^{(K)}(\eta) = \left[Q^{(EPK)}\right]^{-1} H(\gamma^{(EPK)})\, Q^{(EPK)} \tag{A3}$$

of the EPK $L = 1$ limit of any given non-Hermitian but $\mathcal{PT}$-symmetric Hamiltonian of phenomenological relevance. From this point of view one can extend the same transformation (i.e., for matrices with $K < \infty$, a mere choice of the basis in Hilbert space) to a vicinity of the EPK singularity. This yields the Jordan block matrix plus perturbation,

$$\left[Q^{(EPK)}\right]^{-1} H^{(K)}(\gamma)\, Q^{(EPK)} = J^{(K)}(\eta) + \lambda\, V^{(K)}(\gamma)\,. \tag{A4}$$

Such a definition contains a redundant but convenient measure $\lambda = \mathcal{O}(\gamma - \gamma^{(EPK)})$ of the smallness of perturbation.

Due to the conventional postulate of having a specific one-parametric family of Hamiltonians $H^{(K)}(\gamma)$ given in advance, the introduction of the concept of perturbation in (A4) is just a formal step. Nevertheless, the interaction term itself could be also reinterpreted as a model-independent random perturbation which carries the input dynamical information. From such a perspective every preselected perturbation term defines a different Hamiltonian $\widetilde{H^{(K)}})$ which merely coincides with

$H^{(K)}(\gamma)$ in the NHD limit $\gamma \to \gamma^{(EP)}$ (a few exactly solvable samples of such a truly remarkable Hamiltonian matching may be found in [15]). This means that via relation

$$J^{(K)}(\eta) + \lambda\, V^{(K)} = \left[Q^{(EPK)}\right]^{-1} \widetilde{H^{(K)})}\, Q^{(EPK)} \tag{A5}$$

[i.e., via a mere reordering of relation (A4)] we obtain a new picture of physics in which the resulting tilded Hamiltonian with real spectrum need not be $\mathcal{PT}$-symmetric at all.

Appendix B.2. The Description of the Unfolding of the Degeneracy at $L = 1$

A specific constructive extension of the latter observation has been presented in our recent paper [18]. We were able there to prove that in the NHD dynamical regime, the mere boundedness of the norm of matrix $V^{(K)}$ together with the smallness of parameter λ still *do not* guarantee the survival of the unitarity of the system after perturbation. We showed there (cf. also [14]) that one can guarantee the absence of a "quantum catastrophe" (i.e., of an abrupt change of some of the system's observable features, see [31–33]) only via a certain self-consistent revision of the criteria of smallness of matrix $V^{(K)}$.

Having in mind the parameter-independence and invertibility of the transformation matrix $Q^{(EPK)}$ the quantification of the influence of the perturbation is of enormous interest, among others, in the analysis of stability of the system in question [12]. This was the reason we also addressed the $L = 1$ perturbative bound-state problem

$$\left[J^{(K)}(0) + \lambda\, V^{(K)}\right] |\Psi\rangle = \epsilon\, |\Psi\rangle\,, \tag{A6}$$

with $\eta = 0$ in Ref. [17]. With the ket-vector subscripts j in $|\Psi_j\rangle$ running from 1 to K we fixed the norm (by setting $|\Psi_1\rangle = 1$), and we relocated the first column of Equation (A6), viz., vector

$$\vec{r} = \vec{r}(\lambda) = \left(\begin{array}{c} \epsilon - \lambda\, V_{1,1} \\ -\lambda\, V_{2,1} \\ \vdots \\ -\lambda\, V_{K,1} \end{array} \right) \tag{A7}$$

to the right-hand side of the equation. Then we restored the comfortable square-matrix form of the equation via its two further equivalent modifications. First we added a new, temporarily undetermined auxiliary component Ω_K to an "upgrade" of the wave function

$$\vec{y} = \left(\begin{array}{c} y_1 \\ y_2 \\ \vdots \\ y_{K-1} \\ y_K \end{array} \right) = \left(\begin{array}{c} |\Psi_2\rangle \\ |\Psi_3\rangle \\ \vdots \\ |\Psi_K\rangle \\ \Omega_K \end{array} \right). \tag{A8}$$

Subsequently, an introduction of the following auxiliary lower-triangular K by K matrix

$$A = A(K,\epsilon) = \left(\begin{array}{cccccc} 1 & 0 & 0 & \ldots & 0 \\ \epsilon & 1 & 0 & \ddots & \vdots \\ \epsilon^2 & \epsilon & \ddots & \ddots & 0 \\ \vdots & \ddots & \ddots & 1 & 0 \\ \epsilon^{K-1} & \ldots & \epsilon^2 & \epsilon & 1 \end{array} \right) \tag{A9}$$

and of its two-diagonal inverse

$$A^{-1} = \begin{pmatrix} 1 & 0 & 0 & \ldots & 0 \\ -\epsilon & 1 & 0 & \ddots & \vdots \\ 0 & -\epsilon & \ddots & \ddots & 0 \\ \vdots & \ddots & \ddots & 1 & 0 \\ 0 & \ldots & 0 & -\epsilon & 1 \end{pmatrix} \tag{A10}$$

accompanied by a parallel formal upgrade of the interaction matrix,

$$V^{(K)} \to Z = \begin{pmatrix} V_{1,2} & V_{1,3} & \ldots & V_{1,K} & 0 \\ V_{2,2} & V_{2,3} & \ldots & V_{2,K} & 0 \\ \ldots & \ldots & \ldots & \vdots & \vdots \\ V_{K,2} & V_{K,3} & \ldots & V_{K,K} & 0 \end{pmatrix} \tag{A11}$$

enabled us to rewrite our initial homogeneous Schrödinger Equation (A6), in the last step of the construction of its solution, in an equivalent matrix-inversion form

$$(A^{-1} + \lambda\, Z)\vec{y} = \vec{r} \tag{A12}$$

or, better,

$$(I + \lambda\, A\, Z)\vec{y} = A\,\vec{r} \tag{A13}$$

accompanied by the innocent-looking but important self-consistence constraint

$$\Omega_K = 0\,. \tag{A14}$$

In the leading-order application of the recipe we then returned to the slightly vague assumption of the "sufficient smallness" of the perturbation. On these grounds we recalled the formal Taylor-series expansion of the resolvent which yielded the closed formula

$$\vec{y}^{(solution)}(\epsilon) = A\vec{r} - \lambda\, A\; Z\, A\,\vec{r} + \lambda^2\, A\; Z\, A\, Z\, A\,\vec{r} - \ldots \tag{A15}$$

for the modified wave function. It contained a free parameter ϵ which had to be fixed via the supplementary secular Equation (A14). In the light of the Taylor-series formula (A15), such a secular equation now acquires the K-th-vector-component form

$${y_K}^{(solution)}(\epsilon) = 0\,. \tag{A16}$$

of an explicit transcendental equation for the energies ϵ.

Appendix B.3. Unitary-Evolution Process of Unfolding at $L = 1$

The latter constraint (A14) plays the role of an implicit definition of the spectrum. The K-plet of roots $\epsilon_n = \epsilon_n(\lambda)$, $n = 1, 2, \ldots, K$ represents the bound-state energies. After the truncation of the series, just approximate solutions are being obtained. In Ref. [17], incidentally, even the leading-order roots were found complex in general.

This observation was interpreted as indicating that in an immediate EPK vicinity the norm-bounded perturbations $\lambda\, V$ should still be considered, in the unitary theory, "inadmissibly large". The non-unitary, open-quantum system interpretation of the perturbations proved needed forcing the system to perform, at an arbitrarily small but non-vanishing $\lambda \neq 0$, an abrupt quantum phase transition.

Incidentally, qualitatively the same conclusions were also obtained in the above-mentioned more concrete study [11] of the specific Bose–Hubbard model in its EPK dynamical regime. The resolution of an apparent universal-instability paradox was provided in our subsequent study [18] in which we studied the underlying exact as well as approximate secular equations in more detail. Our ultimate conclusion was that the necessary smallness condition specifying the class of the admissible, unitarity non-violating perturbations does not involve their upper-triangular matrix part at all. In contrast, their lower-triangular matrix part must be given the following, matrix-element-dependent form

$$\lambda\, V^{(K)}_{(admissible)} = \left[\begin{array}{cccccc} \lambda^{1/2}\mu_{11} & 0 & \ldots & 0 & 0 & 0 \\ \lambda\,\mu_{21} & \lambda^{1/2}\mu_{22} & \ldots & 0 & 0 & 0 \\ \lambda^{3/2}\,\mu_{31} & \lambda\,\mu_{32} & \ddots & \vdots & \vdots & 0 \\ \lambda^{2}\mu_{41} & \lambda^{3/2}\,\mu_{42} & \ddots & \ddots & 0 & 0 \\ \vdots & \vdots & \ddots & \lambda\,\mu_{K-1K-2} & \lambda^{1/2}\mu_{K-1K-1} & 0 \\ \lambda^{K/2}\mu_{K1} & \lambda^{(K-1)/2}\mu_{K2} & \ldots & \lambda^{3/2}\,\mu_{KK-2} & \lambda\,\mu_{KK-1} & \lambda^{1/2}\mu_{KK} \end{array}\right]. \qquad \text{(A17)}$$

During the decrease of $\lambda \to 0$, all the variable lower-triangle matrix-element parameters must remain bounded, $\mu_{j,k} = \mathcal{O}(1)$. In other words, as long as we are working in a specific, fixed "unperturbed" basis, the matrix structure (A17) may be interpreted as manifesting a characteristic anisotropy and the hierarchically ordered weights of influence of the separate matrix elements. Indeed, we may re-scale

$$\lambda\, V^{(K)}_{(admissible)} = \lambda^{1/2}\, B(\lambda)\, V^{(reduced)}\, B^{-1}(\lambda) \qquad \text{(A18)}$$

where $B(\lambda)$ would be a diagonal matrix with elements $B_{jj}(\lambda) = \lambda^{j/2}$ and where the reduced perturbation matrix would be bounded, $V^{(reduced)}_{jk} = \mathcal{O}(1)$.

On this necessary-condition background valid at all dimensions K, the samples of sufficient conditions retain a purely numerical trial-and-error character, with the small$-K$ non-numerical exceptions discussed, in [18], for the matrix dimensions up to $K = 5$.

References

1. Bender, C.M., Boettcher, S. Real spectra in non-Hermitian Hamiltonians having $\mathcal{PT}$ symmetry. *Phys. Rev. Lett.* **1998**, *80*, 5243.
2. Bender, C.M. Making Sense of Non-Hermitian Hamiltonians. *Rep. Prog. Phys.* **2007**, *70*, 947–1018.
3. Mostafazadeh, A. Pseudo-Hermitian Quantum Mechanics. *Int. J. Geom. Meth. Mod. Phys.* **2010**, *7*, 1191–1306.
4. Bagarello, F.; Gazeau, J.-P.; Szafraniec, F.; Znojil, M. (Eds.) *Non-Selfadjoint Operators in Quantum Physics: Mathematical Aspects*; Wiley: Hoboken, NJ, USA, 2015.
5. Bender, C.M. (Ed.) *PT Symmetry in Quantum and Classical Physics*; World Scientific: Singapore, 2018.
6. Berry, M.V. Physics of non-Hermitian degeneracies. *Czech. J. Phys.* **2004**, *54*, 1039–1047.
7. Kato, T. *Perturbation Theory for Linear Operators*; Springer: Berlin, Germany, 1966.
8. Bloch, I.; Hänsch, T.W.; Esslinger, T. Atom laser with a cw output coupler. *Phys. Rev. Lett.* **1999**, *82*, 3008.
9. Hiller, M.; Kottos, T.; Ossipov, A. Bifurcations in resonance widths of an open Bose-Hubbard dimer. *Phys. Rev. A* **2006**, *73*, 063625.
10. Teimourpour, M.H.; Zhong, Q.; Khajavikhan, M.; El-Ganainy, R. Higher Order EPs in iscrete Photonic Platforms. In *Parity-Time Symmetry and Its Applications*; Christodoulides, D., Yang, J.-K., Eds.; Springer: Singapore, 2018; pp. 261–276.
11. Graefe, E.M.; Günther, U.; Korsch, H.J.; Niederle, A.E. A non-Hermitian PT symmetric Bose-Hubbard model: Eigenvalue rings from unfolding higherorder exceptional points. *J. Phys. A Math. Theor.* **2008**, *41*, 255206.
12. Krejčiřík, D.; Siegl, P.; Tater, M.; Viola, J. Pseudospectra in non-Hermitian quantum mechanics. *J. Math. Phys.* **2015**, *56*, 103513.

13. Moiseyev, N. *Non-Hermitian Quantum Mechanics*; CUP: Cambridge, UK, 2011.
14. Znojil, M.; Růžička, F. Nonlinearity of perturbations in PT-symmetric quantum mechanics. *J. Phys. Conf. Ser.* **2019**, *1194*, 012120.
15. Znojil, M. Passage through exceptional point: Case study. *Proc. R. Soc. A Math. Phys. Eng. Sci.* **2020**, *476*, 20190831.
16. Znojil, M.; Borisov, D.I. Anomalous mechanisms of the loss of observability in non-Hermitian quantum models. *Nucl. Physci. B* **2020**, *957*, 115064.
17. Znojil, M. Admissible perturbations and false instabilities in PT-symmetric quantum systems. *Phys. Rev. A* **2018**, *97*, 032114.
18. Znojil, M. Unitarity corridors to exceptional points. *Phys. Rev. A.* **2019**, *100*, 032124.
19. Seyranian, A.P.; Mailybaev, A.A. *Multiparameter Stability Theory with Mechanical Applications*; World Scientific: Singapore, 2003.
20. Christodoulides, D.; Yang, J.-K. (Eds.) *Parity-Time Symmetry and Its Applications*; Springer: Singapore, 2018.
21. Scholtz, F.G.; Geyer, H.B.; Hahne, F.J.W. Quasi-Hermitian Operators in Quantum Mechanics and the Variational Principle. *Ann. Phys.* **1992**, *213*, 74.
22. Dyson, F.J. General theory of spin-wave interactions. *Phys. Rev.* **1956**, *102*, 1217–1230
23. Jones, H.F. Scattering from localized non-Hermitian potentials. *Phys. Rev. D* **2007**, *76*, 125003.
24. Jones, H.F. Interface between Hermitian and non-Hermitian Hamiltonians in a model calculation. *Phys. Rev. D* **2008**, *78*, 065032.
25. Znojil, M. Discrete PT-symmetric models of scattering. *J. Phys. A Math. Theor.* **2008**, *41*, 292002.
26. Znojil, M. Scattering theory using smeared non-Hermitian potentials. *Phys. Rev. D* **2009**, *80*, 045009.
27. Dieudonné, J. Quasi-Hermitian operators. In Proceedings of the International Symposium on Linear Spaces, Jerusalem, Israel, 5–12 July 1961; Pergamon: Oxford, UK, 1961; pp. 115–122.
28. The On-Line Encyclopedia of Integer Sequences. Available online: http://oeis.org/A083751/ (accessed on 6 November 2019).
29. The On-Line Encyclopedia of Integer Sequences. Available online: http://oeis.org/A000041/ (accessed on 6 November 2019).
30. The On-Line Encyclopedia of Integer Sequences. Available online: http://oeis.org/A000094/ (accessed on 6 November 2019).
31. Znojil, M. Quantum catastrophes: A case study. *J. Phys. A Math. Theor.* **2012**, *45*, 444036.
32. Borisov, D.I. Eigenvalue collision for PT-symmetric waveguide. *Acta Polytech.* **2014**, *54*, 93.
33. Znojil, M.; Borisov, D.I. Two patterns of PT-symmetry breakdown in a non-numerical six-state simulation. *Ann. Phys.* **2018**, *394*, 40.

4

Dynamical Symmetries of the H Atom, One of the Most Important Tools of Modern Physics: SO(4) to SO(4,2), Background, Theory and use in Calculating Radiative Shifts

G. Jordan Maclay

Quantum Fields LLC, St. Charles, IL 60174, USA; jordanmaclay@quantumfields.com

Abstract: Understanding the hydrogen atom has been at the heart of modern physics. Exploring the symmetry of the most fundamental two body system has led to advances in atomic physics, quantum mechanics, quantum electrodynamics, and elementary particle physics. In this pedagogic review, we present an integrated treatment of the symmetries of the Schrodinger hydrogen atom, including the classical atom, the SO(4) degeneracy group, the non-invariance group or spectrum generating group SO(4,1), and the expanded group SO(4,2). After giving a brief history of these discoveries, most of which took place from 1935–1975, we focus on the physics of the hydrogen atom, providing a background discussion of the symmetries, providing explicit expressions for all of the manifestly Hermitian generators in terms of position and momenta operators in a Cartesian space, explaining the action of the generators on the basis states, and giving a unified treatment of the bound and continuum states in terms of eigenfunctions that have the same quantum numbers as the ordinary bound states. We present some new results from SO(4,2) group theory that are useful in a practical application, the computation of the first order Lamb shift in the hydrogen atom. By using SO(4,2) methods, we are able to obtain a generating function for the radiative shift for all levels. Students, non-experts, and the new generation of scientists may find the clearer, integrated presentation of the symmetries of the hydrogen atom helpful and illuminating. Experts will find new perspectives, even some surprises.

Keywords: symmetry; hydrogen atom; group theory; SO(4); SO(4,2); dynamical symmetry; non-invariance group; spectrum generating algebra; Runge-Lenz; Lamb shift

---

1. Introduction

1.1. *Objective of This Paper*

This pedagogic review is focused on the symmetries of the Schrodinger nonrelativistic hydrogen atom exclusively to give it the attention that we believe it deserves. The fundamental results of the early work are known and do not need to be derived again. However, having this knowledge permits us to use the modern language of group theory to do a clearer, more focused presentation, and to use arguments from physics to develop the proper forms for the generators, rather than dealing with detailed, mathematical derivations to prove results we know are correct.

There are numerous articles about the symmetry of the Schrodinger hydrogen atom, particularly the SO(4) group of the degenerate energy eigenstates, including discussions from classical perspectives. The spectrum generating group SO(4,1) and the non-invariance group SO(4,2) have been discussed, but, in many fewer articles, often in appendices, with different bases for the representations. For example, numerous papers employ Schrodinger wave functions in parabolic coordinates, not with the familiar *nlm* quantum numbers, often with the emphasis on the details of the mathematical structures, not the physics, and with the emphasis on the potential role of the symmetry in elementary particle physics. Generators may be expressed in complex and unfamiliar terms, for example, in terms of the raising and lowering operators for quantum numbers characteristic of parabolic coordinates. Other approaches involve regularizing Schrodinger's equation by, for example, multiplying by r. The approach results in generators that are not always Hermitian or manifestly Hermitian, and the need for nonstandard inner products. Indeed, most of the seminal articles do not have the words "hydrogen atom" in the title but are focused on the prize down the road, understanding what was called at the time "the elementary particle zoo".

Yet, these foundational articles and books taken together present the information we have about the symmetry of the H atom, of which experts in the field are aware. On the other hand, to the non-expert, the student, or a researcher new to the field, it does not appear that the relevant information is in a form that is conveniently accessible. Since the hydrogen atom is the most fundamental physical system with an interaction, whose exploration and understanding has led to much of the progress in atomic physics, quantum physics, and quantum electrodynamics, we believe that a comprehensive treatment is warranted and, since most of the relevant papers were published four decades ago, is timely. Many younger physicists may not be acquainted with these results.

Unlike in a number of the foundational papers, here the operators are all Hermitian, and given in terms of the canonical position and momentum variables in the simplest forms. The transformations they generate are clearly explained, and we provide brief explanations for the group theory used in the derivations.

In most papers, a separate treatment for bound and scattering states in needed. In contrast we are able to clarify and simplify the exposition since we use a set of basis states which are eigenfunctions of the inverse of the coupling constant [1], that include both the bound states and the scattering states in a uniform way, and that employ the usual Cartesian position and momenta, with the usual inner product, with the exact same quantum numbers nlm as the ordinary bound states; a separate treatment for bound and scattering states is not required. In addition, we have two equivalent varieties of this uniform basis, one that is more suitable for momentum space calculations and one more suitable for configuration space calculations. This advantage again allows us to simplify the exposition.

We focus on the utility of group theoretic methods using our representation and derive expressions for the unitary transformation of group elements and some new results that allow for us to readily compute the first order radiative shift (Lamb shift) of a spinless electron, which accounts for about 95% of the total shift. This approach allows for us to obtain a generating function for the shifts for all energy levels. For comparison, we derive an expression for the Bethe log.

In summary, we present a unified treatment of the symmetries of the Schrodinger hydrogen atom, from the classical atom to SO(4,2) that focuses on the physics of the hydrogen atom, that gives explicit expressions for all the manifestly Hermitian generators in terms of position and momenta operators in a Cartesian space, that explains the action of the generators on the basis states, which evaluates the Casimir operators characterizing the group representations, and that gives a unified treatment of the bound and continuum states in terms of wave functions that have the same quantum numbers as the ordinary bound states. We give an example of the use of SO(4,2) in a practical application, the computation of the first order radiative shift in the hydrogen atom.

Hopefully, students and non-experts and the new generation of scientists will find this review helpful and illuminating, perhaps motivating some to use these methods in various new contexts. Senior researchers will find new perspectives, even some surprises and encouragements.

1.2. Outline of This Paper

In the remainder of Section 1 we give a brief historical account of the role of symmetry in quantum mechanics and of the work done in order to explore the symmetries of the Schrodinger hydrogen atom.

In Section 2, we provide some general background observations regarding symmetry groups and non-invariance groups and discuss the degeneracy groups for the Schrodinger, Dirac, and Klein-Gordon equations. We also introduce the uncommon $(Z\alpha)^{-1}$ eigenstates that allow us to treat the bound and scattering states in a uniform way, using the usual quantum numbers. In Section 3, the classical equations of motion of the nonrelativistic hydrogen atom in configuration and momentum space are derived from symmetry considerations. The physical meaning of the symmetry transformations and the structure of the degeneracy group SO(4) is discussed. In Section 4, we discuss the symmetries using the language of quantum mechanics. In order to display the symmetries in quantum mechanics in the most elegant and uniform way we use a basis of eigenstates of the inverse of the coupling constant $(Z\alpha)^{-1}$. In Section 5 we discuss these wave functions in momentum and configuration space, how they transform and their classical limit for Rydberg states.

In Section 6, we discuss the noninvariance or spectrum generating group of the hydrogen atom SO(4,l) and relate it to the conformal group in momentum space. In Section 7, the enlarged spectrum generating group SO(4,2) is introduced, with a discussion of the physical meaning of the generators. All of the physical states together form a basis for a unitary irreducible representation of these noninvariance groups. We derive manifestly Hermitian expressions in terms of the momentum and position canonical variables for the generators of the group transformations and obtain the values possible for the Casimir operators. We discuss the important subgroups of SO(4,2).

In Section 8, we use the group theory of SO(4,2) to determine the radiative shifts in energy levels due to the interaction of a spinless electron with its own radiation field, or equivalently with the quantum vacuum. In the nonrelativistic or dipole approximation the level shift contains a matrix element of a rotation operator of an O(1,2) subgroup of the group SO(4,2). We can sum this over all states, obtaining the character of the representation, yielding a single integral that is a generating function for the radiative shift for any level in the nonrelativistic or dipole approximation. A brief conclusion follows.

1.3. Brief History of Symmetry in Quantum Mechanics and Its Role in Understanding the Schrodinger Hydrogen Atom

The hydrogen atom is the fundamental two-body system and perhaps the most important tool of atomic physics and the continual challenge is to continually improve our understanding of the hydrogen atom and to calculate its properties to the highest accuracy possible. The current QED theory is the most precise of any physical theory [2]:

> The study of the hydrogen atom has been at the heart of the development of modern physics...theoretical calculations reach precision up to the 12th decimal place...high resolution laser spectroscopy experiments...reach to the 15th decimal place for the 1S–2S transition...The Rydberg constant is known to six parts in 10^{12} [2,3]. Today, the precision is so great that measurement of the energy levels in the H atom has been used to determine the radius of the proton.

Continual progress in understanding the properties of the hydrogen atom has been central to progress in quantum physics [4]. Understanding the atomic spectra of the hydrogen atom drove the discovery of quantum mechanics in the 1920's. The measurement of the Lamb shift in 1947 and its explanation by Bethe in terms of atom's interaction with the quantum vacuum fluctuations ushered in a revolution in quantum electrodynamics [5–7]. Exploring the symmetries of the hydrogen atom has been an essential part of this progress. Symmetry is a concept that has played a broader role in physics in general; for example, in understanding the dynamics of the planets, atomic, and molecular spectra, and the masses of elementary particles.

When applied to an isolated system, Newton's equations of motion imply the conservation of momentum, angular momentum and energy. But the significance of these conservation laws was not really understood until 1911 when Emily Nother established the connection between symmetry and conservation laws [8]. Rotational invariance in a system results in the conservation of angular momentum; translational invariance in space results in conservation of momentum; and translational invariance in time results in the conservation of energy. We will discuss Nother's Theorem in more detail in Section 2.

Another critical ingredient of knowledge, on which Nother based her proof, was the idea of an infinitesimal transformation, such as a infinitesimal rotation generated by the angular momentum operators in quantum mechanics. These ideas of infinitesimal transformations originated with the Norwegian mathematician Sophus Lie who was studying differential equations in the latter half of the nineteenth century. He studied the collection of infinitesimal transformations that would leave a differential equation invariant [9]. In 1918, German physicist and mathematician Hermann Weyl, in his classic book with the translated title "The Theory of Groups and Quantum Mechanics", would refer to this collection of differential generators leaving an operator invariant as a linear algebra, ushering in a little of the terminology of modern group theory [10]. Still this was a very early stage in understanding the role of symmetry in the language of quantum theory. When he introduced the new idea of a commutator on page 264, he put the word "commutator" in quotes. In the preface Weyl made a prescient observation: ".. the essence of the new Heisenberg-Schrodinger-Dirac quantum mechanics is to be found in the fact that there is associated with each physical system a set of quantities, constituting a non-commutative algebra in the technical mathematical sense, the elements of which are the physical quantities themselves".

A few years later Eugene Wigner published in German, "Group Theory and Its Application to the Quantum Mechanics of Atomic Spectra [11]". One might ask why was this classic not translated into English until 1959. In the preface to the English edition, Prof. Wigner recalled: "When the first edition was published in 1931, there was a great reluctance among physicists toward accepting group theoretical arguments and the group theoretical point of view. It pleases the author that this reluctance has virtually vanished.." It was the application of group theory in particle physics in the early sixties, such as SU(3) and chiral symmetry, which reinvigorated interest in Wigner's book and the field in general. In the 1940's, Wigner and Bargmann developed the representation theory of the Poincare group that later provided an infrastructure for the development of relativistic quantum mechanics [11,12].

The progress in understanding the symmetries of the hydrogen atom, in particular, has some parallels to the history of symmetry in general: there were some decades of interest but after the 1930's interest waned for about three decades in both fields, until stimulated by the work on symmetry in particle physics.

Probably, the first major advance in understanding the role of symmetry in the classical treatment of the Kepler problem after Newton's discovery of universal gravitation, elliptical orbits, and Kepler's Laws, was made two centuries ago by Laplace when he discovered the existence of three new constants of the motion in addition to the components of the angular momentum [13]. These additional conserved quantities are the components of a vector which determines the direction of the perihelion of the motion (point closest to the focus) and whose magnitude is the eccentricity of the orbit. The Laplace vector was later rediscovered by Jacobi and has since been rediscovered numerous times under different names. Today it is generally referred to as the Runge-Lenz vector. However, the significance of this conserved quantity was not well understood until the nineteen thirties.

In 1924, Pauli made the next major step forward in understanding the role of symmetry in the hydrogen atom [14]. He used the conserved Runge-Lenz vector **A** and the conserved angular momentum vector **L** to solve for the energy spectrum of the hydrogen atom by purely algebraic means, a beautiful result, yet he did not explicitly identify that **L** and **A** formed the symmetry group SO(4) corresponding to the degeneracy. At this time, the degree of degeneracy in the hydrogen energy levels was believed to be n^2 for a state with principal quantum number n, clearly greater than the degeneracy due to rotational symmetry which is $(2l+1)$. The n^2 degeneracy arises from the possible values of the angular momentum $l = 0, 1, 2, \ldots n-1$, and the $2l+1$ values of the angular momentum along the azimuthal axis $m = -l, -l+1, ...0, 1, 2, l+1$. The additional degeneracy was referred to as "accidental degeneracy [15]."

Six years after Pauli's paper, Hulthen used the new Heisenberg matrix mechanics to simplify the derivation of the energy eigenvalues of Pauli by showing that the sum of the squares $\mathbf{L}^2 + \mathbf{A}^2$ could be used to express the Hamiltonian and so could be used to find the energy eigenvalues [16]. In a one sentence footnote in this three page paper, Hulthen gives probably the most important information in his paper: Prof. Otto Klein, who had collaborated for years with Sophus Lie, had noticed that the two conserved vectors formed the generators of the Lorentz group, which we can describe as rotations in four dimensions, the fourth dimension being time. This is the non-compact group SO(3,1), the special orthogonal group in four dimension whose transformations leave the magnitude $g_{\mu\nu}z^{\mu}z^{\nu} = -t^2 + x^2 + y^2 + z^2$ unchanged [17]. Klein's perceptive observation triggered the introduction of group theory to understanding the hydrogen atom.

About a decade later, in 1935, the Russian physicist Vladimir Fock published a major article in Zeitshrift fur Physik, the journal in which all the key articles about the hydrogen atom cited were published [18]. He transformed Schrodinger's equation for a given energy eigenvalue from configuration space to momentum space, and did a stereographic projection onto a unit sphere, and showed that the bound state momentum space wave functions were spherical harmonics in four dimensions. He stated that this showed that rotations in four dimensions corresponded to the symmetry of the degenerate bound state energy levels in momentum space, realizing the group SO(4), the group of special orthogonal transformations which leaves the norm of a four-vector $U_0^2 + U_1^2 + U_2^2 + U_3^2$ constant. By counting the number of four-dimensional spherical harmonics Y_{nlm} in momentum space ($m = -l, -l+1 ... 0, 1, ... l$, where the angular momentum l can equal $l = n-1, n-2, ... 0$), he determined that the degree of degeneracy for the energy level characterized by the principal quantum number n was n^2. It is interesting that Fock did not cite the work by Pauli, implying the four dimensional rotational symmetry in configuration space. Fock also presented some ideas about using this symmetry in calculating form factors for atoms.

A year later, the German-American mathematician and physicist Valentine Bargmann showed that for bound states (E < 0) Pauli's conserved operators, the angular momentum **L** and the Runge–Lenz vector **A**, obeyed the commutation rules of the SO(4) [12]. His use of commutators was so early in the field of quantum mechanics, that Bargmann explained the square bracket notation he used for a commutator in a footnote [19]. He gave differential expression for the operators, adapting the approach of Lie generators in the calculation of the commutators. He linked solutions to Schrodinger's equation in parabolic coordinates to the existence of the conserved Runge–Lenz vector and was thereby able to establish the relationship of Fock's results to the algebraic representation of SO(4) for bound states implied by Fock and Pauli [12]. He also pointed out that the scattering states (E > 0) could provide a representation of the group SO(3,1). In a note at the end of the paper, Bargmann, who was at the University in Zurich, thanked Pauli for pointing out the paper of Hulthen and the observation by Klein that the Lie algebra of **L** and **A** was the same as the infinitesimal Lorentz group, which is how he referred to a Lie algebra. Bargmann's work was a milestone demonstrating the relationship of symmetry to conserved quantities and it clearly showed that to fully understand a physical system one needed to go beyond the usual ideas of geometrical symmetry. This work was the birth, in 1936, without much fanfare, of the idea of dynamical symmetry.

Little attention was paid to these developments until the 1960's, when interest arose primarily because of the applications of group theory in particle physics, particularly modeling the mass spectra of hadrons. Particle physicists were faced with the challenge of achieving a quantitative description of hadron properties, particularly the mass spectra and form factors, in terms of quark models. Since little was know about quark dynamics they turned to group-theoretical arguments, exploring groups like SU(3), chiral U(3)xU(3), U(6)xU(6), etc. The success of the eight-fold way of SU(3) (special unitary group in three dimensions) of American physicist Murray Gell-Mann in 1962 brought attention to the use of symmetry considerations and group theory as tools for exploring systems in which one was unsure of the exact dynamics [20].

In 1964, three decades after Fock's work, American physicist Julian Schwinger published a paper using SO(4) symmetry to construct a Green function for the Coulomb potential, which he noted was based on a class he taught at Harvard in 1949 [21]. The publication was a response to the then current emphasis on group theory and symmetry, which led to, as Israeli physicist Yuval Ne'eman described it, "'the great-leap-forward' in particle physics during the years 1961–1966 [22]". Some of the principal researchers leading this effort were Ne'eman [23], Gell-Mann [20,24,25], Israeli physicist Y. Dothan [26], Japanese physicist Yochiri Nambu [27], and English-American Freeman Dyson [28]. Advantage was taken of the mathematical infrastructures of group theory developed years earlier [10–12,29,30].

Interest was particularly strong in systems with wave equations with an infinite number of components, which characterize non-compact groups. In about 1965, this interest in particle physics gave birth to the identification of SO(4,1) and SO(4,2) as Spectrum Generating Algebras that might serve as models for hadronic masses. The hydrogen atom was seen as a model to explore the infinite dimensional representations of non-compact groups. The first mention of SO(4,1) was by Barut, Budini, and Fronsdal [31], where the H atom was presented as an illustration of a system characterized by non-compact representation, and so comprising an infinite number of states. The first mention of a six dimensional symmetry, referred to as the "non-compact group O(6)", appears to be by the Russian physicists I. Malin and V. Man'ko of the Moscow Physico-technical Institute [32]. In a careful three page paper, they showed that all of the bound states of the H atom energy spectrum in Fock coordinates provided a representation of this group, and they calculated the Casimir operators for their symmetric tensor representation in parabolic coordinates.

Very shortly thereafter, Turkish-American theoretical physicist Asim Barut and his student at University of Colorado, German theoretical physicist Hagen Kleinert, showed that including the dipole operator *er* as a generator led to the expansion of SO(4,1) to SO(4,2), and that all the bound states of the H atom formed a representation of SO(4,2) [33]. This allowed them to calculate dipole transition matrix elements algebraically. They give a position representation of the generators based on the use of parabolic coordinates. The generators of the transformations are given in terms of the raising and lowering operators for the quantum numbers for solutions to the H atom in parabolic coordinates. The dilation operator is used to go from one SO(4) subspace with one energy to a SO(4) subspace with different energy and it has a rather complicated form. They also used SO(4,2) symmetry to compute form factors [34].

The papers of the Polish-American physicist Myron Bander and French physicist Claude Itzakson published in 1966, when both were working at SLAC (Stanford Linear Accelerator in California) provide the first mathematically rigorous and "succinct" review of the O(4) symmetry of the H atom and provide an introduction to SO(4,1) [35,36], which is referred to as a spectrum generating algebra SGA, meaning that it includes generators that take the basis states from one energy level to another. They use two approaches in their mathematical analysis, the first is referred to as "the infinitesimal method," based on the two symmetry operators, **L** and **A** and the O(4) group they form, and the other, referred to as the "global method", first done by Fock, converts the Schrodinger equation to an integral equation with a manifest four dimensional symmetry in momentum space. They establish the equivalence of the two approaches by appealing to the solutions of the H atom in parabolic coordinates, and demonstrate that the symmetry operators in the momentum space correspond to the symmetry operators in the configuration space. As they note, the stereographic projection depends on the energy, so the statements for a SO(4) subgroup are valid only in a subspace of constant energy. They then explore the expansion of the SO(4) group to include scale changes so the energy can be changed, transforming between states of different principal quantum number, which correspond to different subspaces of SO(4). To insure that this expansion results in a group, they include other transformations, which results in the the generators forming the conformal group O(4,1). Their mathematical analysis introducing SO(4,1) is based on the projection of a p dimensional space (4 in the case of interest) on a parabaloid in p + 1 dimensions (5 dimensions). In their derivation they treat bound states in their first paper [35] and scattering states in the second paper [36].

As we have indicated the interest in the SO(4,2) symmetry of the Schrodinger equation was driven by a program focused on developing equations for composite systems that had infinite multiplets of energy solutions and ultimately could lead to equations that could be used to predict masses of elementary particles, perhaps using other than four dimensions [27,35–40]. In 1969 Jordan and Pratt showed that one could add spin to the generators $\boldsymbol{A}$ and $\boldsymbol{L}$, and still form a SO(4) degeneracy group. By defining $\boldsymbol{J} = \frac{1}{2}(\boldsymbol{L} + \boldsymbol{A}) + \boldsymbol{S}$, they showed one could obtain a representation of O(4,1) for any spin s [41].

In their review of the symmetry properties of the hydrogen atom, Bander and Itzakson emphasize this purpose for exploring the group theory of the hydrogen atom [35]:

> The construction of unitary representations of non-compact groups that have the property that the irreducible representations of their maximal subgroup appear at most with multiplicity one is of certain interest for physical applications. The method of construction used here in the Coulomb potential case can be extended to various other cases. The geometrical emphasis may help to visualize things and provide a global form of the transformations.

Special attention was also given to solutions for the hydrogen atom from the two body Bethe-Salpeter equation for a proton and electron interacting by a Coulomb potential, since the symmetry was that of a relativistic non-compact group [27,40,42,43].

Finally in 1969, five years after it was published, Schwinger's form of the Coulomb Green's function based on the SO(4) symmetry was used to calculate the Lamb shift by Michael Lieber, one of Schwinger's students at Harvard [21,44]. A year later, Robert Huff, a student of Christian Fronsdal at UCLA, focused on the use of the results from SO(4,2) group theory to compute the Lamb shift [45]. He converted the conventional expression for the Lamb shift into a matrix element containing generators of SO(4,2), and was able to perform rotations and scale changes to simplify and evaluate the matrix elements. After clever mathematical manipulation, he obtained an expression for the Bethe log in terms of a rapidly terminating series for the level shifts. He provided an appendix with a brief discussion of the fundamental of SO(4,2) representations for the H atom, showing the expressions for the three generators needed to express the Schrodinger equation.

In the next few years, the researchers published a few mathematically oriented papers [41,46–50], a short book [51] dealing with the symmetries of the Coulomb problem, and a paper by Barut presenting a SO(4,2) formulation of symmetry breaking in relativistic Kepler problems, with a 1 page summary of the application of SO(4,2) for the non-relativistic hydrogen atom [34,52]. Bednar published a paper applying group theory to a variety of modified Coulomb potentials, which included some matrix elements of SO(4,2) using hydrogen atom basis states with quantum numbers nlm [53]. There also was interest in application of the symmetry methods and dynamical groups in molecular chemistry [54] and atomic spectroscopy [55].

In the 1970's, researchers focused on developing methods of group theory and on understanding dynamical symmetries in diverse systems [56–58]. A book on group theory and its applications appeared in 1971 [59]. Barut and his collaborators published a series of papers dealing with the hydrogen atom as a relativistic elementary particle, leading to an infinite component wave equation and mass formula [60–63].

Papers on the classical Kepler problem, the Runge–Lenz vector, and SO(4) for the hydrogen atom have continued to appear over the years, from 1959 to today. Many were published in the 1970's [64–70] and some since 1980, including [71–75]. Papers dealing with SO(4,2) are much less frequent. In 1986 Barut, A. Bohm, and Ne'eman published a book on dynamical symmetries that included some material on the hydrogen atom [76]. In 1986, Greiner and Muller published the second edition of Quantum Mechanics Symmetries, which had six pages on the Hydrogen atom, covering only the SO(4) symmetry [77]. The 2005 book by Gilmore on Lie algebras has 4 pages of homework problems on the H atom to duplicate results in early papers [78]. The last papers I am aware of that used SO(4,2) were applications in molecular physics [79,80] and more general in scope [81]. Carl Wulfman published a book on dynamical symmetries in 2011, which provides a helpful discussion of dynamical symmetries for the hydrogen atom [82]. He regularizes the Schrodinger equation, essentially multiplying by r, obtaining Sturmian wave functions in parabolic coordinates. This approach allows for him to treat bound and scattering states for SO(4,2) at one time, but requires redefining the inner product, and leads to a non-Hermitian position operator.

1.4. The Dirac Hydrogen Atom

We have focused our discussion on the symmetries of the non-relativistic hydrogen atom described by the Schrodinger equation. Quantum mechanics also describes the hydrogen atom in terms of the relativistic Dirac equation, which we will only discuss briefly in this paper.

The gradual understanding of the dynamical symmetry of the Dirac atom parallels that of the Schrodinger atom, but it has received much less attention, probably because the system has less relevance for particle physics and for other applications. It was known that the rotational symmetry was present and that the equation predicted that the energy depended on the principal quantum number and the quantum number for the total angular momentum j, but not the spin s or orbital angular momentum l separately. This remarkable fact meant that, in some sense, angular momentum contributed the same to the total energy no matter whether it was intrinsic or orbital in origin. This degeneracy is lifted if we include the radiative interactions which leads to the Lamb shift.

In order to understand the symmetry group for the Dirac equation consider that for a given total angular momentum quantum number $J > 0$ there are two degenerate levels for each energy level of the Dirac hydrogen atom: one level has $l = J + 1/2$ and the other has $l = J - 1/2$. Since the l values differ by unity, the two levels have opposite parity. Dirac described a generalized parity operator K, which was conserved. For an operator Λ to transform one degenerate state into the other, it follows that the operator has to commute with J and have parity -1. This means it has to anticommute with K, and so it is a conserved pseudoscalar operator.

The parity $(-1)^{l+J-1/2}$ is conserved in time, so the states are parity eigenstates. Using the two symmetry operators Λ and K, one can build a SU(2) algebra. If we include the O(3) symmetry due to the conservation of angular momentum, we obtain the full symmetry group $SU(2)xO(3)$ which is isomorphic to $SO(4)$ for the degeneracy of the Dirac hydrogen atom.

In 1950, M.Johnson and B. Lippman discovered the operator Λ [83]. Further work was done on understanding Λ by Biedenharn [84]. The Johnson–Lippman operator has been rediscovered and reviewed several times over the decades [85–87]. It has been interpreted in the non-relativistic limit as the projection of the Runge–Lenz vector onto the spin angular momentum [87–89]. The SO(4) group can be expanded to include all states, and then the spectrum generating group is SO(4,1) or SO(4,2) depending on the assumptions regarding relativistic properties and the charges present [33,38]. We will not discuss the symmetries of the Dirac H atom further.

2. Background

2.1. The Relationship between Symmetry and Conserved Quantities

The nature of the relationship between symmetry, degeneracy, and conserved operators is implicit in the equation

$$[H, S] = 0 \tag{1}$$

where H is the Hamiltonian of our system, S is a Hermitian operator, and the brackets signify a commutator if we are discussing a quantum mechanical system, or i times a Poisson bracket if we are discussing a classical system. If S is viewed as the generator of a transformation on H, then Equation (1) says the transformation leaves H unchanged. Therefore, we say S is a symmetry operator of H and leaves the energy invariant. The fact that a non- trivial S exists means that there is a degeneracy. To show this, consider the action of the commutator on an energy eigenstate $|E\rangle$:

$$[H,S]|E\rangle = 0 \tag{2}$$

or

$$H(S|E\rangle) = E(S|E\rangle). \tag{3}$$

If S is nontrivial then $S|E\rangle$ is a different state from $|E\rangle$ but has the same energy eigenvalue. If we label all such degenerate states by

$$|E,m\rangle, m = 1, ..., N \tag{4}$$

then clearly $S|E,m\rangle$ is a linear combination of degenerate states:

$$S|E,m\rangle = S_{ms}|E,s\rangle. \tag{5}$$

S_{ms} is a matrix representation of S in the subspace of degenerate states. In a classical Kepler system S generates an orbit deformation that leaves H invariant. The existence of a nontrivial S therefore implies degeneracy, in which multiple states have the same energy eigenvalue. We can show that the complete set of symmetry operators for H forms a Lie algebra by applying Jacobi's identity to our set of Hermitian operators S_i :

$$[\mathrm{H}, \mathrm{S_i}] = 0, \quad \mathrm{i} = 1, \ldots, \mathrm{L}. \tag{6}$$

$$[S_j, [H, S_i]] + [S_i, [S_j, H]] + [H, [S_i, S_j]] = 0 \tag{7}$$

so

$$[H, [S_i, S_j]] = 0. \tag{8}$$

Therefore, the commutator of S_i and S_j is a symmetry operator of H. Either the commutator is a linear combination of all the symmetry operators $S_i, i = 1, ..., L$:

$$[S_i, S_j] = C_{ij}^k S_k \tag{9}$$

or the commutator defines a new symmetry operator which we label S_{L+1}. We repeat this procedure until the Lie algebra closes as in Equation (9).

By exponentiation, we assume that we can locally associate a group of unitary transformations

$$e^{iS_i a^i} \tag{10}$$

for real a^i with our Lie algebra and so conclude that a group of transformations exists under which the Hamiltonian is invariant [90]. We call this the symmetry or degeneracy group of H. Our energy eigenstates states form a realization of this group.

It is possible to form scalar operators, called Casimir operators, from the generators of the group that commute with all the generators of the group, and, therefore, have numerical values. The values of the Casimir operators characterize the particular representation of the group. For example, for the rotation group in three dimensions, the generators are $\boldsymbol{L} = (L_1, L_2, L_3)$ and the quantity $\boldsymbol{L}^2 = L(L+1)$ commutes with all of the generators. L can have any positive integer value for a particular representation. The Casimir operator for O(3) is $\boldsymbol{L}^2$. The number of Casimir operators that characterize a group is called the rank of the group. O(3) is rank 1 and SO(4,2) is rank 3.

Now let us consider Equation (1) in a different way. If we view H as the generator of translations in time, then we recall that the total time derivative of an operator S_i is

$$\frac{\mathrm{d}S_i}{\mathrm{d}t} = \frac{i}{\hbar}[\mathrm{H}, \ \mathrm{S_i}] + \frac{\partial \mathrm{S_i}}{\partial t} \tag{11}$$

where the commutator and the partial derivative give the implicit and explicit time dependence respectively. Provided that the symmetry operators have no explicit time dependence ($\frac{\partial S_i}{dt} = 0$), then Equation (11) implies that Equation (1) means that the symmetry operators S are conserved in time and $\frac{dS_i}{dt} = 0$. Conversely, we can say that conserved Hermitian operators with no explicit time dependence are symmetry operators of H. This very important relationship between conserved Hermitian operators and symmetry was first discovered by German mathematician Emmy Nother in 1917, and is called Nother's Theorem [8,91–95].

2.2. Non-Invariance Groups and Spectrum Generating Group

As we have discussed, the symmetry algebra contains conserved generators S_i that transform one energy eigenstate into a linear combination of eigenstates all with the same energy. In order to illustrate with hydrogen atom eigenstates:

$$S_i|nlm\rangle = \sum_{l',m'} S_{nlm}^{nl'm'}|nl'm'\rangle \tag{12}$$

where $|nlm\rangle$ refers to a state with energy E_n, angular momentum $l(l+1)$ and $l_z = m$.

A non-invariance algebra contains generators D_i that can be used to transform one energy eigenstate $|nlm\rangle$ into a linear combination of other eigenstates, with the same or a different energy, different angular momentum l, and different azimuthal angular momentum m:

$$D|nlm\rangle = \sum_{n',l',m'} D_{nlm}^{n'l'm'}|n'l'm'\rangle. \tag{13}$$

Because the set of energy eigenstates is complete, the action of the most general operator would be identical to that shown to Equation (13). Therefore this requirement alone is not sufficient to select the generators needed.

The goal is to expand the degeneracy group with its generators S_i into a larger group, so that some or all of the eigenstates form a representation of the larger group with the degeneracy group as a subgroup. Thus, we need to add generators G_i, such that the combined set of generators

$$\{S_i, G_j;\ \text{for all } i,j\} \equiv \{D_k;\ \text{for all } k\}$$

forms an algebra that closes

$$[D_i, D_j] = i\epsilon_{ijk}D_k. \tag{14}$$

This is the Lie algebra for the expanded group. To illustrate with a specific example, consider the O(4) degeneracy group with six generators [96]. One can expand the group to O(5) or O(4,1) which has ten generators by adding a four-vector of generators. One component might be a scalar and the other three a three-vector. The question then is can some or all of the energy eigenstates provide a representation of O(5)? If so, then this would be considered a non-invariance group. The group might be expanded further in order to obtain generators of a certain type or to include all states in the representation. For the H atom the generators D_i can transform between different energy eigenvalues meaning between eigenfunctions with different principal quantum numbers.

Another way to view the expansion of the Lie algebra of the symmetry group is to consider additional generators D_i that are constants in time [97] but do not commute with the Hamiltonian so

$$\frac{\mathrm{d}D_i}{\mathrm{d}t} = 0 = \frac{i}{\hbar}[\mathrm{H}, \quad \mathrm{D_i}] + \frac{\partial \mathrm{D_i}}{\partial t}.$$

If we make the additional assumption that the time dependence of the generators is harmonic

$$\frac{\partial^2 D_i(t)}{\partial t^2} = \omega_{in}D_n(t).$$

then generators D_i, and the first and second partial derivatives with respect to time could close under commutation, forming an algebra. This approach does not tell us what generators to add, but, as we demonstrate in Section 7.5, it does reflect the behavior of the generators that have been added to form the spectrum generating group in the case of the hydrogen atom.

We may look for the largest set of generators D_i, which can transform the set of solutions into itself in an irreducible fashion (meaning no more generators than necessary). These generators form the Lie algebra for the non-invariance or spectrum generating algebra [32,98]. If the generators for the spectrum generating algebra can be exponentiated, then we have a group of transformations for the spectrum generating group. The corresponding wave functions form the basis for a single irreducible representation of this group. This group generates transformations among all the solutions for all energy eigenvalues and it is called the Spectrum Generating Group [99]. For the H atom, SO(4,1) is a spectrum generating group or non-invariance group, which can be reduced to contain one separate SO(4) subgroup for each value of n.

To get a representation of SO(4,1), we need an infinite number of states, which we have for the H atom. This group has been expanded by adding a five vector to form SO(4,2) because the additional generators can be used to express the Hamiltonian and the dipole transition operator. The group SO(p,q) is the group of orthogonal transformations that preserve the quantity $X = x_1^2 + x_2^2 + ... + x_p^2 - ... - x_{p+q}^2$, which may be viewed as the norm or a p+q-dimensional vector in a space that has a metric with p plus signs and q minus signs. The letters SO stand for special orthogonal, meaning the orthogonal transformations have determinant equal to +1.

In terms of group theory, there is a significant difference between a group like SO(4) and SO(4,1). SO(4) and SO(3) are both compact groups, while SO(4,1) and SO(4,2) are non-compact groups. A continuous group G is compact if each function f(g), continuous for all elements g of the group G, is bounded. The rotation group in three dimensions O(3), which conserves the quantity $r^2 = x_1^2 + x_2^2 + x_3^2$, is an example of a compact group.

For a non-compact group, consider the Lorentz group O(3,1) of transformations to a coordinate system moving with a velocity v. The transformations preserve the quantity $r^2 - c^2t^2$. The matrix elements of the Lorentz transformations are proportional to $1/\sqrt{1-\beta^2}$, where $\beta = v/c$, and are not bounded as $\beta \to 1$. Therefore, r and ct may increase without bound, while the difference of the squares remains constant. Unitary representations of non-compact groups are infinite dimensional, for example, the representation of the non-invariance group SO(4,1) has an infinite number of states. Unitary representations of compact groups can be finite dimensional, for example, our representation of SO(4) for an energy level E_n has dimension n^2.

In the nineteen sixties and later, the spectrum generating group was of special interest in particle physics, because it was believed it could provide guidance where the precise particle dynamics were not known. The hydrogen atom provided a physical system as a model. Because the application was in particle physics, there was less interest in exploring representations in terms of the dynamical variables for position and momentum.

The expansion of the group from SO(4,1) to SO(4,2) was motivated by the fact that the additional generators could be used to write Schrodinger's equation entirely in terms of the generators, and to express the dipole transition operator. This allowed for algebraic techniques and group theoretical methods to be used to obtain solutions, calculate matrix elements, and other quantities [33,38].

2.3. Basic Idea of Eigenstates of $(Z\alpha)^{-1}$

We briefly introduce the idea behind these states, since they are unfamiliar [1]. The full derivation is given in Section 4. Schrodinger's equation in momentum space for bound states can be written as

$$\left[p^2 + a^2 - \frac{2mZ\alpha}{r}\right]|a\rangle = 0.$$

where $a^2 = -2mE > 0$ and $Z\alpha$ is the coupling constant, which we will now view as a parameter. This equation has well behaved solutions for certain discrete eigenvalues of the energy or a^2, namely

$$a_n^2 = -2mE_n$$

where $E_n = -\frac{1}{2}\frac{m(Z\alpha)^2}{n^2}$. We can write the eigenvalue condition equivalently as

$$(\frac{a_n}{mZ\alpha}) = \frac{1}{n}.$$

This last equation shows that solutions exist for certain values a_n of the RMS momentum a. To introduce eigenstates of $(Z\alpha)^{-1}$ we simply take a different view of this last equation and say that instead of quantizing a and obtaining a_n, we imagine that we quantize $(Z\alpha)^{-1}$, let a remain unchanged, obtaining

$$\frac{a}{m(Z\alpha)_n} = \frac{1}{n}.$$

Accordingly, now we can interpret Schrodinger's equation as an eigenvalue equation that has solutions for certain values of $(Z\alpha)^{-1}$ namely

$$(Z\alpha)_n^{-1} = \frac{m}{an}.$$

We have the same equation but can view the eigenvalues differently but equivalently. Instead of quantizing a we quantize $(Z\alpha)^{-1}$.

This roughly conveys the basic idea of eigenstates of the inverse of $(Z\alpha)$, but this simplified version does at all reveal the advantages of our reformulation because we have left the Hamiltonian unchanged. In Section 4, we transform Schrodinger's equation to an eigenvalue equation in a, so that the kernel is bounded, which means that there are no states with E > 0, no scattering states, and all states have the usual quantum numbers. Other important advantages to this approach will also be discussed.

2.4. Degeneracy Groups for Schrodinger, Dirac and Klein-Gordon Equations

The degeneracy groups for the bound states described by the different equations of the hydrogen atom are summarized in Table 1. The degeneracy (column 2) is due to the presence of conserved operators which are also symmetry operators (column 3), forming a degeneracy symmetry group (column 4). For example, The symmetry operators for the degeneracy group in the Schrodinger hydrogen atom are the angular momentum **L** and the Runge–Lenz vector **A**. In Section 3, it will be shown that together these are the generators for the direct product SO(3)xSO(3) which is isomorphic to SO(4). Column 5 gives the particular representations present. These numbers are the allowed values of the Casimir operators for the group and they determine the degree of degeneracy (last column) and the corresponding allowed values of the quantum numbers for the degenerate states.

The Casimir operators, which are made from generators of the group, have to commute with all of the members of the group, and the only way this can happen is if they are actually constants for the representation. The generators are formed from the dynamical variables of the H atom, so the Casimir operators are invariants under the group composed of the generators, and their allowed numerical values reflect the underlying physics of the system and determine the appropriate representations of the group [9,82,100]. For example, $\mathbf{L}^2$ is the Casimir operator for the group O(3) and can have the values $l(l+1)$. The relationship between Casimir operators and group representations is true for all irreducible group representations, including the SO(4) degeneracy group, as well as the spectrum generating group SO(4,2) [43,53].

For the Schrodinger equation, there are n^2 states $|nlm\rangle$ that form a representation of the degeneracy group SO(4) formed by $\boldsymbol{L}$ and $\boldsymbol{A}$. These states correspond to the principal quantum number n, the n

different values of the angular momentum quantum number l, and $2l+1$ different values of the z component of the angular momentum $l_Z = m$.

For the Dirac equation, the $2(2J+1)$ dimensional degeneracy group for bound states is realized by the total angular momentum operator **J**, the generalized parity operator K, and the Johnson–Lippman operator Λ, which together form the Lie algebra for SO(4).

For the fully relativistic Klein–Gordon equation, only the symmetry from rotational symmetry survives, leading to the degeneracy group O(3). If the V^2 term, the four-potential term squared is dropped in a semi-relativistic approximation as we describe in Section 4.3, then the equation can be rewritten in the same form as the non-relativistic Schrodinger equation, so a Runge–Lenz vector can be defined and the degeneracy group is again SO(4).

Table 1. In the table **L** = **rxp** is the orbital angular momentum; **A** is the Runge-Lenz vector; **J** = **L** + $\sigma/2$ is the total angular momentum; K is the generalized parity operator; Λ is the conserved pseudoscalar operator.

Degeneracy Groups for Bound States in a Coulomb Potential

Equation	Degeneracy	Conserved Quantities	Degeneracy Group	Representation	Dimension
Schrodinger	E indep. of l, l_z	**A**, **L**	$SO(4)$	$(\frac{n-1}{2}, \frac{n-1}{2})$	n^2
Klein-Gordon	E indep. of l_z	**L**	$O(3)$	Casimir op. is $l(l+1)$	$2l+1$
Klein-Gordon without V^2 term	E indep. of l, l_z	**A**, **L**	$SO(4)$	$(\frac{n-1}{2}, \frac{n-1}{2})$	n^2
Dirac	E depends on J, n only	$\Lambda, K,$ **J**	$SO(4)$	$(1/2, J)$	2(2J+1)

3. Classical Theory of the H Atom

In order to discuss orbital motion and the continuous deformation or orbits we give this discussion in terms of classical mechanics, but much of it is valid in terms of the Heisenberg representation of quantum mechanics if the Poisson brackets are converted to commutators, as will be discussed in Section 4.

For a charged particle in a Coulomb potential, there are two classical conserved vectors: the angular momentum **L**, which is perpendicular to the plane of the orbit, and the Runge–Lenz vector **A**, which goes from the focus corresponding to the center of mass and force along the semi-major axis to the perihelion (closest point) of the elliptical orbit. The conservation of **A** is related to the fact that non-relativistically the orbits do not precess. The Hamiltonian of our bound state classical system with an energy E < 0 is [17]

$$H = \frac{p^2}{2m} - \frac{Z\alpha}{r} = E \tag{15}$$

where m=mass of the electron, r is the location of the electron, p is its momentum, α is the fine structure constant, and E is the total non-relativistic energy.

The Runge–Lenz vector is

$$\mathbf{A} = \frac{1}{\sqrt{-2mH}}(\mathbf{p} \times \mathbf{L} - mZ\alpha\frac{\mathbf{r}}{r}) \tag{16}$$

where **L** is the angular momentum. From Hamilton's equation, $H = E$, so

$$\mathbf{A} = \frac{\mathbf{p} \times \mathbf{L}}{a} - \frac{mZ\alpha}{a}\frac{\mathbf{r}}{r} \tag{17}$$

where a is defined by

$$a = \sqrt{-2mE}. \tag{18}$$

From the virial theorem, the average momentum $\langle p^2 \rangle = -2mE$ so a is the root mean square momentum. We are discussing bound states so E < 0. It is straightforward to verify that **A** is conserved in time:

$$[\mathbf{A}, H] = \frac{d\mathbf{A}}{dt} = 0. \tag{19}$$

From the definition of **A** and the definition of angular momentum

$$\mathbf{L} = \mathbf{r} \times \mathbf{p} \tag{20}$$

if follows that **A** is orthogonal to the angular momentum vector

$$\mathbf{A} \cdot \mathbf{L} = 0. \tag{21}$$

Using the fact that **A** and **L** are conserved, we can easily obtain equations for the orbits in configuration and momentum space and the eccentricity, and other quantities, all usually derived by directly solving the equations of motion.

We will show that $\boldsymbol{A}$ and $\boldsymbol{L}$ are the generators of the group O(4). If we introduce the linear combinations $\boldsymbol{N} = \frac{1}{2}(\boldsymbol{L} + \boldsymbol{A})$ and $\boldsymbol{M} = \frac{1}{2}(\boldsymbol{L} - \boldsymbol{A})$, we find that $\boldsymbol{N}$ and $\boldsymbol{M}$ commute reducing the nonsimple group O(4) to SU(2)x SU(2), which we will discuss in Section 4.2 in the language of quantum mechanics.

3.1. Orbit in Configuration Space

In order to obtain the equation of the orbit one computes

$$\mathbf{r} \cdot \mathbf{A} = rAcos\phi_r = -r\frac{mZ\alpha}{a} + \mathbf{r} \cdot \mathbf{p} \times \mathbf{L}. \tag{22}$$

Noting that $\mathbf{r} \cdot \mathbf{p} \times \mathbf{l} = L^2$, we can solve for r

$$r = \frac{L^2/mZ\alpha}{(a/mZ\alpha)A\ cos\phi_r + 1} \quad . \tag{23}$$

This is the equation of an ellipse with eccentricity $e = (a/mZ\alpha)A$ and a focus at the origin (Figure 1). To find e we calculate $\mathbf{A} \cdot \mathbf{A}$ using the identity $\mathbf{p} \times \mathbf{L} \cdot \mathbf{p} \times \mathbf{L} = p^2L^2$ and obtain

$$A^2 = \frac{p^2L^2}{a^2} - \frac{2mZ\alpha}{a^2}\frac{L^2}{r} + \left(\frac{mZ\alpha}{a}\right)^2. \tag{24}$$

Substituting E for the Hamiltonian Equation (15) gives the usual result

$$\frac{a}{mZ\alpha}A = e = \sqrt{\frac{2EL^2}{m(Z\alpha)^2} + 1} \quad . \tag{25}$$

The length r_c of the semi-major axis is the average of the radii at the turning points

$$r_c = \frac{r_1 + r_2}{2}. \tag{26}$$

Using the orbit equation we find

$$r_c = \frac{L^2}{mZ\alpha}\frac{1}{1-e^2} \tag{27}$$

or

$$r_c = -\frac{Z\alpha}{2E} = \frac{mZ\alpha}{a^2}. \tag{28}$$

The energy depends only on the length of the semi-major axis r_c, not on the eccentricity. This important result is a consequence of the symmetry of the problem. It is convenient to parameterize the eccentricity in terms of the angle ν (see Figure 1) where

$$e = \sin\nu \tag{29}$$

From this definition and from Equations (25), (27), and (28) follow the useful results

$$L = r_c a\cos\nu, \qquad A = r_c a\sin\nu \tag{30}$$

which immediately imply

$$L^2 + A^2 = (r_c a)^2 = \left(\frac{mZ\alpha}{a}\right)^2. \tag{31}$$

This equation is the classical analogue of an important quantum mechanical result first obtained by Pauli and Hulthen allowing us to determine the energy levels from symmetry properties alone [14,16]. From Figure 1, it is apparent that this equation is a statement of Pythagoras's theorem for right triangles.

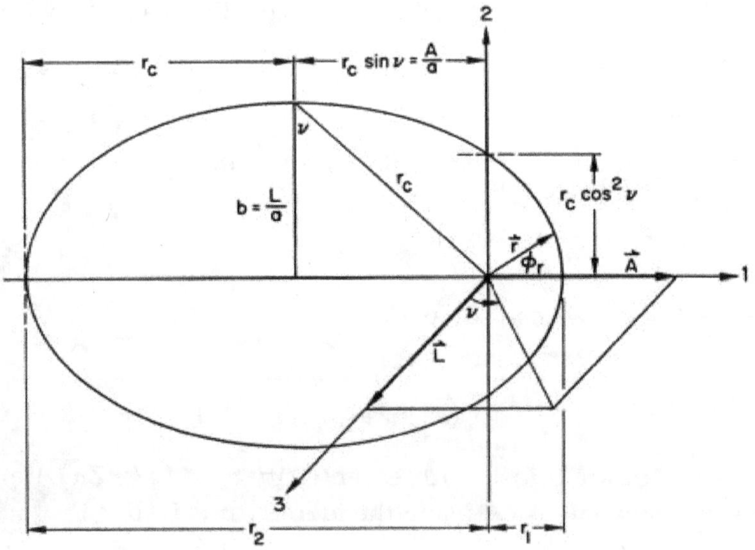

Figure 1. Classical Kepler orbit in configuration space. The orbit is in the 1–2 plane (plane of the paper). One focus, where the proton charge is located, is the origin. The semi-minor axis is $b = r_c\sin\nu$. The semi-major axis is r_c.

The energy equation (Equation (15)) and the orbit equation (Equation (23)) respectively may be rewritten in terms of a, r, and ν:

$$\frac{r_c}{r} = \frac{p^2 + a^2}{2a^2} \tag{32}$$

$$r = \frac{r_c\cos^2\nu}{1 + \sin\nu\cos\phi_r}. \tag{33}$$

3.2. The Period

To obtain the period, we use the geometrical definition of the eccentricity

$$e = \sqrt{1 - (b/r_c)^2} \tag{34}$$

where b is the semi-minor axis. Using $e = \sin\nu$ we find

$$b = r_c \cos\nu \tag{35}$$

so from Equation (30), we obtain

$$L = ab. \tag{36}$$

From classical mechanics, we know the magnitude of the angular momentum is equal to twice the mass times the area swept out by the radius vector per unit time. The area of the ellipse is $\pi b r_c$. It the period of the classical motion is T, then $L = 2m\pi b r_c/T$. Therefore, the classical period is

$$T = 2\pi\frac{mr_c}{a} = 2\pi\sqrt{\frac{m(Z\alpha)^2}{-8E^3}}. \tag{37}$$

and the classical frequency $\omega_{cl} = 2\pi/T$ is

$$\omega_{cl} = \frac{a}{mr_c}. \tag{38}$$

3.3. Group Structure SO(4)

The generators of our symmetry operations form the closed Poisson bracket algebra of $O(4)$:

$$[L_i, L_j] = i\epsilon_{ijk}L_k\ , \quad [L_i, A_j] = i\epsilon_{ijk}A_k\ , \quad [A_i, A_j] = i\epsilon_{ijk}L_k\ . \tag{39}$$

The brackets mean i times the Poisson bracket, which is the classical limit of a commutator. The first bracket says that the angular momentum generates rotations and forms a closed Lie algebra corresponding to O(3). The second bracket says that the Runge-Lenz vector transforms as a vector under rotations generated by the angular momentum. The last commutator says that the multiple transformations generated by the Runge-Lenz vector are equivalent to a rotation. Taken together the commutators form the Lie algebra of O(4). The connected symmetry group for the classical bound state Kepler problem is obtained by exponentiating our algebra giving the symmetry group SO(4). The scattering states with $E > 0$ form an infinite dimensional representation of the non-compact group SO(3,1).

We now want to determine the nature of the transformations generated by A_i and L_i. Clearly, $\boldsymbol{L} \cdot \delta\boldsymbol{\omega}$ generates a rotation of the elliptical orbit about the axis $\delta\boldsymbol{\omega}$ by an amount $\delta\omega$. To investigate the transformations generated by $\boldsymbol{A} \cdot \delta\boldsymbol{v}$ we assume a particular orientation of the orbit, namely that it is in the 1–2 (or x-y) plane and that $\boldsymbol{A}$ is along the 1-axis (see Figure 1). The more general problem is obtained by a rotation generated by L_i. For an example, we choose a transformation with $\delta\boldsymbol{v}$ pointing along the 2-axis, so that $\boldsymbol{A} \cdot \delta\boldsymbol{v} = A_2\delta v$. The change in $\boldsymbol{A}$ is defined by $\delta\boldsymbol{A}$, where

$$\delta\boldsymbol{A} = i[\boldsymbol{A} \cdot \delta\boldsymbol{v}, \boldsymbol{A}]. \tag{40}$$

From the Poisson bracket relations, we find for this particular case:

$$\delta A_1 = L_3\delta v\ , \quad \delta A_2 = 0\ , \quad \delta A_3 = -L_1\delta v \tag{41}$$

For our orbit, $L_1 = 0$ so $\delta A_3 = 0$. We perform a similar computation to find $\delta \boldsymbol{L}$. We find we can characterize the transformation by

$$\begin{array}{lcl} \delta A_1 = L_3 \delta v & \text{or} & \delta A = L \delta \nu \\ \delta L_3 = -A_1 \delta \nu & \text{or} & \delta L = -A \delta \nu \\ \delta e = \sqrt{1-e^2}\delta \nu & \text{or} & \delta(\sin \nu) = \cos \nu \delta \nu \end{array} . \tag{42}$$

Recalling $e = \sin \nu$ and Equation (30), we see that these transformations are equivalent to the substitution

$$\nu \longrightarrow \nu + \delta \nu. \tag{43}$$

In other words, the eccentricity of the orbit, and therefore $\boldsymbol{A}$ and $\boldsymbol{L}$ are all changed in such a way that the energy, a and r_c (length of the semi-major axis) remain constant. In our example, both $\boldsymbol{L}$ and $\boldsymbol{A}$ are changed in length but not direction, so the plane and orientation of the orbit are unchanged. The general transformation $\boldsymbol{A} \cdot \delta \boldsymbol{\mu}$ will also rotate the plane of the orbit or the semi-major axis.

Figure 2 shows a set of orbits in configurations space with different values of the eccentricity $e = \sin \nu$, but the same total energy and the same semimajor axis r_c, which is the bold hypotenuse. The bold vertical and horizontal legs are A/a and L/a and they are related to the hypotenuse r_c by Pythagoras's theorem. The generator $A_2 \nu$ produces a deformation of the circular orbit into the various elliptical orbits shown. This classical degeneracy corresponds to the quantum mechanical degeneracy in energy levels that occurs for different eigenvalues of the angular momentum with a fixed principal quantum

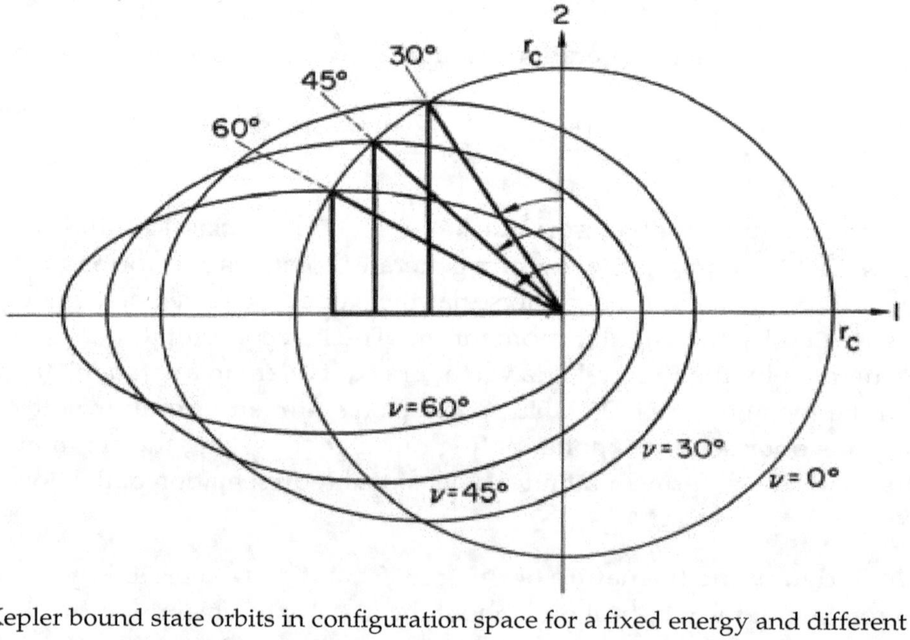

Figure 2. Kepler bound state orbits in configuration space for a fixed energy and different values of the eccentricity $e = \sin \nu$. The bold hypotenuse is the semi-major axis r_c, which makes an angle ν with the vertical 2-axis.

We can visualize all possible elliptical orbits for a fixed total energy or semi-major axis through a simple device. It is possible to produce an elliptical orbit with eccentricity $\sin \nu$ as the shadow of a circle of radius r_c which is rotated an amount ν about an axis perpendicular to the illuminating light. With a complete rotation of the circle, we will see all possible classical elliptical orbits corresponding to a given total energy. In quantum mechanics only certain angles of rotation would be possible corresponding to the quantized values of L. As the circle is rotated, we must imagine that the force center shifts as the sine of the angle of rotation, so that it always remains at the focus [101].

3.4. The Classical Hydrogen Atom in Momentum Space

We can derive the equation for the classical orbit in momentum space of a particle bound in a Coulomb potential using the conserved operators $\boldsymbol{L}$ and $\boldsymbol{A}$. For convenience, we assume that we have rotated our axes, so that $\boldsymbol{L}$ lies along the 3-axis and $\boldsymbol{A}$ the 1-axis, as shown in Figure 1. We compute

$$\boldsymbol{p}\cdot\boldsymbol{A} = p_1 A = \frac{-mZ\alpha}{a}\boldsymbol{p}\cdot\frac{\boldsymbol{r}}{r} \equiv \frac{-mZ\alpha}{a}p_r \tag{44}$$

and we employ Equations (28) and (30) to show $A = \frac{mZ\alpha}{a}\sin\nu$, to obtain [102]

$$p_r = -\sin\nu p_1 \tag{45}$$

which we substitute into the identity

$$p_r^2 + \frac{\boldsymbol{L}^2}{r^2} = p^2 = p_1^2 + p_2^2 \tag{46}$$

Using Equations (30) and (32) we find

$$1 = \left(\frac{2ap_1}{p^2+a^2}\right)^2 + \left(\frac{2ap_2}{p^2+a^2}\right)^2 \frac{1}{\cos^2 v} \tag{47}$$

and

$$p^2 - a^2 = 2ap_2\tan\nu \tag{48}$$

which may also be written as

$$p_1^2 + (p_2 - a\tan\nu)^2 = \frac{a^2}{\cos^2\nu}. \tag{49}$$

From Equation (49), we see the orbit in momentum space is a circle of radius $a/\cos\nu$ with its center displaced from the origin a distance $a\tan\nu$ along the 2-axis. Figure 3 shows the momentum space orbit that corresponds to the configuration space orbit in Figure 1.

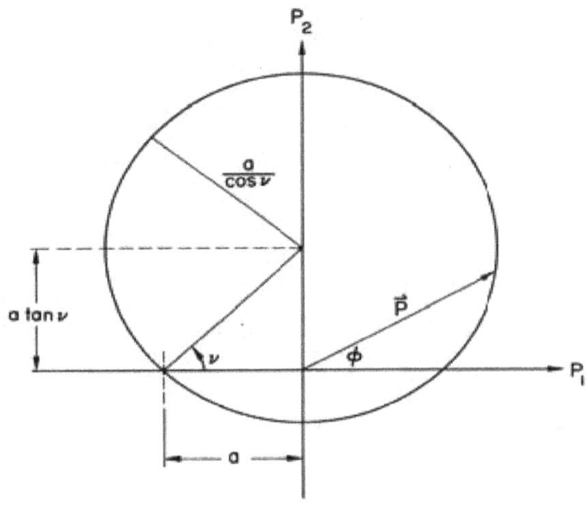

Figure 3. Kepler orbit in momentum space of radius $a/\cos\nu$, with its center at $p_2 = a\tan\nu$, corresponding to the orbit in configuration space shown in Figure 1. A circular orbit in configuration space corresponds to a circular orbit in momentum space centered on the origin with radius a.

As an alternative method of showing the momentum space orbits are circular, we can compute [103]

$$\left(\boldsymbol{p} - a\frac{\boldsymbol{L}\times\boldsymbol{A}}{\boldsymbol{L}^2}\right)^2 = C^2. \tag{50}$$

Using the lemma

$$p \times A = -\frac{L}{2a}(p^2 - a^2), \tag{51}$$

the fact that $L \cdot A = 0$, and Equation (30), we find $C = \frac{a}{\cos\nu}$. The orbit is a circle of radius $\frac{a}{\cos\nu}$ whose center lies at $a\frac{L\times A}{L^2}$, in agreement with the previous result.

We now consider what the generators A_i and L_i do to the orbit in momentum space. Clearly, L generates a rotation of the axes. For an A_i transformation, consider the same situation that we considered in our discussion of the configuration space orbit (see Figures 1 and 3). Because the generator $A_2\delta\nu$ changes ν to $\nu + \delta\nu$, we conclude that in momentum space this shifts the center of the orbit along the 2-axis and changes the radius of the circle. However, the distance a from the 2-axis to the intersection of the orbit with the 1-axis remains unchanged. Figure 4 shows a set of momentum space orbits for a fixed energy that correspond to the set of orbits in configuration space shown in Figure 2.

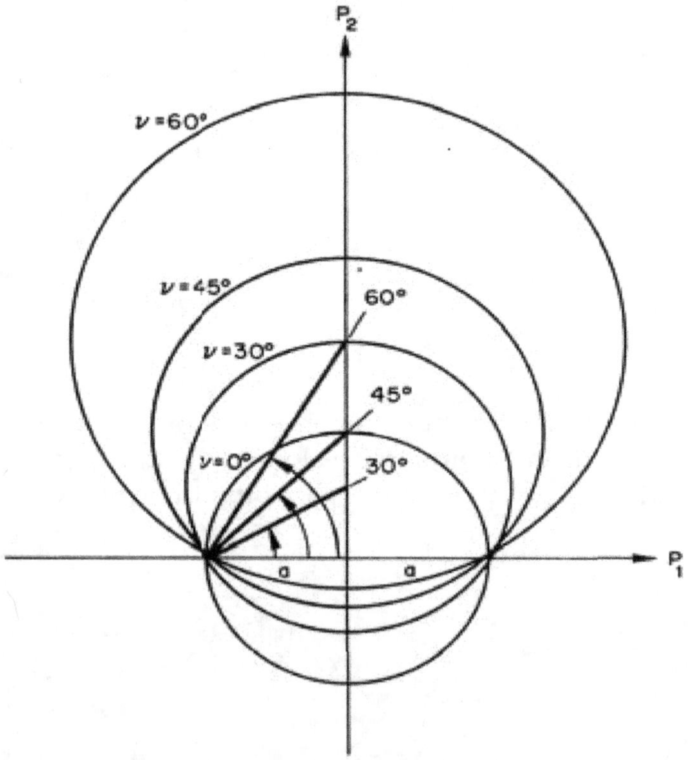

Figure 4. Kepler orbits in momentum space for a fixed energy and RMS momentum a with different values of the eccentricity $e = \sin\nu$ corresponding to the orbits in configuration space shown in Figure 2.

3.5. Four-Dimensional Stereographic Projection in Momentum Space

It is interesting that in classical mechanics the bound state orbits in a Coulomb potential are simpler in momentum space than in configuration space. In quantum mechanics the momentum space wave functions become simply four-dimensional spherical harmonics if one normalizes the momentum p by dividing by the RMS momentum $a = \sqrt{-2mE}$ and performs a stereographic projection onto a unit hyper-sphere in a four-dimensional space [18,35]. We will do the analogous projection procedure for the classical orbits. The three-dimensional momentum space hyperplane passes through the center of the four-dimensional hypersphere, as shown in Figure 5. The unit vector in the fourth direction is $\hat{n} = (1, 0, 0, 0)$. The unit vector $\hat{U}$ goes from the center of the sphere to the surface of the hypersphere, where it is intersected by the line connecting the vector p/a to the north pole of the sphere.

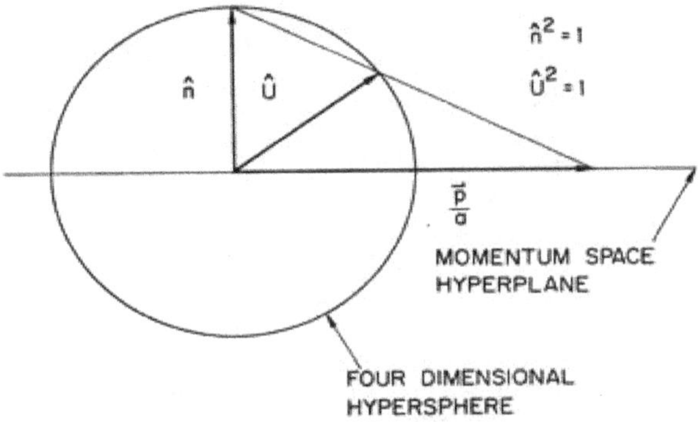

Figure 5. Stereographic projection in momentum space for a fixed energy, mapping $\boldsymbol{p}/a$ into $\hat{U}$. The unit vector in the four direction is $\hat{n}$ and $\hat{n} \cdot \hat{U} = \cos\Theta_4$.

We find

$$U_i = \frac{2ap_i}{p^2 + a^2} \quad i = 1, 2, 3. \quad U_4 = \frac{p^2 - a^2}{p^2 + a^2} \tag{52}$$

or inverting,

$$p_i = \frac{aU_i}{1 - U_4} \qquad p^2 = a^2 \frac{1 + U_4}{1 - U_4}. \tag{53}$$

Momentum space vectors for which $p/a < 1$ are mapped onto the lower hyperhemisphere. The advantage of this projection over one in which the hypersphere is tangent to the hyperplane is that we may have $|\hat{n}| = |\hat{U}| = 1$. At times it is convenient to describe $\hat{U}$ in terms of spherical polar coordinates in four dimensions. Because $\hat{U}$ is a unit vector we define

$$\begin{aligned} U_4 &= \cos\theta_4 \\ U_3 &= \sin\theta_4 \cos\theta \\ U_2 &= \sin\theta_4 \sin\theta \sin\phi \\ U_1 &= \sin\theta_4 \sin\theta \cos\phi \end{aligned} \tag{54}$$

where θ and ϕ are the usual coordinates in three dimensions. By comparison to Equation (52), we have

$$\theta_4 = 2\cot^{-1}\frac{p}{a} \qquad \theta = \cos^{-1}\frac{p_3}{p} \qquad \phi = \tan^{-1}\frac{p_2}{p_1} \tag{55}$$

3.6. Orbit in U space

We want to find the trajectory of the particle on the surface of the hypersphere corresponding to the Kepler orbits in configuration space or the displaced circles in momentum space. We assume we have rotated the axes in configuration space so that $\boldsymbol{L}$ is along the 3-axis and $\boldsymbol{A}$ is along the 1-axis, as shown in Figure 1. The equation for the orbit in three-dimensional momentum space is given by Equations (48) or (50). Dividing Equation (48) by $p^2 + a^2$ immediately gives a parametric equation for the projected orbit in U space:

$$U_4 = U_2 \tan\nu. \tag{56}$$

Because the orbit is in the 1–2 plane in configurations space, $U_3 = 0$. The orbit lies in a hyperplane perpendicular to the 2–4 plane that goes through the origin and makes an angle $\pi/2 - \nu$ with the 4-axis, as shown in Figure 6 [104]. The orbit is the intersection of this plane with the hypersphere and it is therefore a great circle. To derive the exact equation for the projected orbit we express p in Cartesian components p_1 and p_2 in Equation (48) and substitute Equation (52) obtaining

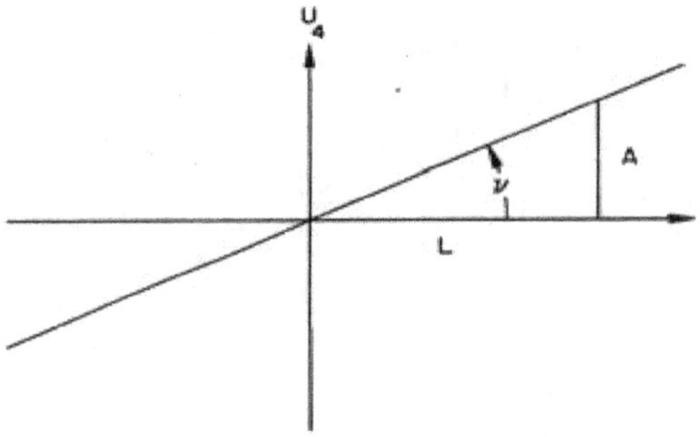

Figure 6. Showing the hyperplane containing the orbit making an angle ν with the U_2 plane. Notice $\tan\nu = A/L$ as required by Equation (30).

$$U_l^2 + U_2^2 - 2\tan\nu\, U_2(l - U_4) = (1 - U_4)^2. \tag{57}$$

To interpret this equation, we consider it in a rotated coordinate system. If we perform a rotation by an amount $\delta\nu$ about the 1–3 plane ($A_2\delta\nu$ is the generator of this rotation), the equations of transformation may be written [105]

$$\begin{aligned} U_2 &= U_2' \cos\delta\nu + U_4' \sin\delta\nu, \quad U_3 = U_3' \\ U_4 &= U_4' \cos\delta\nu - U_2' \sin\delta\nu, \quad U_1 = U_1'. \end{aligned} \tag{58}$$

This transformation is equivalent to making the substitution $\nu \longrightarrow \nu + \delta\nu$ in the equations relating to the orbit. For example, Equation (56) becomes

$$U_4' = U_2' \tan(\nu + \delta\nu). \tag{59}$$

We choose $\delta\nu = -\nu$, which means the orbital plane becomes $U_4' = 0$. Writing Equation (57) in terms of the primed coordinates, we find

$$U_1'^2 + U_2'^2 = 1 \tag{60}$$

which in the original system is

$$U_1^2 + (U_2\cos\nu + U_4\sin\nu)^2 = 1. \tag{61}$$

This is the equation of a great hypercircle $(\nu, 0)$ centered at the origin and lying in a hyperplane making an angle $\pi/2 - \nu$ with the 4-axis and an angle $\pi/2 - 0$ with the 3-axis. If $\boldsymbol{L}$ did not lie along the 3-axis, but, for example, was in the 1–3 plane, at an angle Θ from the 3-axis, then Equation (61) would be modified by the substitution

$$U_1 \longrightarrow U_1 cos\Theta + U_3 \sin\Theta \tag{62}$$

which follows, since U_i transforms as a three-vector. The corresponding great circle (ν, Θ) lies in a hyperplane making an angle $\pi/2 - \nu$ with the 4-axis and $\pi/2 - \Theta$ with the 3-axis.

The motion of the orbiting particle corresponds to a dot moving along the great circle (ν, 0 or Θ) with a period T given by the classical period Equation (37). The velocity in configuration space can be expressed in terms of U_4 by using its definition in terms of p^2 Equation (53) or in terms of θ_4 Equation (55). The particle is moving at maximum velocity when θ_4 is a minimum, which occurs at the perihelion when $\theta_4 = \pi/2 - \nu$:

$$max(\frac{p}{a}) = \sqrt{\frac{1+e}{1-e}} = \sqrt{\frac{r_2}{r_1}} \tag{63}$$

and at a minimum velocity when $\theta_4 = \pi/2 + \nu$:

$$min(\frac{p}{a}) = \sqrt{\frac{1-e}{1+e}}. \tag{64}$$

These values of θ_4 correspond to turning points, at which r and p have extreme values. This is apparent when we use Equation (32) for the total energy to show

$$U_4 = \frac{r_c - r}{r_c}. \tag{65}$$

When $r > r_c$ then $p^2 < a^2$, so the particle is moving more slowly than the RMS velocity. Applying the virial theorem to any orbit we find $\langle p^2 \rangle = a^2$ so as expected a is the RMS momentum and $\langle \frac{1}{1-U_4} \rangle = 1 = \langle r_c/r \rangle$.

Figure 7 is a picture of a simple device illustrating the stereographic projection of the orbit in p/a-space onto the four-dimensional hypersphere in U-space. We assume that the orbit is in the 1–2 plane and that $\boldsymbol{A}$ lies along the 1-axis, so $p_3 = 0$, $U_3 = 0$. Because of this trivial dependence on p_3, we have omitted the 3-axis. The vertical pin or rod represents the unit vector $\hat{n}$ lying along the 4-axis. The circumference of the larger circle perpendicular to the 4-axis represents the orbit in p/a-space. One can see that it is displaced from the origin along the 2-axis. Centered at the origin, we must imagine a hypersphere of unit radius $\hat{U}^2 = 1$. The stereographic projection $\hat{U}$ of the vector $\boldsymbol{p}/a$ is obtained by placing the string coming from the top of n directly at the head of the vector $\boldsymbol{p}/a$. The intersection of the string with the unit hypersphere defines $\hat{U}$. As the string is moved along the orbit in p/a-space, it intersects the hypersphere along a great circle shown by the circumference of the unit circle making an angle ν with the 1–2 plane. We can see for example, that at the closest approach, θ_4 is a minimum and U4 is a maximum, $\hat{U} \cdot \boldsymbol{A} = U_1$ is a minimum, and p/a is a maximum.

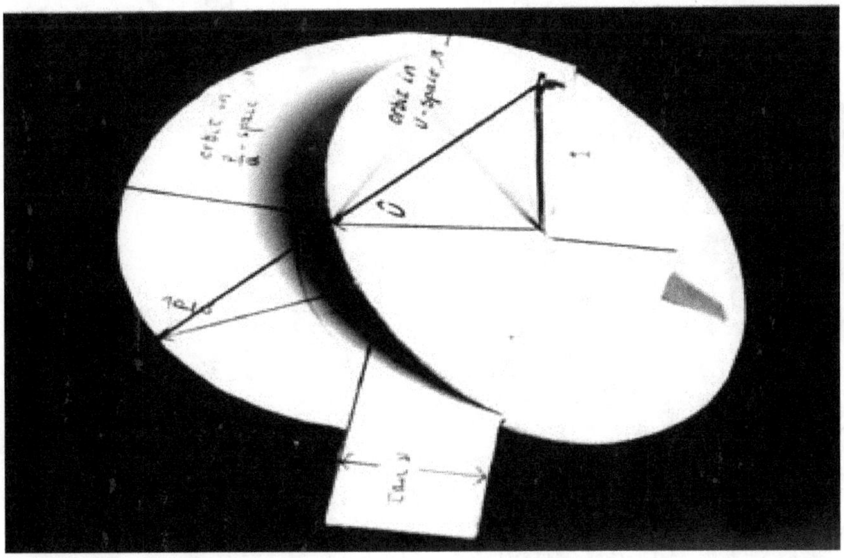

Figure 7. Model illustrating the stereographic projection from the 1–2 plane to a 4-d hypersphere. The pin represents the unit vector $\hat{n}$ along the 4-axis, normal to the 1–2 plane.

3.7. Classical Time Dependence of Orbital Motion

We can determine the time dependence of the orbital motion by integrating the expression for the angular momentum $L = mr^2 d\phi_r/dt$

$$\int dt = \int \frac{mr^2}{L} d\phi_r \tag{66}$$

where r is given by the orbit equation Equation (33) and we are assuming the orbit is in the 1–2 plane. After integrating, we can use the equations relating the momentum space and configuration space variables to obtain the time dependence in p-space and U-space. We obtain

$$\frac{1}{r_c^2}\frac{1}{\cos\nu}\frac{L}{m}\int_0^t dt = \cos^3\nu\int_0^{\phi_r(t)}\frac{d\phi_r}{(1+\sin\nu\cos\phi_r)^2}. \tag{67}$$

The left-hand side of this equation is equal $\omega_{cl}t$, where ω_{cl} is the classical frequency. This follows by substituting Equations (30) and (38)

$$L = ar_c\cos\nu \qquad \omega_{cl} = \frac{a}{mr_c}. \tag{68}$$

The integral on the right side gives [106]

$$\omega_{cl}t = -\frac{\sin\nu\cos\nu\sin\phi_r}{1+\sin v\cos\phi_r} - \tan^{-1}\frac{\cos\nu\sin\phi_r}{\sin\nu+\cos\phi_r} \tag{69}$$

which may be simplified as

$$\omega_{cl}t = -\frac{A}{L}\frac{y}{r_c} - \tan^{-1}\frac{A}{L}\frac{y}{r_c - r} \tag{70}$$

where $y = r\sin\phi_r$.

The relationship between the angle $\phi \equiv \phi_p$ in momentum space and ϕ_r in configuration space follows by either differentiating the orbit equation Equation (33) with respect to time and using $L = mr^2\dot{\phi}_r$ or by simultaneously solving the configuration space orbit, the momentum space orbit equation Equation (48) and the energy equation Equation (32). We find

$$\sin\phi_r = -p\cos\phi\frac{\cos\nu}{a} \qquad \cos\phi_r = p\sin\phi\frac{\cos v}{a} - \sin\nu. \tag{71}$$

From these equations, the definitions of the U_i, Equation (52), and the orbit and energy equations, it follows that for the classical orbit in the 1–2 plane

$$\begin{aligned} U_1 &= -\frac{r\sin\phi_r}{r_c\cos\nu} \\ U_2 &= \frac{r}{r_c}\frac{\sin\nu+\cos\phi_r}{\cos v} = U_4\cot\nu \\ U_3 &= 0 \\ U_4 &= \frac{r_c - r}{r_c}. \end{aligned} \tag{72}$$

Using these results in Equation (69) gives

$$\omega_{cl}t = U_1\sin\nu + \tan^{-1}\left(\frac{U_2}{U_1\cos\nu}\right) \tag{73}$$

which gives the time dependence in U space, and it agrees with the results of [97,107]. We could do rotations to generalize this result [102]. We can also rewrite the inverse tangent as $\cos^{-1}U_1$ using

$$U_1^2 + \frac{U_2^2}{cos^2\nu} = 1. \tag{74}$$

If we consider the last two equation for circular orbits with $e = \sin\nu = 0$ we obtain $\tan^{-1}(U_2/U_1) = \phi(t) = \omega_{cl}t = \phi_r(t) + \pi/2$, and our familiar circle $U_1^2 + U_2^2 = 1$.

Remark on Harmonic Oscillator

We can find a conserved Runge–Lenz vector for the non-relativistic hydrogen atom because the elliptical orbit does not precess, as it does for the relativistic atom. The only central force laws that yield classical elliptical orbits that do not precess are the inverse Kepler force and the linear harmonic oscillator force [108]. Thus, it seems reasonable that one could construct a constant vector similar to $\boldsymbol{A}$ for the oscillator, although the force center for the atom is at a focus and for the oscillator it is at the center of the ellipse. However, it is not readily possible [109]. Instead, one can construct a constant Hermitian second rank tensor T_{ij}:

$$T_{ij} = \frac{1}{m\omega_0} p_i p_j + m\omega_0 x_i x_j. \tag{75}$$

This constant tensor is analogous to the moment of inertia tensor for rigid body motion. The eigenvectors of the tensor will be constant vectors along the principal axes for the particular orbit being considered. The existence of the conserved tensor leads to the U(3) symmetry algebra of the oscillator. The generators are $\lambda^a_{ij} T_{ij}$, where the λ's are are the usual $U(3)$ matrices [22]. The spectrum generating algebra is SU(3,1).

In another approach, the Schrodinger equation for the hydrogen atom has been transformed into an equation for a four dimensional harmonic oscillator or two two dimensional harmonic oscillators. This approach which fits well with parabolic coordinates was used especially in the 1980's to analyze the group structure of the atom and relate it to SU(3) [110–119]. We will not discuss this approach further.

4. The Hydrogenlike Atom in Quantum Mechanics; Eigenstates of the Inverse of the Coupling Constant

In this section we switch from classical dynamics to quantum mechanics and discuss the group structure and exploit it to determine the bound state energy spectrum directly, as Pauli and followers did almost a century ago [14,16]. In Section 4.3 we introduce a new set of basis states for the hydrogenlike atom, eigenstates of the coupling constant. Using these states allows us to display the symmetries in the most convenient manner and to treat bound and scattering states uniformly.

4.1. The Degeneracy Group SO(4)

The quantum mechanical Hamiltonian is

$$H = \frac{p^2}{2m} - \frac{Z\alpha}{r} = E. \tag{76}$$

The classical expression for the Runge–Lenz vector needs to be symmetrized to insure the corresponding quantum mechanical operator $\boldsymbol{A}$ is Hermitian:

$$\boldsymbol{A} = \frac{1}{\sqrt{-2mH}} \left(\frac{\boldsymbol{p} \times \boldsymbol{L} - \boldsymbol{L} \times \boldsymbol{p}}{2} - mZ\alpha \frac{\boldsymbol{r}}{r} \right). \tag{77}$$

We may verify that $\boldsymbol{A}$ and $\boldsymbol{L} = \boldsymbol{r} \times \boldsymbol{p}$ both commute with the Hamiltonian H. The commutation relations of L_i and A_i are the same as the corresponding classical Poisson bracket relations for bound states:

$$[L_i, L_j] = i\epsilon_{ijk} L_k\,, \quad [L_i, A_j] = i\epsilon_{ijk} A_k\,, \quad [A_i, A_j] = i\epsilon_{ijk} L_k \tag{78}$$

and form the algebra of O(4) [35]. We can write the commutation relations in a single equation which makes the 0(4) symmetry explicit. If we define

$$S_{ij} = \epsilon_{ijk} L_k \qquad S_{i4} = A_i \tag{79}$$

then

$$[S_{ab}, S_{cd}] = i(\delta_{ac}S_{bd} + \delta_{bd}S_{ac} - \delta_{ad}S_{bc} - \delta_{bc}S_{ad}) \quad a, b = 1,2,3,4. \tag{80}$$

The Kronecker delta function δ_{ab} acts like a metric tensor.

4.2. Derivation of the Energy Levels

We can obtain the energy levels by determining which representations of the group SO(4) are realized by the degenerate eigenstates of the hydrogenlike atom [12,14,35]. The representations of SO(4) may be characterized by the numerical values of the two Casimir operators for SO(4):

$$C_1 = \boldsymbol{L} \cdot \boldsymbol{A} \qquad C_2 = \boldsymbol{L}^2 + \boldsymbol{A}^2 \tag{81}$$

Once we know the value of C_2, then the eigenvalues of H follow from the quantum mechanical form of Equation (31), namely

$$\boldsymbol{L}^2 + \boldsymbol{A}^2 + 1 = \frac{(mZ\alpha)^2}{-2mH}. \tag{82}$$

In order to determine the possible values of C_2, we factor the 0(4) algebra into two disjoint SU(2) algebras [120], each of which has the same commutation relations as the ordinary angular momentum operators,

$$\boldsymbol{N} = \tfrac{1}{2}(\boldsymbol{L} + \boldsymbol{A}) \qquad \boldsymbol{M} = \tfrac{1}{2}(\boldsymbol{L} - \boldsymbol{A}). \tag{83}$$

The commutation relations are

$$[M_i, N_j] = 0 \quad [M_i, M_j] = i\epsilon_{ijk}M_k \quad [N_i, N_j] = i\epsilon_{ijk}N_k. \tag{84}$$

In analogy with the results for the ordinary angular momentum operators, the Casimir operators are

$$\begin{aligned} M^2 &= j_1\,(j_2+1)\,, \quad j_1 = 0, \tfrac{1}{2}, 1, \ldots \\ N^2 &= j_2\,(j_2+1)\,, \quad j_2 = 0, \tfrac{1}{2}, 1, \ldots \end{aligned} \tag{85}$$

The numbers j_1 and j_2, which may have half-integral values for SU(2) but not O(3), define the (j_1, j_2) representation of SO(4). From the definitions of A and L in terms of the canonical variables, it follows that $C_1 = \boldsymbol{L} \cdot \boldsymbol{A} = 0$ which means $j_1 = j_2 = j$, as in the classical case. For our representations, we find

$$\boldsymbol{M}^2 = \boldsymbol{N}^2 = \tfrac{1}{4}(\boldsymbol{L}^2 + \boldsymbol{A}^2) = j(j+1), j = 0, \tfrac{1}{2}, 1, .. \tag{86}$$

and therefore

$$\boldsymbol{L}^2 + \boldsymbol{A}^2 + 1 = (2j+1)^2. \tag{87}$$

Substituting this result into Equation (82) gives the usual formula for the bound state energy levels of the hydrogen atom:

$$H' = -\frac{m(Z\alpha)^2}{2n^2} = E_n \tag{88}$$

where the principal quantum number n = 2j+1 = 1,2,.. and the prime on H signifies an eigenvalue of the operator H.

Within a subspace of energy E_n, the Runge–Lenz vector is

$$\boldsymbol{A} = \frac{1}{a_n}\left(\frac{\boldsymbol{p} \times \boldsymbol{L} - \boldsymbol{L} \times \boldsymbol{p}}{2} - mZ\alpha\frac{\boldsymbol{r}}{r}\right). \tag{89}$$

where $a_n = \sqrt{-2mE_n} = \frac{mZ\alpha}{n}$.

Our considerations of the Casimir operators have shown that the hydrogen atom provides completely symmetrical tensor representations of SO(4), namely, $(j, j) = (\frac{n-1}{2}, \frac{n-1}{2}), \quad n = 1, 2, \ldots$ The dimensionality is n^2, corresponding to the n^2 degenerate states. The appearance of only symmetrical tensor representations ($j_1 = j_2$) may be traced to $\boldsymbol{L} \cdot \boldsymbol{A}$ vanishing, which is a consequence of the structure of $\boldsymbol{L}$ and $\boldsymbol{A}$ in terms of the dynamical variables for the hydrogenlike atom. For systems other than the hydrogenlike atom, it is not generally possible to find the expression for the energy levels in terms of all the different quantum numbers alone. It worked here, since we could express the Hamiltonian as a function of the Casimir operators that contained all quantum numbers explicitly.

There are a variety of possible basis states. We could choose basis states for the SO(4) representation that reflect the SU(2) decomposition, namely eigenstates of $\boldsymbol{M}^2$, $\boldsymbol{N}^2$, M_3 and N_3 [120]. Another possibility is to have a basis with eigenstates of the Casimir operator C_2, and A_3, and M_3. This choice fits well with the use of parabolic coordinates [60]. A more physically understandable choice is to choose the common basis states $|nlm\rangle$ that are eigenstates of C_2, $\boldsymbol{L}^2$, and L_3. For this set of basis states, we have

$$\sqrt{(\boldsymbol{L}^2 + \boldsymbol{A}^2 + 1)}|nlm\rangle = n|nlm\rangle \qquad \boldsymbol{L}^2|nlm\rangle = l(l+1)|nlm\rangle \quad L_3|nlm\rangle = m|nlm\rangle \tag{90}$$

We can define raising and lowering operators for m:

$$L_\pm = L_1 \pm iL_2 \tag{91}$$

which obeys the commutation relations

$$[\boldsymbol{L}^2, L_\pm] = 0] \qquad [L_3, L_\pm] = \pm L_\pm. \tag{92}$$

Therefore, we can use $L_\pm$ to change the value of m for the basis states

$$L_\pm|nlm\rangle = \sqrt{(l(l+1) - m(m \pm 1)}|nl\ m \pm 1\rangle \tag{93}$$

for $l \geq 1$. We can also use the generators $\boldsymbol{A}$ to change the angular momentum. A general SO(4) transformation can be expressed as a rotation induced by $\boldsymbol{L}$, followed by a rotation induced by A_3, followed by another rotation generated by $\boldsymbol{L}$ [121]. Our interest is primarily in changing the angular momentum l, which is most directly done while using A_3, which commutes with L_3 and C_2, and so only changes l:

$$A_3|nlm\rangle = \left(\frac{(n^2 - (l+1)^2)((l+1)^2 - m^2)}{4(l+1)^2 - 1}\right)^{\frac{1}{2}} |n\,l+1\,m\rangle + \left(\frac{(n^2 - l^2)(l^2 - m^2)}{4l^2 - 1}\right)^{\frac{1}{2}} |n\ l-1\ m\rangle. \tag{94}$$

for $l \geq 1$.

4.3. *Relativistic and Semi-Relativistic Spinless Particles in the Coulomb Potential and Klein–Gordon Equation*

The Klein–Gordon equation

$$\left(p^2 - (\tilde{E} - V)^2 + m^2\right)\tilde{\psi} = 0$$

where $\tilde{E}$ is the relativistic total energy, may be solved exactly for a Coulomb potential, $V = -(Z\alpha)/r$ [122]. The energy levels depend on a principal quantum number and on the magnitude of the angular momentum but not its direction. The only degeneracy present is associated with the O(3) symmetry of the Hamilton. For a relativistic scalar particle, there is no degeneracy to be lifted by a Lamb shift.

If we neglect the V^2 term the resulting equation can be written in the form

$$\left(\frac{p^2}{2\tilde{E}} + V - \frac{\tilde{E}^2 - m^2}{2\tilde{E}}\right)\tilde{\psi} = 0$$

This is exactly the same as the nonrelativistic Schrodinger equation with the substitutions

$$m \to \tilde{E} \qquad E \to \frac{\tilde{E}^2 - m^2}{2\tilde{E}}.$$

Thus, we regain the O(4) symmetry of the nonrelativistic hydrogen atom, and can define two conserved vectors, as indicated in Table 1. It is possible to take the "square root" of this approximate Klein-Gordon equation (in the same sense that the Dirac equation is the square root of the Klein–Gordon equation) and get an approximate Dirac equation whose energy eigenvalues are independent of the orbital angular momentum [123].

4.4. *Eigenstates of the Inverse Coupling Constant* $(Z\alpha)^{-1}$

Solutions to Schrodinger's equation for a particle of energy $E = -\frac{a^2}{2m}$ in a Coulomb potential

$$\left[p^2 + a^2 - \frac{2mZ\alpha}{r}\right] |a\rangle = 0 \tag{95}$$

may be found for certain critical values of the energy $E_n = -\frac{a_n^2}{2m}$ where $a_n = \frac{mZ\alpha}{n}$. The corresponding eigenstates of the Hamiltonian are $|nlm\rangle$ which satisfy Equation (95) with a replaced by a_n. In addition to the bound states, because there is no upper bound on p^2 in the Hamiltonian, we also have the continuum of scattering states that have E > 0.

Because the quantity that must have discrete values for a solution to exist is actually $\frac{a}{mZ\alpha}$, as noted in Section 2.3, we might ask if eigensolutions to Equation (95) exist for certain critical values of $Z\alpha$ while keeping a and the energy fixed [1]. To investigate such solutions it is convenient to algebraically transform Equation (95):

$$\left[\frac{1}{\sqrt{\rho(a)}}\left(\frac{p^2+a^2}{a}\right)\frac{1}{\sqrt{\rho(a)}} - \frac{1}{\sqrt{\rho(a)}}\left(\frac{2mZ\alpha}{ar}\right)\frac{1}{\sqrt{\rho(a)}}\right]\sqrt{\rho(a)}|a\rangle = 0$$

where

$$\rho(a) = \frac{p^2 + a^2}{2a^2}. \tag{96}$$

Because $\rho(a)$ commutes with p^2, we obtain the eigenvalue equation

$$\left[\left(\frac{a}{mZ\alpha}\right) - K(a)\right] |a) = 0 \tag{97}$$

where the totally symmetric and real kernel is

$$K(a) = \sqrt{\frac{2a^2}{p^2+a^2}}\frac{1}{ar}\sqrt{\frac{2a^2}{p^2+a^2}} \tag{98}$$

and

$$|a) = (\rho(a))^{\frac{1}{2}}|a\rangle. \tag{99}$$

As before, solutions to this transformed equation may found for the eigenvalues

$$K'(a) = \left(\frac{a}{mZ\alpha}\right)' = \frac{1}{n}. \tag{100}$$

If we hold $Z\alpha$ constant and let a vary, we obtain the usual spectrum $a_n = \sqrt{-2mE_n} = \frac{mZ\alpha}{n}$. For $a = a_n$, Equation (99) reduces to the equation for the eigenstates

$$\sqrt{\rho(a_n)}|nlm\rangle = |nlm; a_n). \tag{101}$$

Alternatively, if we hold a constant, then $Z\alpha$ has the spectrum

$$(Z\alpha)_n = \frac{na}{m} \tag{102}$$

with the corresponding eigenstates of $(Z\alpha)^{-1}$ being

$$|nlm;a). \tag{103}$$

The relationship between the usual energy eigenstates of the energy $|nlm\rangle$ and the eigenstates $|nlm;a)$ of $(Z\alpha)^{-1}$ is

$$|nlm\rangle = \frac{1}{\sqrt{\rho(a_n)}}|nlm;a_n), \tag{104}$$

which requires that both sets of states have the same quantum numbers. Note that the magnitude $(a|K(a)|a)$ is proportional to $\langle 1/ar\rangle_n = (1/a_n)(1/n^2 a_0)$ where a_0 is the Bohr radius for the ground state, and so is positive and bounded. The kernel K(a) is real and symmetric in p and r and manifestly Hermitean. Because the kernel K in Equation (98) is bounded, definite, and Hermitian with respect to the eigenstates $|nlm;a_n)$ [124] the set of normalized eigenstates

$$|nlm;a) \quad n = 0,1,2....; \quad l = 0,1,..n-2,n-1; \quad m = -l,-l+1,...l-1,l. \tag{105}$$

where

$$\left(\frac{1}{n} - K(a)\right) \; |nlm;a) = 0 \tag{106}$$

is a complete orthonormal basis for the hydrogenlike atom:

$$(nlm;a|n'l'm';a) = \delta_{nn'}\delta_{ll'}\delta_{mm'} \tag{107}$$

$$\sum_{nlm} |nlm;a)(nlm;a| = 1. \tag{108}$$

There are several important points to notice with regard to these eigenstates of the inverse of the coupling constant:

(1) **Because of the boundedness of K, there is no continuum portion in the eigenvalue spectrum of $(Z\alpha)^{-1}$, the eigenvalues are discrete**. Because K is a positive definite Hermitian operator, all eigenvalues are positive, real numbers. This feature leads to a unified treatment of all states of the hydrogenlike atom as opposed to the treatment in terms of energy eigenstates in which we must consider separately the bound states and the continuum of scattering states.

(2) It follows from Equation (104) that **the quantum numbers, multiplicities, and degeneracies of these states $|nlm;a)$ are precisely the same as those of the usual bound energy eigenstates.** For example, there are n^2 eigenstates of $(Z\alpha)^{-1}$ that have the principal quantum number equal to n or $(Z\alpha)$ equal to $\frac{na}{m}$.

(3) **A single value of the RMS momentum a or the energy $E = \frac{-a^2}{2m}$ applies to all the states in our complete basis**, as opposed to the usual energy eigenstates where each nondegenerate state has a different value of a. We have made this explicit by including a in the notation for the states: $|nlm;a)$. Sometimes we will write the states as $|nlm)$, provided the value of a has been specified. This behavior in which a single value of a applies to all states will prove to be very useful. In essence, it permits us to generalize from statements applicable in a subspace of Hilbert space with energy E_n or energy parameter a_n to the entire Hilbert space.

(4) **By a suitable scale change or dilation, we may give the quantity a any positive value we desire**. This is affected by the unitary operator

$$D(\lambda) = e^{i\frac{1}{2}(p\cdot r + r\cdot p)\lambda} \tag{109}$$

which transforms the canonical variables

$$D(\lambda)\boldsymbol{p}D^{-1}(\lambda) = e^{-\lambda}\boldsymbol{p} \quad D(\lambda)\boldsymbol{r}D^{-1}(\lambda) = e^{\lambda}\boldsymbol{r} \tag{110}$$

and the kernel K(a)

$$D(\lambda)K(a)D^{-1}(\lambda) = K(ae^{\lambda}). \tag{111}$$

and the eigenvalue equation

$$\left(\frac{1}{n} - K(ae^{\lambda})\right) D(\lambda)|nlm;a\rangle = 0. \tag{112}$$

Therefore the states transform as

$$D(\lambda)|nlm;a\rangle = |nlm;ae^{\lambda}\rangle. \tag{113}$$

These transformed states form a new basis corresponding to the new value $e^{\lambda}a$ of the RMS momentum.

The relationship between the energy eigenstates and the $(Z\alpha)^{-1}$ eigenstates can be written while using the dilation operator:

$$|nlm\rangle = \frac{1}{\sqrt{\rho\,(a_n)}}D\,(\lambda_n)\,|n\ell m;a\rangle \quad \text{where } e^{\lambda_n} = \frac{a_n}{a}. \tag{114}$$

The usual energy eigenstates $|nlm\rangle$ are obtained from the eigenstates of $(Z\alpha)^{-1}$ by first performing a scale change to ensure that the energy parameter a has the value a_n and then multiplying by a factor $\rho^{-1/2}$. The need for the scale change is apparent from dimensional considerations: from the $(Z\alpha)^{-1}$ eigenvalue equation, we see that the eigenfunctions are functions of p/a or ar while from the energy eigenvalue equation the eigenfunctions are functions of p/a_n and $a_n r$. The factor $\rho^{-1/2}$ was required in order to convert Schrodinger's equation to one involving a bounded Hermitean operator.

Using the eigenstates of $(Z\alpha)^{-1}$ as our basis allows for us to analyze the mathematical and physical structure of the hydrogenlike atom in the easiest and clearest way.

4.5. Another Set of Eigenstates of $(Z\alpha)^{-1}$

We may transform Schrodinger's equation Equation (95) to an eigenvalue equation for $(Z\alpha)^{-1}$ that differs from Equation (96) by similar methods:

$$\left(\frac{1}{n} - K_1(a)\right) \; |nlm;a\rangle = 0 \tag{115}$$

where

$$K_1(a) = \frac{1}{\sqrt{ar}}\frac{2a^2}{p^2+a^2}\frac{1}{\sqrt{ar}} \qquad \rho(a) = n/ar \quad \sqrt{\rho(a_n)}|nlm\rangle = |nlm;a_n\rangle \tag{116}$$

This kernel, like K(a), is a bounded, positive definite, Hermitian operator so the eigenstates form a complete basis. The relationship of these basis states to the energy eigenstates is the same as that of the previously discussed eigenstates of $(Z\alpha)^{-1}$ Equation (114) but with $\rho(a) = n/ar$. The n insures that the two sets of eigenstates have consistent normalization, which may be checked by means of the virial theorem. The n cancels out when similarity transforming from the basis of energy eigenstates to the basis of $(Z\alpha)^{-1}$. Note that, classically, both kernels equal $1/ar_c$.

The first set of basis states of $(Z\alpha)^{-1}$ with $\rho(a) = \frac{p^2+a^2}{2a^2}$ is more convenient to use when working in momentum space and the second set with $\rho(a) = n/ar$ is more convenient in configuration space.

Other researchers have used other approaches to secure a bounded kernel for the Schrodinger hydrogen atom, for example, by multiplying the equation from the left by r to regularize it [82].

However, the methods used have not symmetrized the kernels to make them Hermitian, nor are all the generators of the corresponding groups Hermitian, and they have to redefine the inner product [38,82].

4.6. Transformation of A and L to the New Basis States

We must transform $\boldsymbol{A}$ as given in Equation (89) and $\boldsymbol{L} = \boldsymbol{r} \times \boldsymbol{p}$ when we change our basis states from eigenstates of the energy to eigenstates of the inverse coupling constant. The correct transformation may be derived by requiring that the transformed generators produce the same linear combination of new states as the original generators produced of the old states. Thus, because

$$\boldsymbol{A}|nlm\rangle = \sum_{l',m'} |nl'm'\rangle A^{lm}_{l'm'} \tag{117}$$

where the coefficients $A^{lm}_{l'm'}$ are the matrix elements of $\boldsymbol{A}$, we require that the transformed generator $\boldsymbol{a}$ satisfies the equation

$$\boldsymbol{a}|nlm) = \sum_{l'm'} |nl'm')A^{lm}_{l'm'}. \tag{118}$$

In other words, since the Runge–Lenz vector $\boldsymbol{A}$ is a symmetry operator of the original energy eigenstates, $\boldsymbol{a}$ will be a symmetry operator of the new states with precisely the same properties and matrix elements. Because $\boldsymbol{A}$ is Hermitian, $\boldsymbol{a}$ is Hermitian.

To obtain a differential expression for $\boldsymbol{a}$ acting on the new states, we need to transform the generator using Equation (114):

$$\boldsymbol{a} = D^{-1}(\lambda_n)\left(\sqrt{\rho(a_n)}\boldsymbol{A}\frac{1}{\sqrt{\rho(a_n)}}\right)D(\lambda_n). \tag{119}$$

The effect of the scale change on the quantity in large parenthesis is to replace a_n everywhere by a. By explicit calculation, we find

$$\boldsymbol{a} = \frac{1}{2a}\left(\frac{\boldsymbol{r}p^2 + p^2\boldsymbol{r}}{2} - \boldsymbol{r}\cdot\boldsymbol{p}\boldsymbol{p} - \boldsymbol{p}\boldsymbol{p}\cdot\boldsymbol{r}\right) - \frac{a\boldsymbol{r}}{2} \tag{120}$$

for $\rho(a) = (p^2 + a^2)/2a^2$. And we obtain

$$\boldsymbol{a} = \frac{1}{2a}\left(\frac{\boldsymbol{r}p^2 + p^2\boldsymbol{r}}{2} - \boldsymbol{r}\cdot\boldsymbol{p}\boldsymbol{p} - \boldsymbol{p}\boldsymbol{p}\cdot\boldsymbol{r} - \frac{\boldsymbol{r}}{4r^2}\right) - \frac{a\boldsymbol{r}}{2} \tag{121}$$

for $\rho(a) = n/ar$.

Both of these expressions for $\boldsymbol{a}$ are manifestly Hermitian. In addition, since there is no dependence on the principal quantum number these expressions are valid in the entire Hilbert space, and not just in a subspace spanned by the degenerate states, as was the case when we used the energy eigenstates as a basis (Equation (89)).

The angular momentum operator is invariant under scale changes and it commutes with scalar operators. Therefore $\boldsymbol{L}$ is invariant under the similarity transformation $\frac{1}{\sqrt{\rho(a_n)}}D(\lambda_n)$ and the expression for the angular momentum operator with respect to the eigenstates of $(Z\alpha)^{-1}$ is the same as the expression with respect to the eigenstates of the energy.

4.7. The $\langle U'|$ Representation

The U' coordinates provide the natural representation for the investigation of the symmetries of the hydrogenlike atom in quantum mechanics, as in classical mechanics [125]. Therefore, we briefly consider the relevant features of this representation and, in particular, its relationship to the momentum representation. The eigenstate $< U'|$ of $U_b, b = 1,2,3,4$, is defined by

$$\langle U'|\ U_b = U'_b\ \langle U'| \tag{122}$$

These states are complete on the unit hypersphere in four dimensions:

$$\int |U'\rangle\langle U'|\ d^3\Omega' = 1 \tag{123}$$

where Ω refers to the angles (θ_4, θ, ϕ) defined in Equation (54). The U variables are defined in terms of the momentum variables and the quantity a in Equation (52). Therefore, the momentum and the U operators commute

$$[p_i, U_b] = 0 \tag{124}$$

and the state $\langle U'|$ is proportional to a momentum eigenstate $\langle p'|$:

$$\langle U'| = \langle p'|\ \sqrt{J(p)} \tag{125}$$

where the momentum eigenstate is defined by $\langle p'|\boldsymbol{p} = \boldsymbol{p'}\langle p|$ and

$$\int d^3p'|p'\rangle\langle p'| = 1. \tag{126}$$

The function $J(p')$ may be determined by equating the completeness conditions and substituting Equation (125):

$$1 = \int d^3p'|p'\rangle\langle p'| = \int d^3\Omega'|U'\rangle\langle U'| = \int d^3\Omega' J(p')|p'\rangle\langle p'| \tag{127}$$

which leads to the identification of the differential quantities

$$d^3p' = d^3\Omega' J(p') \tag{128}$$

demonstrating that $J(p')$ is the Jacobian of the transformation from the p- to the U- space. Noting that on the unit sphere

$$U_4^2 = 1 - U_i U_i \tag{129}$$

we can compute the Jacobian

$$J(p') = \left[\frac{p'^2 + a^2}{2a}\right]^3. \tag{130}$$

Therefore from Equation (125) we have the important result

$$\langle U'| = \langle p'|\left[\frac{p^2 + a^2}{2a}\right]^{3/2}. \tag{131}$$

We can use this result to compute the action of $\boldsymbol{r}$ on $\langle U'|$ in terms of the differential operators. Using the equation

$$\langle p'|\boldsymbol{r} = i\boldsymbol{\nabla}_{p'}\langle p'| \tag{132}$$

we obtain

$$\langle U'|\boldsymbol{r} = \left(i\boldsymbol{\nabla}_{p'} - \frac{3i\boldsymbol{p'}}{p'^2 + a^2}\right)\langle U'|. \tag{133}$$

Action of $\boldsymbol{a}$ and $\boldsymbol{L}$ on $\langle U'|$

Using Equation (133) for the action of $\boldsymbol{r}$ on $\langle U'|$ and using the expression Equation (120) for $\boldsymbol{a}$ since $\langle U'|$ is proportional to $\langle p'|$, we immediately find that when acting on $\langle U'|$, $\boldsymbol{a}$ has the differential representation

$$\boldsymbol{a}' = \frac{i}{2a}\left((p'^2 - a^2)\boldsymbol{\nabla}_{p'} - 2\boldsymbol{p}'\boldsymbol{p}'\cdot\boldsymbol{\nabla}_{p'}\right) \tag{134}$$

where

$$\langle U'|\boldsymbol{a} = \boldsymbol{a}'\langle U'|. \tag{135}$$

We can also write $\boldsymbol{a}'$ in terms of the U' variables by using the relationship Equation (52) between the p and U variables:

$$\boldsymbol{a}' = U_4' i\boldsymbol{\nabla}_{U'} - \boldsymbol{U}' i\partial/\partial_4' \tag{136}$$

where the spatial part of the four vector U' is $\boldsymbol{U} = (U_1, U_2, U_3)$ and U_4 is the fourth component. This is the differential representation of a rotation operator mixing the spatial and the fourth components of U_a'. When acting on the state $\langle U'|$, clearly $e^{i\boldsymbol{a}\cdot\boldsymbol{v}}$ generates a four-dimensional rotation that produces a new eigenstate $\langle U''|$. To explicitly derive the form of the finite transformation, we compute

$$[a_j', U_j'] = iU_4'\delta_{ij} \qquad [a_j', U_4'] = -iU_i'. \tag{137}$$

For a finite transformation $a^{i\boldsymbol{a}\cdot\boldsymbol{n}v}$ with $\boldsymbol{n}^2 = 1$, we have

$$\begin{aligned}\boldsymbol{U}'' &= e^{i\boldsymbol{a}'\cdot\boldsymbol{n}v}\boldsymbol{U}'e^{-i\boldsymbol{a}'\cdot\boldsymbol{n}v} \\ &= \boldsymbol{U}' - \boldsymbol{n}\boldsymbol{n}\cdot\boldsymbol{U}' + \boldsymbol{n}\boldsymbol{n}\cdot\boldsymbol{U}'\cos v - \boldsymbol{n}U_4'\sin v\end{aligned} \tag{138}$$

and

$$\begin{aligned}U_4'' &= e^{i\boldsymbol{a}'\cdot\boldsymbol{n}v}U_4'e^{-i\boldsymbol{a}'\cdot\boldsymbol{n}v} \\ &= U_4'\cos v + \boldsymbol{n}\cdot\boldsymbol{U}'\sin v\end{aligned} \tag{139}$$

These equations of transformation are like those for a Lorentz transformation of a four-vector (r, it). We can illustrate the equations for e^{ia_2v} (cf Equation (58)), which mixes the two and four components of U':

$$\begin{array}{ll} U_1'' = U_1' & U_3'' = U_3' \\ U_2'' = U_2'\cos v - U_4'\sin v & U_4'' = U_2'\sin v + U_4'\cos v \end{array}. \tag{140}$$

When $\boldsymbol{L}$ acts on $\langle U'|$, it has the differential representation

$$\boldsymbol{L}' = \boldsymbol{U}' \times i\boldsymbol{\nabla}_{U'} \tag{141}$$

This result follows directly, since U_i' equals p_i' times a factor that is a scalar under rotations in three dimensions. When $e^{i\boldsymbol{L}\cdot\boldsymbol{\omega}}$ acts on $\langle U'|$ it produces a new state $\langle U''|$, where the spatial components of U' have been rotated to produce U''.

In summary, we see that U' is a four-vector under rotations generated by $\boldsymbol{a}'$ and $\boldsymbol{L}'$. Therefore the states $\langle U'|$ provide a vector representation of the group of rotations in four dimensions SO(4), with the generators $\boldsymbol{a}$ and $\boldsymbol{L}$.

5. Wave Functions for the Hydrogenlike Atom

In this section, we analyze the wave functions of the hydrogenlike atom, working primarily in the $\langle U'|$ representation and using eigenstates of the inverse of the coupling constant $(Z\alpha)^{-1}$ for the basis states. In this representation the wave functions are spherical harmonics in four dimensions. We derive the relationship of the usual energy eigenfunctions in momentum space to the spherical harmonics and discuss the classical limits in momentum and configuration space.

5.1. Transformation Properties of the Wave Functions under the Symmetry Operations

We can show that the wave functions $Y_{nlm}(U')$ in the $\langle U'|$ representation with respect to the eigenstates of $(Z\alpha)^{-1}$

$$Y_{nlm}(U') \equiv \langle U'|nlm) \tag{142}$$

transform as four-dimensional spherical harmonics under the four-dimensional rotations generated by the Runge-Lenz vector $\boldsymbol{a}$ and the angular momentum $\boldsymbol{L}$. We note that the quantity a is implicit in both the bra and the ket in Equation (142). For our basis states we employ the set of $(Z\alpha)^{-1}$ eigenstates $|nlm)$ of the inverse coupling constant that are convenient for momentum space calculations ($\rho = \frac{p^2+a^2}{2a^2}$). We choose these states rather than those convenient for configuration space calculations, because the $\langle U'|$ eigenstates are proportional to the $\langle p'|$ eigenstates.

If we transform our system by the unitary operator $e^{i\theta}$ where $\theta = \boldsymbol{L}\cdot\boldsymbol{\omega} + \boldsymbol{a}\cdot\boldsymbol{v}$, then the wave function in the new system is

$$Y'_{nlm}(U') = \langle U'|e^{i\theta}|nlm). \tag{143}$$

There are two ways in which we may interpret this transformation, corresponding to what have been called the active and the passive interpretations. In the passive interpretation we let $e^{i\theta}$ act on the coordinate eigenstate $\langle U'|$. As we have seen in Section 3.6, this produces a new eigenstate $\langle U''|$, where the four-vector U'' is obtained by a four-dimensional rotation of U' (Equations (138) and (139)). Thus, we have

$$Y'_{nlm}(U') = \langle U''|nlm) = Y_{nlm}(U''). \tag{144}$$

In the active interpretation, we let $e^{i\theta}$ act on the basis state $|nlm)$. Because $\boldsymbol{L}$ and $\boldsymbol{a}$ are symmetry operators of the system, transforming degenerate states into each other, it follows that $e^{i\theta}|nlm)$ must be a linear combination of states with principal quantum number equal to n. Therefore, we have

$$Y'_{nlm}(U') = \sum_{l'm'}\langle U'|R^{nlm}_{nl'm'}|nl'm') \quad = \quad \sum_{l'm'} R^{nlm}_{nl'm'}Y_{nl'm'}(U'). \tag{145}$$

The wave functions for degenerate states with a given n transform irreducibly among themselves under the four-dimensional rotations, forming a basis for an irreducible representation of SO(4) of dimensions n^2 . Equating the results of the two different interpretations gives

$$Y_{nlm}(U'') = \sum_{l'm'} R^{nlm}_{nl'm'}Y_{nl'm'}(U'). \tag{146}$$

The transformation properties Equation (146) of Y_{nlm} are precisely analogous to those of the three-dimensional spherical harmonic functions. It follows that the Y_{nlm} are four-dimensional spherical harmonics [35,124].

5.2. Differential Equation for the Four Dimensional Spherical Harmonics $Y_{nlm}(U')$

The differential equation for the $Y_{nlm}(U')$ may be obtained from the equation

$$(\boldsymbol{L}'^2 + \boldsymbol{a}'^2)Y_{nlm}(U') = (n^2-1)Y_{nlm}(U') \tag{147}$$

which follows from $C_2 = n^2 - 1$ and the definition of C_2, Equation (81). Substituting in the differential expressions Equations (134) and (141) for $\boldsymbol{a}'$ and $\boldsymbol{L}'$ we find that $\boldsymbol{L}'^2 + \boldsymbol{a}'^2$ equals $\nabla^2_{U'} - (\boldsymbol{U}'\cdot\boldsymbol{\nabla}_{U'})^2$, which is the angular part of the Laplacian operator in four dimensions (cf in three dimensions, $\boldsymbol{L}^2/r^2 = p^2 - p_r^2$). Thus, Equation (147) is the differential equation for four-dimensional spherical harmonics with the degree of homogeneity equal to $n-1$, which means n^2 such functions exist, in agreement with the know degree of degeneracy.

5.3. *Energy Eigenfunctions in Momentum Space*

We want to determine the relationship between the usual energy eigenfunctions in momentum space $\psi_{nlm}(p') \equiv \langle p'|nlm\rangle$ (with $a = a_n$) and the four-dimensional spherical harmonic eigenfunctions $Y_{nlm}(U';a) = \langle U'|nlm;a)$.

We choose the RMS momentum a to have the value a_n. If we use the expression Equation (131) for $\langle U'|$ in terms of $\langle p'|$

$$\langle U'| = \langle p'| \left(\frac{p^2 + a_n^2}{2a_n}\right)^{3/2} \tag{148}$$

and the expression Equation (104) for the eigenstates of $(Z\alpha)^{-1}$ in terms of the energy eigenstates

$$|nlm; a_n) = \sqrt{\frac{p^2 + a_n^2}{2a_n^2}} |nlm\rangle \tag{149}$$

we find the desired result

$$Y_{nlm}(U'; a_n) = \left(\frac{p^2 + a_n^2}{2a_n}\right)^2 \frac{1}{\sqrt{a_n}} \psi_{nlm}(p'). \tag{150}$$

The usual method of deriving this relationship between the wave function in momentum space and the corresponding spherical harmonics in four dimensions involves transforming the Schrodinger wave equation to an integral equation in momentum space [18,35]. As in the classical case, to do this we first replace $\boldsymbol{p}$ by $\boldsymbol{p}/a$ and perform a stereographic projection from the hyperplane corresponding to the three- dimensional momentum space to a unit hypersphere in a four-dimensional space. The resulting integral equation manifests a four-dimensional invariance. When the wave functions are normalized as in Equation (150), the solutions are spherical harmonics in four dimensions. As another alternative to this procedure, we can Fourier transform the configuration space wave functions directly [126].

5.4. *Explicit Form for the Spherical Harmonics*

The spherical harmonics in four dimensions can be expressed as [127]:

$$Y_{nlm}(\Omega) = N_1(n,l)(\sin\theta_4)^l C_{n-1-l}^{l+1}(\cos\theta_4) \cdot \left[N_2(l,m)(\sin\theta)^m C_{l-m}^{m+1/2}(\cos\theta)\frac{e^{im\phi}}{\sqrt{2\pi}}\right]. \tag{151}$$

The factor in brackets is equal to $Y_l^m(\theta,\phi)$, the usual spherical harmonic in three-dimensions [127]. The Gegenbauer polynomials C_n^λ of degree n and order λ are defined in terms of a generating function:

$$\frac{1}{(1 - 2tx + t^2)^\lambda} = \sum_{n=0} t^n C_n^\lambda. \tag{152}$$

$N_1(n,l)$ and $N_2(l,m)$ are chosen to normalize the Y_{nlm} on the surface of the unit sphere:

$$\int |Y_{nlm}(\Omega)|^2 d^3\Omega_U = 1 \tag{153}$$

where $d^3\Omega_U = \sin^2\theta_4 \sin\theta d\theta d\phi$. We find

$$N_1(n,l) = \sqrt{\frac{2^{2l+1}}{\pi}\frac{n(n-l-1)!(l!)^2}{(n+l)!}} \tag{154}$$

$$N_2(l,M) = \sqrt{\frac{2^{2m}}{\pi}(l+\frac{1}{2})\frac{(l-m)!}{(l+m)!}[\Gamma(m+\frac{1}{2})]^2}\ . \tag{155}$$

In the next section, we discuss the asymptotic behavior of Y_{nlm} for large quantum numbers and compare it to the classical results of Section 3.

5.5. Wave Functions in the Classical Limit

5.5.1. Rydberg Atoms

Advances in quantum optics, such as the development of ultra short laser pulses, microwave spectroscopy, and atom inteferometery, have opened new possibilities for experiments with atoms and Rydberg states, meaning hydrogenlike atoms in states with very large principal quantum numbers and correspondingly large diameter electron orbits. The pulsed electromagnetic fields can be used to modify the behavior of the orbital electrons. Semi-classical electron wave packets in hydrogenlike atoms were first generated in 1988 by ultrashort laser pulses, and today are often generated by unipolar teraherz pulses [128–130]. Over the last few decades, there has been interest in the classical limit of the hydrogenlike atom for n very large, Rydberg states, for a number of reasons [131]: 1. Rydberg states are at the border between bound states and the continuum, and any process which leads to excited bound states, ions or free electrons usually leads to the production of Rydberg states. This includes, for example, photo-ionization or the application of microwave fields. The very large cross section for scattering is unique. 2. Rydberg states can be used to model atoms with a higher atomic number that have an excited valence electron that orbits beyond the core. 3. In Rydberg states, the application of electric and magnetic fields breaks the symmetry of the atom and allows for the study of different phenomena, including the transition from classical to quantum chaos [132]. 4. Rydberg atoms can be used to study coherent transient excitation and relaxation, for example, the response to short laser pulses creating coherent quantum wave packets that behave like a classical particle.

The square of the wave function for a given quantum state gives a probability distribution for the electron that is independent of time. If we want to describe the movement of an electron in a semiclassical state, with a large radius, going around the nucleus with a classical time dependence, then we need to form a wave packet. The wave packet is built as a superposition of many wave functions with a band of principal quantum numbers.

A variety of theoretical methods have been used to derive expressions for the hydrogen atom wave functions and wave packets for highly excited states. There is general agreement on the wave functions for large n, and that the wave functions display the expected classical behavior, elliptical orbits in configurations space, and great circles in the four dimensional momentum space [133–136].

Researchers have proposed a variety of wave packets to describe Rydberg states. There are general similarities in the wave packets that describe electrons going in circular or elliptical orbits with a classical time dependence for some characteristic number of orbits, and it is maintained that the quantum mechanical wave packets provide results that agree with the classical results [129,130,133–139]. Most of the approaches exploit the SO(4) or SO(4,2) symmetry of the hydrogen atom which is used to rotate a circular orbital to an elliptical orbit. The starting orbital is often taken as a coherent state, which is usually considered a classical like state with minimum uncertainty. The most familiar example of a coherent state is for a one dimensional harmonic oscillator characterized by creation and annihilation operators $a^\dagger$ and a. The coherent state $|\alpha\rangle$ is a superposition of energy eigenstates that is an eigenstate of a where $a|\alpha\rangle = \alpha|\alpha\rangle$ for a complex α. This coherent state will execute harmonic motion like a classical particle [140]. To obtain a coherent state for the hydrogenlike atom, eigenstates of the operator than lowers the principal quantum number n (which will be discussed in Section 7.4) have been used [141], as well as lowering operators based on the equivalence of the four dimensional harmonic oscillator representation of the hydrogen atom [134,135,142].

In either case, this coherent eigenstate is characterized by a complex eigenvalue, which needs to be specified. Several constraints have been used to obtain the classical wave packet that presumably obeys Kepler's Laws, such as requiring that the orbit lie in a plane so $\langle z\rangle = 0$ for the orbital, or that $\langle r - r_{classical}\rangle$ be a minimum, or that some minimum uncertainty relationship is obtained. In addition, there are issues regarding the approximations used, in particular, those that relate to time. For times characteristic of the classical hydrogen atom, the wavepackets act like a classical system. For longer times, the wavepacket spreads in the azimuthal direction and after some number of classical revolutions of order 10 to 100 the spread is 2π, so the electron is uniformly spread over the entire orbit. The spread arises because the component wave functions forming the wavepacket have different momenta. In two derivations, still longer times were considered, and recoherence was predicted to occur after about $n/3$ (where n is the approximate principal quantum number) revolutions, although there is some difference in the predicted amount of recoherence [131,136]. Because of the conservation of $\boldsymbol{L}$ and $\boldsymbol{A}$, the spread of the wave packets is inhibited, except in the azimuthal direction.

This is a system with very interesting physics. For example, one can view the transition from a well defined wave packet representing the electron to a 2π spread and back again as a dynamical illustration of the wave-particle duality.

Brown took a different approach to develop a wavepacket for a circular orbit [139]. He first developed the asymptotic wave function for large n and then optimized the coefficients in a Gaussian superposition to minimize the spread in ϕ, obtaining a predicted characteristic decoherence time of about 10 minutes, considerable longer than any other predicted decoherence time.

Other authors have explored the problem from the perspective of classical physics and the correspondence principle [133,137,143,144]. The results from the different methods are similar with the basic conclusion that the wave functions are peaked on the corresponding classical Kepler trajectories: "atomic elliptic states sew the wave flesh on the classical bones" [129].

With the variety of experimental methods used to generate Rydberg states, a variety of Rydberg wave packets are created, and it is not clear which theoretical model, if any, is preferred [131]. We take a very simple approach to forming a wavepacket and simply use a Gaussian weight for the different frequency components. This does not give an intentionally optimized wave packet, but it is a much simpler approach and the result has all of the expected classical behavior that is very similar to that obtained from much more complicated derivations. We start with a circular orbit and then do a SO(4) rotation to secure an elliptical orbit. We show that it has the classical period of rotation.

5.5.2. Wave Functions in the Semi-Classical Limit

We need to derive the semiclassical limit of the wave functions that correspond to circular orbits in configuration space. For this case, $\sin\nu$, which we interpret as the expectation value of the eccentricity, vanishes. We derive expressions for the wave functions in momentum space and then form a wavepacket. To obtain corresponding expressions for elliptical orbits, we perform a rotation by $e^{ia\cdot v}$, which does not alter the energy but changes the eccentricity and the angular momentum.

<u>Case 1: Circular orbits, $\sin\nu = e = 0$</u>

We derive the asymptotic form of Y_{nlm} for large quantum numbers, where for simplicity we choose the quantum numbers $n-1 = l = m$ corresponding to a circular orbit in the 1–2 plane. From Equation (151) we see we encounter Gegenbauer polynomials of the form C_0^λ, which, by Equation (152), are unity. For a very large l, $\sin^l\theta$ will have a very strong peak at $\theta = \pi/2$, so we make the expansion [103]

$$\sin\theta = \sin\left(\frac{\pi}{2} + \left(\theta - \frac{\pi}{2}\right)\right) = 1 - \frac{1}{2}\left(\theta - \frac{\pi}{2}\right)^2 + \cdots \quad \approx e^{-(1/2)(\theta-\pi/2)^2} \tag{156}$$

to obtain

$$\sin^l\theta \approx e^{-(1/2)l(\theta-\pi/2)^2}. \tag{157}$$

The asymptotic forms for N_1 and N_2 can be computed using the properties of Γ functions:

$$\begin{aligned}\Gamma(2z) &= \left(\frac{1}{\sqrt{2\pi}}\right)2^{2z-\frac{1}{2}}\Gamma(z)\Gamma\left(z-\frac{1}{2}\right)\\ \lim_{z\to\infty}\Gamma(az+b) &\simeq \sqrt{2\pi}e^{-az}(az)^{az+b-\frac{1}{2}}.\end{aligned} \tag{158}$$

We finally obtain [145]

$$Y_{n,n-1,n-1}(\Omega) = \sqrt{\frac{n}{2\pi^2}} e^{-\frac{1}{2}n\left(\theta_4 - \frac{\pi}{2}\right)^2} \cdot e^{-\frac{1}{2}n\left(\theta - \frac{\pi}{2}\right)^2} e^{i(n-1)\phi}, \tag{159}$$

which gives the probability density

$$|Y_{n,n-1,n-1}(\Omega)|^2 = \frac{n}{(2\pi^2)} e^{-n\left(\theta_4 - \frac{\pi}{2}\right)^2} \cdot e^{-n\left(\theta - \frac{\pi}{2}\right)^2}. \tag{160}$$

We have Gaussian probability distributions in θ_4 and θ about the value $\pi/2$. The distributions are quite narrow with widths $\Delta\theta_4 \approx \Delta\theta \approx 1/\sqrt{n}$ and the spherical harmonic essentially describes a circle ($\theta_4 = \theta = \pi/2$) on the unit sphere in the 1–2 plane. As n becomes very large, $U_4 = \cos\theta_4 \approx (r - r_c)/r$ (Equations (54) and (72)) and $U_3 = \sin\theta_4 \cos\theta$, which is proportional to p_3, both go to zero as $1/\sqrt{n}$. The distribution approaches the great circle $U_1^2 + U_2^2 = 1$ that we found in Section 3.6 for a classical particle moving in a circular orbit in the 1–2 plane in configuration space. Note that this state is a stationary state with a constant probability density. To get the classical time dependence we need to form a wavepacket.

Forming a Wavepacket

We form a time dependent wavepacket for circular orbits by superposing circular energy eigenstates:

$$\chi(\Omega, t) = \sum_n e^{itE_n} Y_{n,n-1,n-1} A_{n-N} \tag{161}$$

where A_{n-N} is an amplitude peaked about $n = N >> 1$. For $n >> 1$ we expand E_n about E_N:

$$E_n = E_N + \left.\frac{\partial E}{\partial n}\right|_N s + \left.\frac{\partial^2 E}{\partial n^2}\right|_N s^2 + \ldots \tag{162}$$

where $s = n - N$. From the equation for the energy levels, $E = -m(Z\alpha)^2/(2n^2)$ we compute

$$\left.\frac{\partial E_n}{\partial n}\right|_N = \frac{m(Z\alpha)^2}{N^3} = \sqrt{\frac{-8E_N^3}{m(Z\alpha)^2}}. \tag{163}$$

In agreement with the Bohr Correspondence Principle, the right-hand side of this equation is just the classical frequency ω_{cl} as given in Equation (38). For the second order derivative, we have

$$\left.\frac{\partial^2 E}{\partial n^2}\right|_N = -\frac{3}{N}\omega_{cl} \equiv \beta \tag{164}$$

which gives

$$\chi(\Omega, t) = e^{-itE_N} e^{i\phi(N-1)} \sum_{s=-N+1}^{\infty} e^{-i\left(\omega_{cl} st - \left(\frac{\beta}{2}\right)s^2 t - s\phi\right)} \cdot A_s |Y_{N+s,N+s-1,N+s-1}|. \tag{165}$$

We choose a Gaussian form for A_s

$$A_s = \frac{1}{\sqrt{2\pi N}} e^{-s^2/(2N)}. \tag{166}$$

Brown used $A_s = Ce^{-s^2 3\omega_{cl} t/N}$ which minimizes the diffusion in ϕ at time t [139]. Since $|Y_{N+s,N+s-1,N+s-1}|$ varies slowly with s for $N >> 1$ we can take it outside the summation in Equation (165). We now replace the sum by an integral over s. Because A_s is peaked about N

we can integrate from $s = -\infty$ to $s = +\infty$. We perform the integral by completing the square in the usual way. The final result for the probability amplitude for a circular orbital wave packet is

$$|x(\Omega,t)|^2 = |Y_{N,N-1,N-1}|^2 \left(1+\beta^2 t^2 N^2\right)^{-1} \cdot \exp\left[-\left(\phi-\omega_{cl}t\right)^2 \frac{N}{1+(\beta t N)^2}\right]. \tag{167}$$

This represents a Gaussian distribution in ϕ that is centered about the classical value $\phi = \omega_{cl}t$, meaning that the wavepacket is traveling in the classical trajectory with the classical time dependence. The width of the ϕ distribution is

$$\Delta\phi = (N)^{(-1/2)}(1+\beta^2 t^2 N^2)^{1/2} = (N)^{(-1/2)}(1+9\omega_{cl}^2 t^2)^{1/2}. \tag{168}$$

The distribution in ϕ at $t = 0$ is very narrow, proportional to $1/\sqrt{N}$, but after several orbits $\Delta\phi$ is increasing linearly with time.

The distributions in θ_4 and θ are Gaussian and centered about $\pi/2$ in each case as for the circular wave function (cf. Equation (160)) with widths equal to $(N)^{-1/2}$. The spreading of these distributions in time is inhibited because of the conservation of angular momentum and energy. The detailed behavior of the widths depend on our use of the Gaussian distribution. Other distributions will give different widths, although the general behavior is expected to be similar.

As a numerical example, consider a hydrogen atom that is in the semiclassical region when the orbital diameter is about 1 cm. The corresponding principal quantum number is about 10^4, the mean velocity is about 2.2×10^4 cm/s and the period about 1.5×10^{-4} s. After about 34 revolutions or 5×10^{-3} s, the spread in ϕ is about 2π, meaning the electron is spread uniformly about the entire circular orbit. This characteristic spreading time can be compared to 1.6×10^{-3} s for a fully optimized wave packets formed from coherent SO(4,2) states [136,146]. In order to make predictions about significantly longer times, we would need to retain more terms in the expansion Equation (162) of E_n.

Case 2: Elliptical orbits $\sin\nu = e \neq 0$

We can obtain the classical limit of the wave function for elliptical orbits by first writing our asymptotic form Equation (159) for $Y_{n,n-1,n-1}$ in terms of the U variables instead of the angular variables by using definitions Equation (52), and setting $a = a_n$. Retaining only the lowest order terms in $(\theta_4 - \pi/2)$ and $(\theta - \pi/2)$, we find

$$Y_{n,n-1,n-1}(U) = \left(\frac{n}{2\pi^2}\right)^{\frac{1}{2}} e^{i(n-1)\tan^{-1}\left(\frac{U_2}{U_1}\right)} \cdot e^{-\frac{1}{2}n(U_4)^2} e^{-\frac{1}{2}n(U_3)^2} \tag{169}$$

For large n, this represents a circular orbit in the 1–2 plane. We now perform a rotation by $A_2\nu$. which will change the eccentricity to $\sin\nu$, and change the angular momentum, but will not change the energy or the orbital plane. Using Equation (140) to express the old coordinates in terms of the new coordinates, we find to lowest order

$$Y'_{n,n-1,n-1}(U) = \left(\frac{n}{2\pi^2}\right)^{1/2} e^{i(n-1)\tan^{-1}\left(\frac{U_2}{U_1\cos v}\right)} \cdot e^{-\frac{1}{2}n\{U_2\sin\nu - U_4\cos\nu\}^2} \cdot e^{-\frac{1}{2}n(U_3)^2}. \tag{170}$$

In Section 3.6, we found that the vanishing of the term in braces $0 = U_2\sin\nu - U_4\cos\nu$ specifies the classical great hypercircle orbit (Equation (56)) corresponding to an ellipse in configuration space with eccentricity $e = \sin\nu$ and lying in the 1–2 plane. The probability density $|Y'_{n,n-1,n-1}(U)|^2$ vanishes except within a hypertorus with a narrow cross section of radius approximately $\frac{1}{\sqrt{n}}$, which is centered about the classical distribution. Because the width $\frac{1}{\sqrt{n}}$ of the distribution is constant in U space, it will not be constant when projected onto p space.

In terms of the original momentum space variables, the asymptotic spherical harmonic is

$$\begin{aligned} Y_{n,n-1,n-1}(\boldsymbol{p}) &= \left(\tfrac{n}{2\pi^2}\right)^{\frac{1}{2}} e^{i(n-1)\tan^{-1}\left(\frac{p_2}{p_1 \cos\nu}\right)} \\ \cdot \exp&\left\{-\left(\tfrac{n}{2}\right)\left[p_1^2 + (p_2 - a\tan\nu)^2 - \tfrac{a^2}{\cos^2\nu}\right]^2 \left(\tfrac{\cos\nu}{p^2+a^2}\right)\right\}^2 \\ &\cdot \exp\left\{-\left(\tfrac{n}{2}\right)\left(\tfrac{2p_3 a}{p^2+a^2}\right)\right\}^2 \Big|a = a_n \end{aligned} \quad . \tag{171}$$

The expression in brackets corresponds to the momentum space classical orbit equation we found previously (Equation (49)). As we expect, p_3 is Gaussian about zero since the classical orbit is in the 1–2 plane. We can simplify the expressions for the widths by observing that to lowest order we can use Equation (48), which implies $p^2 + a^2 = 2a^2 + 2ap_2\tan\nu$ in the exponentials. The widths of both distributions therefore increase linearly with p_2. We also note that, since classically there exists a one-to-one correspondence between each point of the trajectory in momentum space and each point in configuration space, we may interpret the widths of the distributions using Equation (32) $\frac{p^2+a^2}{a^2} = \frac{2r_c}{r}$. Accordingly, the widths increase as the momentum increases or as the distance to the force center decreases (Figure 8).

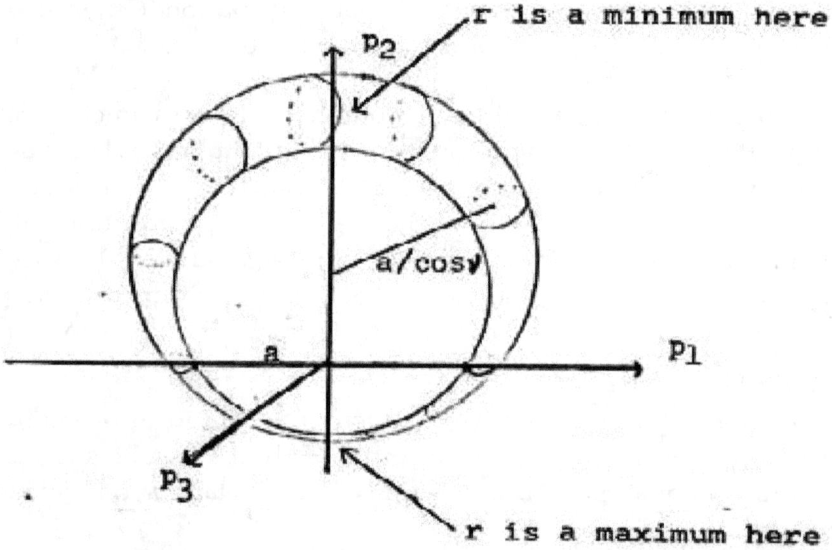

Figure 8. Wave function probability distribution $|Y'_{n,n-1,n-1}(\boldsymbol{p})|^2$ in momentum space for large n, showing the variation in the width of the momentum distribution about the classical circular orbit. The center of the distribution is at $p_2 = a\tan\nu$. The classical orbit is in the 1–2 plane.

Forming a Wave packet for Elliptical Motion

We may form a time dependent wavepacket superposing the wave functions of Equation (170). Care must be taken to include the first order dependence (through a_n) of $\tan^{-1}(U_2/U_1\cos\nu)$ on the principal quantum number when integrating over the Gaussian weight function. The result for the probability density is the same as before (Equation (167)), except $|Y'_{N,N-1,N-1}|^2$ (given in Equation (170)) replaces $|Y_{N,N-1,N-1}|^2$ and

$$\omega_{cl}t = \tan^{-1}\left(\frac{U_2}{U_1\cos\nu}\right) + \sin\nu\,(U_1) \tag{172}$$

where replaces $\omega_{cl}t = \phi$. The result is exactly the same as the classical time dependence Equation (73). The spreading of the wave packet will be controlled by the same factor as for the circular wave packet.

Remark on the Semiclassical Limit in Configuration Space

The time dependent quantum mechanical probability density follows the classical trajectory in momentum space meaning that the probability is greatest at the classical location of the particle in momentum space. Because the configuration space wave function is the Fourier transform of the momentum space wave function, the classical limit must also be obtained in configuration space. That this limit is obtained is made explicit by observing that the momentum space probability density is large when

$$(U_2 \sin\nu - U_4 \cos\nu)^2 \approx 0. \tag{173}$$

However from Section 3 we can show that

$$U_2 \sin\nu - U_4 \cos\nu = \cos\nu\left(\frac{r - r_{classical}}{r_{classical}}\right) \tag{174}$$

where $r_{classical}$ is given by the classical orbit Equation (33). Accordingly, we see that the configurations space probability will be large when

$$\left[\frac{r - r_{classical}}{r_{classical}}\right]^2 \approx 0. \tag{175}$$

5.6. Quantized Semiclassical Orbits

It is convenient at times to have a semiclassical model for the orbitals of the hydrogenlike atom. Historically this was first done by Pauling and Wilson [147]. We can obtain a model by interpreting the classical formulae for the geometrical properties of the orbits as corresponding to the expectation values of the appropriate quantum mechanical expressions. Thus, when the energy $E = -a^2/2m$ appears in a classical formula, we employ the expression for a for the quantized energy levels $a = \frac{1}{nr_0}$ where $r_0 = (mZ\alpha)^{-1}$, which is the radius 0.53 Angstrom of the ground state. Similarly if $\boldsymbol{L}^2$ appears in a classical formula, we substitute $l(l+1)$, where l is quantized $l = 0, 1, 2..n-2, n-1$; and m , the component of $\boldsymbol{L}$ along the 3-axis is quantized: $m = -l, -l+1, -l+2, ..l$.

Orbits in Configuration Space

Recalling Equation (28) $r_c = \frac{mZ\alpha}{a^2}$, we see $ar_c = n$, which gives a semimajor axis of length $r_c = n^2 r_0$, where r_0 is the radius for the circular orbit of the ground state and r_c is for a circular orbit for a state with principal quantum number n. The equations for the magnitude of $\boldsymbol{L}$ and $\boldsymbol{A}$ are

$$L = r_c a \cos v = n \cos v = \sqrt{l(l+1)} \tag{176}$$

$$A = r_c a \sin v = n \sin v = \sqrt{n^2 - l(l+1)} \tag{177}$$

This gives an eccentricity $\sin\nu$ equal to

$$e = \sin v = \sqrt{1 - \frac{l(l+1)}{n^2}} \tag{178}$$

and a semi-major axis equal to

$$b = r_c \sin\nu = n\sqrt{l(l+1)}. \tag{179}$$

Note that the expression for e is limited in its meaning. For an s state, it always gives $e = 1$, and for states with $l = n-1$ it give $e = \sqrt{1/n}$, not the classically expected 0 for a circular orbit.

Orbits in U-Space

The corresponding great hypercircle orbits (ν, Θ) in U-space are described by giving the quantized angle ν, between the three-dimensional hyperplane of the orbit and the 4-axis, and the quantized angle Θ, between the hyperplane of the orbit and the 3-axis:

$$\cos\nu = \sqrt{\frac{l(l+1)}{n^2}} \tag{180}$$

$$\cos\Theta = \sqrt{\frac{m^2}{l(l+1)}}. \tag{181}$$

Note the similarity in these two equations, suggesting that m relates to l the same way that l relates to n, which suggests a generalization of the usual vector model of the atom which only describes the precession of $\boldsymbol{L}$ about the z axis.

The results for orbits in configuration and momentum space illustrate some interesting features:

1. The equation $ar = n$ illustrates that the characteristic dimensions of an orbit in configuration space and the corresponding orbit in momentum space are inversely proportional, as expected, since they are related by a Fourier transform, consistent with the Heisenberg Uncertainty Principle.

2. If $l = 0$, then no classical state exists. The orbit in configuration space degenerates into a line passing through the origin while the corresponding circular orbit in momentum space attains an infinite radius and an infinite displacement from the origin. Although this seems peculiar from the pure classical viewpoint, quantum mechanically it follows, since for S states there is a nonvanishing probability of finding the electron within the nucleus.

In order to interpret these statements about quantized semiclassical elliptical orbits we observe that for quantum mechanical state of the hydrogenlike atom with definite n, l, m, the probability density is (1) independent of ϕ_r or ϕ_p and (2) it does not confine the electron to some orbital plane. Because the quantum mechanical distribution for such a state specifies no preferred direction in the 1–2 plane, we must imagine this distribution as corresponding in some way to an average over all possible orientations of the semiclassical elliptical orbit. This interpretation is supported by the fact that the region within which the quantum mechanical radial distribution function differs largely from zero is included between the values of r corresponding to the semiclassical turning points $r_c(1 \pm \sin\nu)$.

5.7. Four-Dimensional Vector Model of the Atom

In configuration space or momentum space,, the angle between the classical plane of the orbit and the 3-axis is Θ, which is usually interpreted in terms of the vector model of the atom in which we imagine $\boldsymbol{L}$ to be a vector of magnitude $\sqrt{l(l+1)}$ precessing about the 3-axis, with m as the component along the 3-axis. This precession may be linked to the ϕ_r independence of the probability and the absence of an orbital plane as mentioned at the end of the preceding section. The precession constitutes a classical mechanism which yields the desired average over all possible orientations of the semiclassical elliptical orbit. Because the angle Θ is restricted to have only certain discrete values one can say that there is a quantization of space.

The expression for $\cos\nu = \sqrt{l(l+1)/n^2}$ is quite analogous to that for Θ and so suggests a generalization of the vector model of the atom to four dimensions. The projection of the four-dimensional vector model onto the physical three-dimensional subspace must give the usual vector model. We can achieve this by imagining that a four-dimensional vector of length n, where n is the principal quantum number, is precessing in such a way that its three and four components are constants, while the one and two components vary periodically. The projection onto the 1–2–3 hyperplane is a vector of constant magnitude $\sqrt{l(l+1)}$ precessing about the 3-axis. The component along the 3-axis is m. The component along the 4-axis is $A = \sqrt{n^2 - l(l+1)}$ the magnitude of the vector $\boldsymbol{A}$. The vectors $\boldsymbol{L}$ and $\boldsymbol{A}$ are perpendicular to each other. Thus, the precessing n vector makes a constant angle Θ with the 3-axis and a constant angle $\pi/2 - \nu$ with the 4-axis. Because both angles are restricted to certain values, we may say that we have a quantization of four-dimensional space.

6. The Spectrum Generating Group SO(4,1) for the Hydrogenlike Atom

We consider the Schrodinger hydrogen atom and its unitary "noninvariance" or spectrum generating operators $e^{iD_i\beta_i}$ where D_i is a generator and β_i is a real parameter, using eigenstates of $(Z\alpha)^{-1}$ for our basis of our representation. These operators transform an eigenstate of the kernel K with a definite value of the coupling constant (or principal quantum number) into a linear combination of eigenstates with different values of the coupling constant (or different principal quantum numbers), and different l and m. Unlike the invariance generators $\boldsymbol{L}$ and $\boldsymbol{A}$, the noninvariance generators clearly do not generally commute with the kernel K, $[D_i, K] \neq 0$, so they change the principal quantum number.

The set of all invariance and noninvariance operators forms a group with which we may generate all eigenstates in our complete set from a given eigenstate. We show that this group, called the Spectrum Generating Group of the hydrogenlike atom, is SO(4,1), the group of orthogonal transformations in a 5-dimensional space with a metric $g_{AB} = (-1,1,1,1,1)$, where $A, B = 0,1,2,3,4$. The complete set of eigenstates of $(Z\alpha)^{-1}$ for the hydrogenlike atom forms a unitary, irreducible, infinite-dimensional representation of SO(4,1) which, we shall find, can be decomposed into an infinite sum of irreducible representations of SO(4), each corresponding to the degeneracy group for a particular principal quantum number. A unitary representation means all generators are unitary operators. An irreducible representation does not contain lower dimensional representations of the same group. In Section 6.3, we discuss the isomorphism between the spectrum generating group SO(4,l) and the group of conformal transformations in momentum space. An isomorphism means the groups have the same structure and can be mapped into each other.

6.1. Motivation for Introducing the Spectrum Generating Group Group SO(4,1)

We have examined the group structure for the degenerate eigenstates of $(Z\alpha)^{-1}$ for the Schrodinger hydrogenlike atom: the n^2 degenerate states form an irreducible representation of SO(4). The next question we might ask is: Do all or some of the states with different principal quantum numbers form an irreducible representation of some larger group which is reducible into SO(4) subgroups? If such a group exists then it clearly is not an invariance group of the kernel K (Equation (106)). If we want our noninvariance group to include just some of the states then it will be a compact group, since unitary representations of compact groups can be finite dimensional. If we want to include all states then it will be a noncompact group since there are an infinite number of eigenstates of $(Z\alpha)^{-1}$ and all unitary representations of noncompact groups are infinite dimensional [9].

We can find a compact noninvariance group for the first N levels of the coupling constant, $n = 1,2,...N$. The dimensionality of our representation is

$$\sum_{n=1}^{N} n^2 = \frac{N(N+1)(2N+1)}{6}. \tag{182}$$

Mathematical analysis of the group SO(5) shows that this is the dimensionality of the irreducible symmetrical tensor representation of SO(5), given by the tensor with five upper indices $T^{abc...}$, where $a, b, .. = 1,2,3,4$ or 5 [11]. Reducing this representation of SO(5) into its SO(4) components gives

$$\begin{aligned}
(\text{ symm } \cdot \text{ tensor } N)_{SO(5)} =&(0,0) \oplus \left(\frac{1}{2}, \frac{1}{2}\right) \oplus \ldots \quad \oplus \left(\frac{N-1}{2}, \frac{N-1}{2}\right) \\
=&(\text{ symm } \cdot \text{ tensor } n = 1)_{SO(4)} \oplus \\
&(\text{ symm } \cdot \text{ tensor } n = 2)_{SO(4)} \oplus \\
&\ldots \oplus \quad (\text{ symm } \cdot \text{ tensor } n = N)_{SO(4)}
\end{aligned} \tag{183}$$

which is precisely the structure of the first N levels of a hydrogenlike atom. If we want to include all levels then we guess that the appropriate noncompact group is SO(4,1), whose maximal compact subgroup is SO(4). Thus, we conjecture that all states form a representation of SO(4,1).

Consider the Lie algebra of O(4,1) and the general structure of its generators in terms of the canonical variables. The algebra of O(n) has $\frac{n(n-1)}{2}$ generators so to extend the algebra of O(n) to O(n+1) takes n generators, which can be taken as the components of a n-vector. To extend the Lie algebra from O(4) to O(5) or O(4,1) we can choose the additional generators G_a to be components of a four-vector G under O(4):

$$[S_{ab}, G_c] = i(G_b\delta_{ac} - G_a\delta_{bc}) \qquad a,b,c = 1,2,3,4. \tag{184}$$

If we apply Jacobi's identity to S_{ab}, G_a, and G_b and use Equation (184) we find

$$[S_{ab}, [G_a, G_b]] = 0. \tag{185}$$

We require that the Lie algebra closes, so $[G_a, G_b]$ must be a linear combination of the generators, clearly proportional to S_{ab} and we choose the normalization, such that

$$[G_a, G_b] = -iS_{ab}. \tag{186}$$

If we define

$$G_4 = S_{40} = S; \quad G_i = S_{i0} = B_i \tag{187}$$

and recall Equation (79)

$$L_i = e_{ijk}S_{jk} \quad A_i = S_{i4}$$

then the additional commutation relations that realize SO(4,1) may be written in terms of $\boldsymbol{L}, \boldsymbol{A}, \boldsymbol{B}$, and S:

$$\begin{aligned} [L_i, B_j] &= i\epsilon_{ijk}B_k & [L_i\,, S] &= 0 \\ [S\,, A_j] &= iB_j & [S\,, B_j] &= iA_j \\ [A_j, B_k] &= i\delta_{jk}S & [B_i, B_j] &= -i\epsilon_{ijk}L_k. \end{aligned} \tag{188}$$

The top two commutators show that $\boldsymbol{B}$ transforms as a three-vector under O(3) rotations and that S is a scalar under rotations. Alternatively, we can write the commutation relations in terms of the generators S_{AB} , $A, B = 0,1,2,3,4$:

$$[S_{AB}, S_{CD}] = i\,(g_{AC}S_{BD} + g_{BD}S_{AC} - g_{AD}S_{BC} - g_{BC}S_{AD}) \tag{189}$$

where $g_{00} = -1, g_{aa} = 1$.

The commutators above follow directly from the mathematical theory of SO(4,1), but the theory does not tell us what these generators represent, just their commutation properties. We now investigate the general features of the representations of SO(4,1) provided by the hydrogenlike atom and how to represent the generators in terms of the canonical variables.

6.2. Casimir Operators

The two Casimir operators of SO(4,1) are [120]

$$Q_2 = -\frac{1}{2}S_{AB}S^{AB} = S^2 + \boldsymbol{B}^2 - \boldsymbol{A}^2 - \boldsymbol{L}^2 \tag{190}$$

and

$$Q_4 = -w_Aw^A = (S\boldsymbol{L} - \boldsymbol{A}x\boldsymbol{B})^2 - \frac{1}{4}[\boldsymbol{L}\cdot(\boldsymbol{A}+\boldsymbol{B}) - (\boldsymbol{A}+\boldsymbol{B})\cdot\boldsymbol{L}]^2 \tag{191}$$

where $w_A = \frac{1}{8}\epsilon_{ABCDE}S^{BC}S^{DE}$.

For SO(4), we recall that for the SO(4) representations the structure of the generators in terms of the canonical variables led to the vanishing of one Casimir operator $C_1 = \boldsymbol{L} \cdot \boldsymbol{A}$ and, consequently, the appearance of only symmetrical tensor representations. We will find Q_4 vanishes for analogous reasons.

If $\boldsymbol{B}$ is a pseudovector, it is proportional to $\boldsymbol{L}$, which is the only independent pseudovector that can be constructed from the dynamical variables. The coefficient of proportionality, a scalar, X need not commute with H:

$$\boldsymbol{B} = X\boldsymbol{L} \quad [X, \boldsymbol{L}] = 0 \quad [X, H] \neq 0 \tag{192}$$

Because $[B_i, B_j] = -ie_{ijk}L_k$ it follows that $X^2 = -1$ and $\boldsymbol{B}$ would therefore be a constant multiple of $\boldsymbol{L}$ and not an independent generator. Thus $\boldsymbol{B}$ must be a vector and expressible as

$$\boldsymbol{B} = f\boldsymbol{r} + h\boldsymbol{p} \tag{193}$$

where f and h are scalar functions of r, p^2, and $\boldsymbol{r} \cdot \boldsymbol{p}$. Accordingly we find

$$\boldsymbol{B} \cdot \boldsymbol{L} = \boldsymbol{L} \cdot \boldsymbol{B} = 0 \tag{194}$$

Further, since $\boldsymbol{B}$ is a vector and $\boldsymbol{A}$ is a vector, $\boldsymbol{A} \times \boldsymbol{B}$ is a pseudovector and therefore is proportional to $\boldsymbol{L}$:

$$\boldsymbol{A} \times \boldsymbol{B} = Y\boldsymbol{L} \ , \qquad [Y, \boldsymbol{L}] = 0 \tag{195}$$

For this equation to be consistent with the SO(4,1) commutation relations we find $Y = S$ and, therefore

$$\boldsymbol{A} \times \boldsymbol{B} = S\boldsymbol{L}. \tag{196}$$

It follows from substituting $\boldsymbol{L} \cdot \boldsymbol{A} = 0$ and Equations (194) and (196) into Equation (191) that for the SO(4,1) representations realized by the hydrogenlike atom

$$Q_4 = 0. \tag{197}$$

As with the SO(4) symmetry, the dynamics of the hydrogen atom require that only certain representations of SO(4,1) appear. From the mathematical theory of irreducible infinite dimensional unitary representations of SO(4,1), we have the following results:

$$\begin{gathered} \text{Class I:} \quad Q_4 = 0; \quad Q_2 \text{ real, } > 0 \\ \text{SU(2)} \times \text{SU(2) content:} \\ (Q)^I = (0,0) \oplus \left(\frac{1}{2}, \frac{1}{2}\right) \oplus (1,1) \oplus \ldots \end{gathered} \tag{198}$$

$$\begin{gathered} \text{Class II: } Q_4 = 0, Q_2 = -(s-1)(s+2), s = \text{ integer } > 0 \\ SU(2) \times SU(2) \text{ content:} \\ (Q)^{II} = \left(\frac{s}{2}, \frac{s}{2}\right) \oplus \left(\frac{s+1}{2}, \frac{s+1}{2}\right) \oplus \cdots \end{gathered} \tag{199}$$

The class I representations are realized by the complete set of eigenstates of $(Z\alpha)^{-1}$ for the hydrogenlike atom. Note, however, that we have an infinite number of such class I representations since Q_2 may have any positive real value. We shall find that for $Q_2 = 2$ we may extend our group from SO(4,l) to SO(4,2). The class II representations are realized by the eigenstates of $(Z\alpha)^{-1}$ with principal quantum numbers from $n = s + 1$ to n becomes infinite. The first s levels could, if we desire, be described by SO(5).

In this section, we have analyzed the group structure and the representations using the complete set of eigenstates of $(Z\alpha)^{-1}$ for our basis. We might ask: what if we used energy eigenstates instead as a basis for the representations? From Section 4.3 we know that the quantum numbers and multiplicities of the $(Z\alpha)^{-1}$ eigenstates are precisely the same as those of the bound energy eigenstates. Thus, with the energy eigenstates as our basis, we would reach the same conclusions about the group structure as before but we would be including only the bound states in our representations and we would be ignoring all scattering states.

6.3. Relationship of the Dynamical Group SO(4,1) to the Conformal Group in Momentum Space

We can give a more complete analysis of the hydrogenlike atom in terms of SO(4,1) by considering the relationship between the four-dimensional rotations of the four-vector U'_a, with $a = l,2,3,4$, which we discussed in Section 4.6, and the group of conformal transformations in momentum space. Conformal transformations preserve the angles between directed curves, but not necessarily lengths. The rotations generated by the Runge-Lenz vector $\boldsymbol{a}$ and the angular momentum $\boldsymbol{L}$ leave the scalar product $U_a V^a$ of four-vectors invariant and, therefore, are conformal transformations. The stereographic projection we employed is also a conformal transformation. Since the product of two conformal transformations is itself a conformal transformation, we must conclude that $\boldsymbol{a}$ generates a conformal transformation of the momentum three-vector $\boldsymbol{p}$.

We must introduce two additional operators that correspond to the operators $\boldsymbol{B}$ and S introduced in Section 6.1 in order to express the most general conformal transformation. By employing the isomorphism between the generators $\boldsymbol{L}$, $\boldsymbol{a}$, $\boldsymbol{B}$, and S of SO(4,1) and the generators of conformal transformations in momentum space we can immediately obtain expressions for the additional generators $\boldsymbol{B}$ and S in terms of the canonical variables, which is our objective. We need these additional generators to complete our SO(4,1) group for the hydrogen atom.

To derive the isomorphism we use the most convenient representation, namely that based on eigenstates of $(Z\alpha)^{-1}$ convenient for momentum space calculations ($\rho = \frac{(p^2+a^2)}{2a^2}$). Once established, the isomorphism becomes a group theoretical statement and it is independent of the particular representation.

The Conformal Group in Momentum Space

An arbitrary infinitesimal conformal transformation in momentum three-space may be written as

$$\delta p_j = \delta a_j + \delta\omega_{jk} p_k + \delta\rho p_j + \left(p^2 \delta c_j - 2p_j \boldsymbol{p} \cdot \delta \boldsymbol{c}\right) \tag{200}$$

where $\delta\omega_{jk} = -\delta\omega_{kj}$.

The terms in δp_j arise as follows:

$$\begin{aligned}
&\delta a_j \text{ translation generated by } \boldsymbol{R} \cdot \delta \boldsymbol{a} \\
&\delta\omega_{jk} \text{ rotation generated by } \boldsymbol{J} \cdot \delta\boldsymbol{\omega},\, J_{ij} = \epsilon_{ijk} J_k \\
&\delta\rho \text{ dilation generated by } D\delta\rho \\
&\delta c_j \text{ special conformal transformation generated by } \boldsymbol{K} \cdot \delta \boldsymbol{c}
\end{aligned} \tag{201}$$

This is a ten parameter group with the generators $(\boldsymbol{R}, \boldsymbol{J}, D, \boldsymbol{K})$ which obey the following commutation relations:

$$\begin{aligned}
&[D, R_j] \cdot = iR_j && [D, J_i] = 0 \\
&[D, K_j] = -iK_j && [R_i, J_k] = i\epsilon_{ijk} R_k \\
&[K_n, R_m] = 2i\epsilon_{nmr} J_r - 2i\delta_{mn} D && [J_i, J_k] = i\epsilon_{ikm} \mathrm{J_m} \\
&[R_i, R_j] = 0 && [K_i, J_k] = i\epsilon_{ikm} K_m \\
&[K_j, K_j] = 0
\end{aligned} \tag{202}$$

There is an isomorphism between the algebra of the generators of conformal transformations and the dynamical noninvariance algebra of SO(4,1) of the hydrogen atom. Because J_i is the generator of spatial rotations we make the association $L_i = J_i$. Comparing the differential change in p_i from a transformation generated by $\boldsymbol{a} \cdot \delta \boldsymbol{v}$ (in the representation with $\rho = \frac{(p^2+A^2)}{2a^2}$)

$$\begin{aligned} \delta p_i &= i\left[\boldsymbol{a} \cdot \delta \boldsymbol{v}, p_i\right] \\ &= -\frac{1}{2a}\left[\left(p^2 - a^2\right)\delta v_i - 2\boldsymbol{p} \cdot \delta \boldsymbol{v} p_i\right] \end{aligned} \tag{203}$$

to the differential change in p_i from a conformal transformation leads to the association

$$a_i = \frac{1}{2}\left(\frac{K_i}{a} - aR_i\right). \tag{204}$$

To confirm the identification we can use the commutation relations of the conformal group to show that the O(4) algebra of $\boldsymbol{L}$ and $\boldsymbol{a}$ corresponds precisely to that of $\boldsymbol{J}$ and $\frac{1}{2}(\frac{\boldsymbol{K}}{a} - a\boldsymbol{R})$. This result alone suggests that our SO(4) degeneracy group should be considered as a subgroup of the larger group SO(4,1). It suggests introducing the operators

$$\boldsymbol{B} = \frac{1}{2}\left(\frac{\boldsymbol{K}}{a} + a\boldsymbol{R}\right) \qquad S = D. \tag{205}$$

The commutation relations of S and $\boldsymbol{B}$ which follow from Equations (205) and the commutation relations Equation (202) are identical to the commutation relations given for S and $\boldsymbol{B}$ in Section 6.1. Thus, by considering the $\boldsymbol{a}$ and $\boldsymbol{L}$ transformations in momentum space as conformal transformations, we were led to introduce the generators $\boldsymbol{B}$ and S and obtain the dynamical algebra SO(4,l). Further, we are led to the expressions for these generators in terms of the canonical variables.

By comparing the expression for $\boldsymbol{a}$ in terms of the conformal generators with our known expressions for $\boldsymbol{a}$, Equation (120) or Equation (121), we obtain expressions for K_i and R_i in terms of the canonical variables. If we use the eigenstates convenient for configuration space calculations ($\rho = n/ar$) we make the identifications

$$\begin{aligned} \boldsymbol{K} &= \tfrac{1}{2}(\boldsymbol{r}p^2 + p^2\boldsymbol{r}) - \boldsymbol{r} \cdot \boldsymbol{p}\boldsymbol{p} - \boldsymbol{p}\boldsymbol{p} \cdot \boldsymbol{r} - \boldsymbol{r}\tfrac{1}{4r^2} \\ \boldsymbol{R} &= \boldsymbol{r}. \end{aligned} \tag{206}$$

Substituting these results in the equation for $\boldsymbol{B}$ we find

$$\boldsymbol{B} = \frac{1}{2a}\left(\frac{p^2\boldsymbol{r} + \boldsymbol{r}p^2}{2} - \boldsymbol{r} \cdot \boldsymbol{p}\boldsymbol{p} - \boldsymbol{p}\boldsymbol{p} \cdot \boldsymbol{r} - \frac{\boldsymbol{r}}{4r^2}\right) + \frac{a\boldsymbol{r}}{2} \tag{207}$$

which is a manifestly Hermitean operator valid throughout Hilbert space. From Equations (204) and (205) we note that

$$\boldsymbol{B} - \boldsymbol{a} = a\boldsymbol{r}. \tag{208}$$

To compute D, we substitute the expressions for $\boldsymbol{K}$ and $\boldsymbol{R}$ into the commutation relation

$$D = \frac{i}{2}[K_i, R_i]$$

obtaining the result

$$S = \frac{1}{2}(\boldsymbol{p} \cdot \boldsymbol{r} + \boldsymbol{r} \cdot \boldsymbol{p}) = D \tag{209}$$

which is identical to the generator of the scale change transformation $D(\lambda)$ defined in Equation (109) in Section 4.3.

The significance of the generator $D = S$ of the scale change in terms of SO(4,1) is apparent if we compute

$$e^{i\lambda D}\left(\frac{\boldsymbol{K}}{a} \pm a\boldsymbol{R}\right)e^{-i\lambda D} = \frac{\boldsymbol{K}}{a'} \pm a'\boldsymbol{R} \tag{210}$$

where $a' = e^{\lambda}a$.

The unitary transformation $e^{i\lambda D}$ may be viewed as generating an inner automorphism of SO(4,1) which is an equivalent representation of SO(4) that is characterized by a different value of the quantity a or the energy. In other words, under the scale change $e^{i\lambda D}$, the basis states for our representation of SO(4,1), $|nlm;a)$, transform to a new set, $|nlm;e^{\lambda}a)$ in agreement with our discussion in Section 4.3.

Because the algebra of our generators closes, we may also view $e^{i\lambda D}$ as transforming a given generator into a linear combination of the generators. With the definitions of $\boldsymbol{a}$ and $\boldsymbol{B}$ (Equation (204) and (205)), we can easily show that Equation (210), with the upper sign, may also be written

$$e^{i\lambda D}\boldsymbol{B}e^{-i\lambda D} = \boldsymbol{B}\cosh\lambda + \boldsymbol{a}\sinh\lambda. \tag{211}$$

7. The Group SO(4,2)

7.1. Motivation for Introducing SO(4,2)

We would like to express Schrodinger's equation as an algebraic equation in the generators of some group [37,38]. As we are unable to do this with our SO(4,l) generators S_{AB} we again expand the group. To guide us, we recall that to expand SO(3) to SO(4) we added a three-vector of generators $\boldsymbol{A}$, and to expand SO(4) to SO(4,1), we added a four-vector of generators $(S, \boldsymbol{B})$. In both cases, this type of expansion produced a set of generators convenient for the study of the hydrogenlike atom. We guess that the appropriate expansion of SO(4,1) is obtained by adding a five-vector (under SO(4,1)) of generators Γ_A to obtain SO(4,2) [37,38]. We can provide additional motivation for this choice by considering Schrodinger's equation. The generators in terms of which we want to express this equation must be scalars under L_i rotations. Additionally we know $S = S_{40}$ (Equation (187)) generates scale changes of Schrodinger's equation. The fact that S_{40} mixes the zero and four components of a five-vector suggests that Schrodinger's equation may be expressed in terms of the components Γ_0 and Γ_4 which are scalars under L_i, of the five vector Γ_A. Since Γ_A is a five-vector under SO(4,1), it must satisfy the equation

$$[S_{AB}, \Gamma_C] = i\left(\Gamma_B g_{AC} - \Gamma_A g_{BC}\right). \tag{212}$$

The spatial components of Γ_A which are $(\Gamma_1, \Gamma_2, \Gamma_3) = \boldsymbol{\Gamma}$ transform as a vector under rotations generated by $\boldsymbol{L}$.

To construct the Lie algebra of SO(4,2) we require that the set of operators $\{\Gamma_A, S_{AB}; A, B = 0,1,2,3,4\}$ must close under the operations of commutation. By applying Jacobi's identity to Γ_A, Γ_B, and S_{AB}, and requiring that Γ_A and Γ_B do not commute, we find

$$[S_{AB}, [\Gamma_A, \Gamma_B]] = 0 \quad A, B = 0,1,2,3,4.$$

Because we require that our Lie algebra closes, the commutator of Γ_A and Γ_B must be proportional to S_{AB}. We normalize Γ, so

$$[\Gamma_A, \Gamma_B] = -iS_{AB} \quad A, B = 0,1,2,3,4. \tag{213}$$

If we define

$$S_{A5} = \Gamma_A = -S_{5A} \quad A = 0,1,2,3,4. \tag{214}$$

and recall

$$A_i = S_{i4} \quad B_i = S_{io} \quad L_i = e_{ijk}S_{jk} \quad S = S_{40}$$

then we may unite all the commutations relations of Γ_A and S_{AB} in the single equation :

$$[S_{\mathcal{AB}}, S_{\mathcal{CD}}] = i(g_{\mathcal{AC}}S_{\mathcal{BD}} + g_{\mathcal{BD}}S_{\mathcal{AC}} - g_{\mathcal{AD}}S_{\mathcal{BC}} - g_{\mathcal{BC}}S_{\mathcal{AD}}) \tag{215}$$

where $\mathcal{A}, \mathcal{B}, .. = 0,1,2,3,4,5$ and $g_{00} = g_{55} = -1$; $g_{aa} = 1, a = 1,2,3,4$.

These are the commutation relation for the Lie algebra of SO(4,2). In terms of $\boldsymbol{A}, \boldsymbol{B}, \boldsymbol{L}, S$ and Γ_A the additional commutation relations for the noncommuting generators are [53]:

$$\begin{array}{ll} [B_i, \Gamma_j] = i\Gamma_0\delta_{ij} & [\Gamma_i, \Gamma_j] = -i\epsilon_{ijk}L_k \\ [A_i, \Gamma_j] = i\Gamma_4\delta_{ij} & [\Gamma_i, \Gamma_0] = -iB_i \\ [L_i, \Gamma_j] = i\epsilon_{ijk}\Gamma_k & [\Gamma_i, \Gamma_4] = -iA_i \\ & [\Gamma_4, \Gamma_0] = -iS \\ [B_i, \Gamma_0] = i\Gamma_i & [A_i, \Gamma_4] = -i\Gamma_i \\ [S, \Gamma_0] = i\Gamma_4 & [S, \Gamma_4] = i\Gamma_0 \end{array} \tag{216}$$

7.2. Casimir Operators

The Lie algebra of SO(4,2) is rank three, so it has three Casimir operators W_2, W_3, and W_4 [43]:

$$W_2 = -\frac{1}{2}S_{\mathcal{AB}}S^{\mathcal{AB}} = Q_2 + \Gamma_A\Gamma^A \tag{217}$$

where Q_2 is the nonvanishing SO(4,1) Casimir operator Equation (190) and

$$W_3 = \epsilon^{\mathcal{ABCDEF}}S_{\mathcal{AB}}S_{\mathcal{CD}}S_{\mathcal{EF}} \tag{218}$$

$$W_4 = S_{\mathcal{AB}}S^{\mathcal{BC}}S_{\mathcal{CD}}S^{\mathcal{DA}}. \tag{219}$$

Computation of W_3

We can show that $W_3 = 0$ from dynamical considerations similar to those used in the discussion of SO(4,1) Casimir operators. The only terms that can be included in W_3 are scalars that are formed from products of three generators with different indices

$$\begin{array}{lll} \boldsymbol{B}\cdot\boldsymbol{A}\times\boldsymbol{\Gamma}, & \boldsymbol{A}\cdot\boldsymbol{\Gamma}\times\boldsymbol{B}, & \boldsymbol{\Gamma}\cdot\boldsymbol{B}\times\boldsymbol{A} \\ \Gamma_4\boldsymbol{L}\cdot\boldsymbol{B}, & \Gamma_0\boldsymbol{L}\cdot\boldsymbol{A}, & S\boldsymbol{\Gamma}\cdot\boldsymbol{L} \end{array} \tag{220}$$

It is interesting that these terms are actually all pseudoscalars. Terms like $\boldsymbol{B}\cdot\boldsymbol{A}\times\boldsymbol{L}$ are simply not possible because of the structure of W_3. We know that $\boldsymbol{\Gamma} = (\Gamma_1, \Gamma_2, \Gamma_3)$ must not be pseudovector, otherwise it would be proportional to $\boldsymbol{L}$. Since it is a vector, it must equal a linear combination of $\boldsymbol{r}$ and $\boldsymbol{p}$. Therefore, we conclude

$$\boldsymbol{\Gamma}\cdot\boldsymbol{L} = \boldsymbol{L}\cdot\boldsymbol{\Gamma} = 0. \tag{221}$$

Because $\boldsymbol{\Gamma}$ and $\boldsymbol{B}$ are both vectors and $\boldsymbol{L}$ is the only pseudovector we have $\boldsymbol{\Gamma}\times\boldsymbol{B} = \Lambda\boldsymbol{L}$. In order to determine the scalar Λ we evaluate the commutators

$$[B_k, (\boldsymbol{\Gamma}\times\boldsymbol{B})_k], \quad \text{and } [\Gamma_k, (\boldsymbol{\Gamma}\times\boldsymbol{B})_k] \tag{222}$$

and find

$$\boldsymbol{\Gamma}\times\boldsymbol{B} = \Gamma_0\boldsymbol{L} = -\boldsymbol{B}\times\boldsymbol{L}. \tag{223}$$

The analogous equations for $\boldsymbol{A}$ and $\boldsymbol{\Gamma}$, and for $\boldsymbol{A}$ and $\boldsymbol{B}$, are

$$\begin{array}{l} \boldsymbol{\Gamma}\times\boldsymbol{A} = -\Gamma_4\boldsymbol{L} = -\boldsymbol{A}\times\boldsymbol{\Gamma} \\ \boldsymbol{A}\times\boldsymbol{B} = S\boldsymbol{L} = -\boldsymbol{B}\times\boldsymbol{A}. \end{array} \tag{224}$$

From Equations (223) and (224), we see that because of the dynamical structure of the generators each of the quantities in the first line of Equation (220) is proportional to the quantity directly below in the second line. We also have shown that (Equations (194) and (221))

$$\boldsymbol{L} \cdot \boldsymbol{B} = \boldsymbol{L} \cdot \boldsymbol{A} = \boldsymbol{\Gamma} \cdot \boldsymbol{L} = 0. \tag{225}$$

Accordingly, each scalar in our list vanishes and

$$W_3 = 0. \tag{226}$$

Computation of W_2

In order to compute W_2 we need to evaluate

$$\Gamma^2 \equiv \Gamma_A \Gamma^A = \Gamma_4^2 + \Gamma_i \Gamma^i - \Gamma_0^2. \tag{227}$$

From the structure of W_2 as shown in Equation (217), we see that Γ^2 must be a number since W_2 and Q_2 are both Casimir operators and therefore equal numbers for a particular representation. Accordingly, we have

$$[\Gamma^2, \Gamma_A] = 0. \tag{228}$$

From this equation, we can deduce a lemma allowing us to easily evaluate W_2 and W_4 in terms of the number Γ^2. Using Equation (227) and the definition of S_{AB} Equation (213) we find

$$\Gamma^A S_{AB} + S_{AB} \Gamma^A = 0.$$

Contracting Equation (212) with g_{AC} gives

$$S_{AB} \Gamma^A - \Gamma^A S_{AB} = 4i \Gamma_B.$$

Consequently, it must follow that

$$S_{AB} \Gamma^A = 2i \Gamma_B = -\Gamma^A S_{AB}. \tag{229}$$

We are now able to evaluate the quantity

$$S_{AB} S^B{}_C = i S_{AB} [\Gamma^B, \Gamma_C] = i(S_{AB} \Gamma^B \Gamma_C - S_{AB} \Gamma_C \Gamma^B).$$

Using Equation (212) for the commutator of S_{AB} with Γ_C and Equation (229) for the contraction $S_{AB} \Gamma^B$ we prove the lemma

$$S_{AB} S^B{}_C = 2i S_{CA} - \Gamma_A \Gamma_C + \Gamma^2 g_{AC}. \tag{230}$$

The value of the SO(4,1) Casimir operator $Q_2 = \frac{1}{2} g^{AC} S_{AB} S^B{}_C$ follows directly from the lemma:

$$Q_2 = 2\Gamma^2. \tag{231}$$

Accordingly, we have from Equation (217)

$$W_2 = 3\Gamma^2. \tag{232}$$

Computation of W_4

The Casimir operator W_4 can be written as

$$W_4 = S_{AB} S^{BC} S_{CD} S^{DA} + S_{AB} S^{B5} S_{5D} S^{DA} + S_{5B} S^{BC} S_{CD} S^{D5} + S_{5B} S^{B5} S_{5D} S^{D5}. \tag{233}$$

where $\mathcal{B}, \mathcal{D} = 0, 1, 2, 3, 4, 5$ and $A, C = 0, 1, 2, 3, 4$.

In order to evaluate W_4 in terms of Γ^2 we compute $S_{AB}S^{BC}$. Recalling $\Gamma_A = S_{A5}$ we see

$$S_{A\mathcal{B}}S^{\mathcal{B}C} = \Gamma_A\Gamma^C + S_{AB}S^{BC}. \tag{234}$$

Substituting the lemma Equation (230), we find

$$S_{A\mathcal{B}}S^{\mathcal{B}C} = 2iS^C_{\ A} - \Gamma^2 g_A^{\ C}. \tag{235}$$

From Equation (229), it follows that

$$S_{5\mathcal{B}}S^{\mathcal{B}C} = 2i\Gamma^C. \tag{236}$$

Substituting Equations (231), (232), (235), and (236) into Equation (233) for W_4 we find

$$W_4 = 6(\Gamma^2)^2 - 24\Gamma^2. \tag{237}$$

The fact that the nonvanishing Casimir operators (Q_2, W_2, and W_4) for SO(4,1) and SO(4,2) are given in terms of Γ^2 implies that the representation of SO(4,2) determines the particular representation of SO(4,l) appropriate to the hydrogenlike atom. In turn the value of Γ^2 is determined by the structure of the Γs in terms of the canonical variables. In Section 7.4, we derive these structures and find that

$$\Gamma^2 = 1.$$

Therefore, the quadratic SO(4,1) Casimir operator Q_2 has the value

$$Q_2 = 2$$

and the SO(4,2) Casimir operators have the values:

$$W_2 = 3 \qquad W_3 = 0 \qquad W_4 = -18.$$

The researchers that have published different representations of SO(4,2) based on the hydrogen atom that give their Casimir operators all have $W_2 = 3$ (or its equivalent) and W_3=0 [34,79,82], however, two authors have representations with $W_4 = 0$ [34,82] and one [79] has $W_4 = -12$, as compared to our value of -18.

From the mathematical theory of representations, it follows that our representations of SO(4,1) and SO(4,2) are both unitary and irreducible. This means there is no subset of basis vectors that transform among themselves as either SO(4,1) or as SO(4,2).

7.3. Some Group Theoretical Results

In this section, we derive the transformation properties of the generators of SO(4,2) and then a novel contraction formula that will prove useful for situations in which we want to employ perturbation theory, for example, in our calculation of the radiative shift for the hydrogen atom in Section 8. We will work primarily with the SO(4,2) generators expressed as the combination of the SO(4,1) generators S_{AB} and the five-vector Γ, with $g_{AB} = (-1,1,1,1,1)$ where $A, B = 0,1,2,3,4$.

Transformation Properties of the Generators

We can evaluate quantities like

$$^{AB}\Gamma_B(\theta) \equiv e^{iS_{AB}\theta}\Gamma_B e^{-iS_{AB}\theta} \quad \text{no sum over A or B} \tag{238}$$

by expanding the exponentials in an infinite series and then using the commutation relations Equations (212) and (213) of the generators S_{AB} and Γ_A, Γ_B repeatedly. However, it is easier to solve the differential equations satisfied by ${}^{AB}\Gamma_B$ and to use the appropriate boundary conditions. Differentiating Equation (238) and using the commutation relations, we obtain the equations

$$\frac{d}{d\theta}\ {}^{AB}\Gamma_B = -g_{BB}\ {}^{AB}\Gamma_A \qquad \frac{d^2}{d\theta^2}\ {}^{AB}\Gamma_B = -g_{AA}g_{BB}\ {}^{AB}\Gamma_B \tag{239}$$

which have the solution

$${}^{AB}\Gamma_B = \Gamma_B \cos\sqrt{g_{AA}g_{BB}}\theta + \frac{g_{BB}}{\sqrt{g_{AA}g_{BB}}}\Gamma_A \sin\sqrt{g_{AA}g_{BB}}\theta. \tag{240}$$

Using a similar procedure we find

$$e^{i\Gamma_A\theta}S_{AB}e^{-i\Gamma_A\theta} = S_{AB}\cosh\sqrt{g_{AA}}\theta + \sqrt{g_{AA}}\Gamma_B \sinh\sqrt{g_{AA}}\theta \tag{241}$$

$$e^{i\Gamma_A\theta}\Gamma_B e^{-i\Gamma_A\theta} = \Gamma_B\cosh\sqrt{g_{AA}}\theta + \frac{1}{\sqrt{g_{AA}}}S_{AB}\sinh\sqrt{g_{AA}}\theta \tag{242}$$

where no summation over A or B is implied.

These formulae, Equation (240)–(242), give the SO(4,2) transformation properties of the SO(4,2) generators.

<u>The Contraction Formula</u>

If we multiply Equation (241) from the right by $e^{i\Gamma_A\theta}$ and then contract from the left with Γ_B, we obtain

$$\sum_B \Gamma^B e^{i\Gamma_A\theta}\Gamma_B = \left[(1-g_{AA}\Gamma_A^2)\cosh\sqrt{g_{AA}}\theta + \frac{2i\Gamma_A}{\sqrt{g_{AA}}}\sinh\sqrt{g_{AA}}\theta\right]e^{i\Gamma_A\theta} + g_{AA}\Gamma_A^2 e^{i\Gamma_A\theta} \tag{243}$$

where we have used $\Gamma^2 = 1$ and Equation (229). Expanding the hyperbolic functions in terms of exponentials and collecting terms gives

$$\sum_B \Gamma^B e^{i\Gamma_A\theta}\Gamma_B = \frac{1}{2}(1+\frac{i\Gamma_A}{\sqrt{g_{AA}}})^2 e^{i(\Gamma_A - i\sqrt{g_{AA}})\theta} + \frac{1}{2}(1-\frac{i\Gamma_A}{\sqrt{g_{AA}}})^2 e^{i(\Gamma_A + i\sqrt{g_{AA}})\theta} + g_{AA}\Gamma_A^2 e^{i\Gamma_A\theta}. \tag{244}$$

A Fourier decomposition of a function Γ_A may be written

$$f(\Gamma_A) = \frac{1}{2\pi}\int d\theta\, h(\theta)e^{i\Gamma_A\theta}. \tag{245}$$

Consequently we have

$$\sum_B \Gamma^B f(\Gamma_A)\Gamma_B = \frac{1}{2}(1+\frac{i\Gamma_A}{\sqrt{g_{AA}}})^2 f(\Gamma_A - i\sqrt{g_{AA}}) + \frac{1}{2}(1-\frac{i\Gamma_A}{\sqrt{g_{AA}}})^2 f(\Gamma_A + i\sqrt{g_{AA}}) + g_{AA}\Gamma_A^2 f(\Gamma_A). \tag{246}$$

By performing a suitable rotation we can generalize this formula from functions of Γ_A to functions of $\Gamma_A n^A$ where $n_A n^A = \pm 1$. For $n^2 = -1$ we start with $\Gamma_A = \Gamma_0$ and rotate to obtain a very general result

$$\sum_B \Gamma_B f(n\Gamma)\Gamma^B = \frac{1}{2}(n\Gamma+1)^2 f(n\Gamma+1) + \frac{1}{2}(n\Gamma-1)^2 f(n\Gamma-1) - (n\Gamma)^2 f(n\Gamma). \tag{247}$$

We will have occasion to apply this formula for the special case

$$f(n\Gamma) = \frac{1}{\Gamma n - \nu}. \tag{248}$$

Using the representation

$$\frac{1}{\Gamma n - \nu} = \int_0^\infty ds e^{\nu s} e^{-\Gamma n s} \tag{249}$$

we obtain the result

$$\Gamma_A \frac{1}{\Gamma n - \nu} \Gamma^A = -2\nu \int_0^\infty ds\, e^{\nu s} \frac{d}{ds} (\sinh^2 \frac{s}{2}\, e^{-\Gamma n\, s}) \tag{250}$$

which is in a form convenient for perturbation calculations.

Derivation of the Γ_A in terms of the Canonical Variables

For our basis states, we shall use eigenstates of $(Z\alpha)^{-1}$ convenient for configuration space calculations ($\rho = na/r$). We choose these states rather than those convenient for momentum space calculations because they lead to simpler expressions for the Γ_A in terms of the canonical variables, although the expression for $\boldsymbol{a}$ is slightly more complicated. Thus, our states obey the equation

$$\left[\frac{1}{K_1(a)} - n\right] |nlm) = 0 \tag{251}$$

where $K_1(a)$ is given by Equation (116). We know that K_1^{-1} must commute with the generators of the SO(4) symmetry group $(a)_i = S_{i4}$ and $S_{ij} = \epsilon_{ijk} L_k$. This suggests that we choose

$$\Gamma_0 = [K_1(a)]^{-1} = \sqrt{ar} \frac{p^2 + a^2}{2a^2} \sqrt{ar} = \frac{1}{2} \left(\frac{\sqrt{r} p^2 \sqrt{r}}{a} + ar \right) \tag{252}$$

so that

$$(\Gamma_0 - n) |nlm) = 0. \tag{253}$$

This last equation is the Schrodinger equation expressed in our language of SO(4,2): our states $|nlm)$ are eigenstates of Γ_0 with eigenvalue n.

To find Γ_4, we calculate $\Gamma_4 = -i[S, \Gamma_0]$, using Equation (209) for S,

$$\Gamma_4 = \sqrt{ar} \frac{p^2 - a^2}{2a^2} \sqrt{ar} = \frac{1}{2} \left(\frac{\sqrt{r} p^2 \sqrt{r}}{a} - ar \right). \tag{254}$$

Sometimes it is convenient to use the linear combinations

$$\Gamma_0 - \Gamma_4 = ar \qquad \Gamma_0 + \Gamma_4 = \frac{\sqrt{r} p^2 \sqrt{r}}{a} \tag{255}$$

which can be used to express the dipole transition operator [33]. We can find Γ_i from Equation (216), $\Gamma_i = -i[B_i, \Gamma_0]$

$$\Gamma_i = \sqrt{r} p_i \sqrt{r} \tag{256}$$

which we might have guessed initially since $[rp_i, rp_j] \sim L_k$. Every component of Γ_A is Hermitean, consequently the generators S_{AB} given by the commutators Equation (213) are also Hermitian. We may explicitly verify that these expressions for Γ_A lead to a consistent representation of all generators in the SO (4,2) Lie algebra.

Under a scale change generated by S, Γ_i is invariant and Γ_4 and Γ_0 transform in the same manner as $\boldsymbol{a}$ and $\boldsymbol{B}$ (Equation (210)): they retain their form but a is transformed into $e^\lambda a$:

$$e^{i\lambda S} \begin{Bmatrix} \Gamma_0 \\ \Gamma_4 \end{Bmatrix} e^{-i\lambda S} = \frac{1}{2} \left(\frac{\sqrt{r} p^2 \sqrt{r}}{e^\lambda a} \pm e^\lambda ar \right) \tag{257}$$

The scale change generates an inner automorphism of SO(4,2) characterized by a different value of the parameter a.

7.4. Subgroups of SO(4,2)

The two most significant subgroups are [53]:

1. L_i, a_i or S_{jk}, S_{i4}, forming an SO(4) subgroup. These generators commute with Γ_0 and therefore constitute the degeneracy group for states of energy $-a^2/(2m)$ and fixed principal quantum number n (or fixed coupling constant na/m). The Casimir operator for this subgroup is

$$\boldsymbol{a}^2 + \boldsymbol{L}^2 = n^2 - 1 = \Gamma_0^2 - 1. \tag{258}$$

We discussed this subgroup in Section 4.2 in terms of $\boldsymbol{L}$ and $\boldsymbol{A}$ and the states $|nlm\rangle$. The same results are obtained with the generators $\boldsymbol{L}$ and $\boldsymbol{a}$ with the states $|nlm)$. For example, we have the raising and lowering operators for m and l (Equations (93) and (94)). With the definition

$$L_\pm = L_1 \pm iL_2 \tag{259}$$

it follows that

$$[L_3, L_\pm] = \pm L_\pm \tag{260}$$

which gives

$$L_\pm|nlm) = \sqrt{(l(l+1) - m(m\pm 1)}|nl\ m\pm 1) \qquad L_3|nlm) = m|nlm). \tag{261}$$

for $l \geq 1$. In analogy to $L_\pm$ one can define

$$a_\pm = a_1 \pm ia_2 \tag{262}$$

which obey the relations

$$[a_3, a_\pm] = \pm L_3 \qquad [L_3, a_\pm] = \pm a_\pm \tag{263}$$

and

$$\begin{aligned} a_\pm|nlm) = &\mp \left(\frac{(n^2-(l+1)^2)(l+2\pm m)(l+1\pm m)}{4(l+1)^2-1}\right)^{\frac{1}{2}} |n\ l+1\ m\pm 1) \\ &\pm \left(\frac{(n^2-l^2)(l\mp m)(l-1\mp m)}{4l^2-1}\right)^{\frac{1}{2}} |n\ l-1\ m\pm 1) \end{aligned} \tag{264}$$

for $l \geq 1$. The action of $a_\pm$ is not directly analogous to that of $L_\pm$, because we are using $|nlm)$ as basis states. If we used $|na_3l_3 = m)$ as basis states, the action would be similar. An operator that only changes the angular momentum is a_3

$$a_3|nlm\rangle = \left(\frac{(n^2-(l+1)^2)((l+1)^2-m^2)}{4(l+1)^2-1}\right)^{\frac{1}{2}} |n\,l+1\,m\rangle + \left(\frac{(n^2-l^2)(l^2-m^2)}{4l^2-1}\right)^{\frac{1}{2}} |n\ l-1\ m\rangle. \tag{265}$$

for $l \geq 1$. Since a_3 commutes with L_3 and Γ_0, it does not change n or m.

2. Γ_4, $S = S_{40}$, Γ_0, forming a SO(2,1) subgroup. These operators commute with $\boldsymbol{L}$ but not with Γ_0, hence then can change n but not $\boldsymbol{L}$ or m. The Casimir operator for this subgroup is

$$\Gamma_0^2 - \Gamma_4^2 - S^2 = \boldsymbol{L}^2 = l(l+1). \tag{266}$$

We can define the operators [53]

$$j_1 = \Gamma_4 \qquad j_2 = S \qquad j_3 = \Gamma_0 \tag{267}$$

with commutators

$$[j_1, j_2] = -ij_3 \quad [j_2, j_3] = ij_1 \quad [j_3, j_1] = ij_2 \tag{268}$$

We can define the raising and lowering operators

$$j_{\pm} = j_1 \pm i j_2 = \Gamma_4 \pm iS \tag{269}$$

which obey the commutation relations

$$[j_{\pm}, j_3] = \mp j_{\pm}. \tag{270}$$

We find (in analogy to Equation (261))

$$\Gamma_0|nlm) = n|nlm) \qquad (\Gamma_4 \pm iS)|nlm) = \sqrt{n(n \pm 1) - l(l+1)}|n \pm 1\ lm) \tag{271}$$

We can express the action of $\Gamma_0 - \Gamma_4 = ar$ on our states

$$ar|nlm) = \tfrac{1}{2}\Big((n)(n-l) - l(l+1)\Big)^{\frac{1}{2}}|n-1\ lm) \ + n|n\ lm) \ + \tfrac{1}{2}\Big((n)(n+l) - l(l+1)\Big)^{\frac{1}{2}}|n+1\ lm) \tag{272}$$

As previously mentioned, the operator S generates scale changes as shown in Equation (257), where the value of a is changed. We can also express the action of S equivalently as transforming Γ_0 into Γ_4

$$e^{iS\lambda}\Gamma_0 e^{\bar{\ } iS\lambda} = \Gamma_0 \cosh\lambda - \Gamma_4 \sinh\lambda \qquad e^{iS\lambda}\Gamma_4 e^{\bar{\ } iS\lambda} = \Gamma_4 \cosh\lambda - \Gamma_0 \sinh\lambda. \tag{273}$$

7.5. Time Dependence of SO(4,2) Generators

For a generator to be a constant it must commute with the Hamiltonian as discussed in Section 2.1. Because the SO(4,2) group is the non-invariance or spectrum generating group, the additional generators do not all commute with the Hamiltonian and may have a harmonic time dependence as discussed in Section 2.2. It is notable that as far as we know only one paper considers the time dependence of the generators of non-invariance groups in general and one considers SO(4,2) specifically [97,136]. Our results certainly clarify and make explicit the time dependence, and show that it is just a particular aspect of the SO(4,2) transformations. In our representation with basis states $|nlm; a)$, the Hamiltonian, which is the generator of translations in time, has been transformed into Γ_0 and the Schrodinger energy eigenvalue equation has become $\Gamma_0|nlm) = n|nlm)$. Accordingly, all of the generators that commute with Γ_0 are constants of the motion, which includes $\boldsymbol{a}$, $\boldsymbol{L}$. The other operators, $\boldsymbol{B}, \boldsymbol{\Gamma}, S, \Gamma_4$ have a time dependence given by Equations (241) and (242), for example

$$S(t) = e^{iHt}S(0)E^{-iHt} = e^{i\Gamma_0 t}Se^{-i\Gamma_0 t} = S\cos t + \Gamma_4 \sin t. \tag{274}$$

$$\Gamma_4(t) = e^{iHt}\Gamma_4(0)E^{-iHt} = e^{i\Gamma_0 t}\Gamma_4 e^{-i\Gamma_0 t} = \Gamma_4 \cos t - S\sin t. \tag{275}$$

Consequently, terms like $j_{\pm}$ have a simple exponential time dependence

$$j_{\pm}(t) = j_{\pm}(0)e^{\pm it}. \tag{276}$$

Similarly $\boldsymbol{\Gamma} \pm i\boldsymbol{B}$ has an exponential time dependence.

7.6. Expressing the Schrodinger Equation in Terms of the Generators of SO(4,2)

We can write the Schrodinger equation for the energy eigenstate $E_n = -a_n^2/2m$ of a particle in a Coulomb potential in terms of the SO(4,2) generators, which are expressed in terms of the energy $-a^2/2m$, by making a scale change. From Section 4.3, Equation (114), the relationship between the Schrodinger energy eigenstate $|nlm\rangle$ and the eigenstate of $(Z\alpha)^{-1}$ is:

$$|nlm; a) = e^{-iS\lambda_n}\sqrt{\rho(a_n)}|nlm\rangle \tag{277}$$

where

$$e^{\lambda_n} = \frac{a_n}{a} \qquad \rho(a_n) = \frac{n}{a_n r}. \tag{278}$$

Substituting Equation (277) into the eigenvalue equation Equation (253) for $|nlm;a\rangle$ and employing the transformation Equation (273), we find the usual Schrodinger equation can be expressed in SO(4,2) terms as

$$(\Gamma n - n)\sqrt{\rho(a_n)}|nlm\rangle = 0 \tag{279}$$

where

$$\Gamma n \equiv \Gamma_A n^A = \Gamma_0 n^0 + \Gamma_i n^i + \Gamma_4 n^4 \tag{280}$$

$$n^o = \cosh\lambda_n = \frac{a^2 + a_n^2}{2aa_n}, \quad n^i = 0, \quad n^4 = -\sinh\lambda_n = \frac{a^2 - a_n^2}{2aa_n} \tag{281}$$

and $n_A n^A = n_4^2 - n_0^2 = -1$.

Equation (279) expresses Schrodinger's equation for an ordinary energy eigenstate $|nlm\rangle$ with energy $E_N = -a_n^2/2m$ in the language of SO(4,2). It shows the relationship between these energy eigenstates and the basis states of $(Z\alpha)^{-1}$ used for the SO(4,2) representation,

8. SO(4,2) Calculation of the Radiative Shift for the Schrodinger Hydrogen Atom

In the 1930's, it was generally believed that the Dirac equation predicted the energy levels of the hydrogen atom with excellent accuracy, but there were some questions about the prediction that the energy levels for a given principal quantum number and given total angular momentum were independent of the orbital angular momentum. To finally resolve this issue, in 1947, Willis Lamb and his student Robert Retherford at Columbia University in New York City employed rf spectroscopy and exploited the metastability of the hydrogen $2s_{1/2}$ level in a beautiful experiment and determined that the $2s_{1/2}$ and $2p_{1/2}$ levels were not degenerate and that the energy difference between the levels was about 1050 MHz, or 1 part in 10^6 of the $2s_{1/2}$ level [5,148]. Shortly thereafter Hans Bethe [6] published a ground breaking nonrelativistic quantum theoretical calculation of the shift assuming it was due to the interaction of the electron with the ground state electromagnetic field of the quantum vacuum field. This radiative shift accounted for about 96% of the measured shift. The insight that one needed to include the interaction of the atom with the vacuum fluctuations and how one could actually do it ushered in the modern world of quantum electrodynamics [7]. Here, we compute in the non-relativistic dipole approximation and to first order in the radiation field, as did Bethe, the radiative shift, but we use group theoretical methods based on the SO(4,2) symmetry of the non-relativistic hydrogen atom as developed in this paper. Bethe's calculation required the numerical sum over intermediate states to obtain the average value of the energy of the states contributing to the shift. In our calculation, we do not use intermediate states, and we derive an integral equivalent to Bethe's log, and more generally derive the shift for all levels in terms of a double integral.

An expression for the radiative shift Δ_{NL} for energy level E_N of a hydrogen atom in a state $|NL\rangle$ can be readily obtained using second order perturbation theory (to first order in α the radiation field) [6,149–151]

$$\Delta^{NL} = \frac{2\alpha c}{3\pi m^2}\sum_n^S \int_0^{\omega_c} d\omega \frac{(E_n - E_N)\langle NL|p_i|n\rangle\langle n|p_i|NL\rangle}{E_n - E_N + \omega - i\epsilon}, \tag{282}$$

where ω_C is a cutoff frequency for the integration that we will take as $\omega_c = m$.

This expression, which is the same as Bethe's, has been derived by inserting a complete set of states $|n\rangle\langle n|$, a step that we eliminate with our group theoretical approach:

$$\Delta_{NL} = \frac{2\alpha}{3\pi m^2}\int_0^{\omega_c} d\omega \langle NL|p_i \frac{H-E}{H-(E-\omega)-i\epsilon} p_i|NL\rangle \tag{283}$$

If we add and subtract ω from the numerator, we find the real part of the shift is

$$Re\Delta_{NL} = \frac{2\alpha}{3\pi m^2} Re \int_0^{\omega_c} d\omega[\langle NL|p^2|NL\rangle - \omega\Omega_{NL}] \tag{284}$$

where

$$\Omega_{NL} = \langle NL|p_i \frac{1}{H - E_N + \omega - i\epsilon} p_i|NL\rangle \tag{285}$$

and

$$H = \frac{p^2}{2m} - \frac{Z\alpha}{r}. \tag{286}$$

The imaginary part of the shift gives the width of the level [7].

The matrix element Ω_{NL} can be converted to a matrix element of a function of the generators Γ_A taken between eigenstates $|nlm)$ of $(Z\alpha)^{-1}$. To do this we insert factors of $1 = \sqrt{r}\frac{1}{\sqrt{r}}$ and use the definitions of the Γ_A in terms of the canonical variables, Equations (254)–(256). Letting the parameter a take the value a_N, we obtain the result

$$\Omega_{NL} = \frac{m\nu}{N^2}(NL|\Gamma_i \frac{1}{\Gamma n(\xi) - \nu}\Gamma_i|NL) \tag{287}$$

where

$$n^0(\xi) = \frac{2+\xi}{2\sqrt{1+\xi}} = \cosh\phi \qquad n^i = 0 \qquad n^4(\xi) = -\frac{\xi}{2\sqrt{1+\xi}} = -\sinh\phi \tag{288}$$

and

$$\xi = \frac{\omega}{|E_N|} \qquad \nu = \frac{N}{\sqrt{1+\xi}} = Ne^{-\phi}. \tag{289}$$

From the definitions we see $\phi = \frac{1}{2}ln(1+\xi) > 0$ and $n_A(\xi)n^A(\xi) = -1$. The quantity

$$\nu = \frac{mZ\alpha}{\sqrt{-2m(E_N - \omega)}}$$

may be considered the effective principal quantum number for a state of energy $E_N - \omega$. The contraction over i in Ω_{NL} may be evaluated using the group theoretical formula Equation (250):

$$\begin{aligned}\Omega_{NL} = &-2\frac{m\nu^2}{N^2}\int_0^\infty ds e^{\nu s}\frac{d}{ds}\left(\sinh^2\frac{s}{2}M_{NL}(s)\right) \\ &- m\frac{\nu}{N^2}(NL|\Gamma_4\frac{1}{\Gamma n(\xi)-\nu}\Gamma_4|NL) + m\frac{\nu}{N^2}(NL|\Gamma_0\frac{1}{\Gamma n(\xi)-\nu}\Gamma_0|NL)\end{aligned} \tag{290}$$

where

$$M_{NL}(s) = (NL|e^{-\Gamma n(\xi)s}|NL). \tag{291}$$

In order to evaluate the last-two terms in Ω_{NL}, we can express the action of Γ_4 on our states in terms of $\Gamma n(\xi) - \nu$. Substituting the equation

$$\Gamma_0|NL) = N|NL) \tag{292}$$

into the expression for $\Gamma n(\xi) - \nu$, with $n(\xi)$ given by Equation (288), gives

$$\Gamma_4 = N - \left(\frac{1}{\sinh\phi}\right)(\Gamma n(\xi) - \nu) \tag{293}$$

when acting on the state $|NL)$. If we substitute Equation (293) into the expression for the $Re\Delta E_{NL}$ Equation (284) and simplify using the virial theorem

$$(NL|p^2|NL) = a_N^2$$

we find that the term in p^2 exactly cancels the last two terms in Ω_{NL}, yielding the result

$$Re\Delta E_{NL} = \frac{4m\alpha(Z\alpha)^4}{3\pi N^4}\int_0^{\phi_c} d\phi \sinh\phi e^{\phi}\int_0^{\infty} ds\, e^{\nu s}\frac{d}{ds}\left(\sinh^2\frac{s}{2}M_{NL}(s)\right) \tag{294}$$

where

$$\phi_c = \frac{1}{2}ln\left(1+\frac{\omega_c}{|E_N|}\right) = \frac{1}{2}ln\left(1+\frac{2N^2}{(Z\alpha)^2}\right) \tag{295}$$

and $\omega_c = m$.

Comparison to the Bethe Logarithm

The first order non-relativistic radiative shift is commonly given in terms of the Bethe logarithm $\gamma(N,L)$, which is interpreted as the average over all states, including scattering states, of $ln\frac{|E_n-E_N|}{\frac{1}{2}m(Z\alpha)^2}$ [149] :

$$\begin{aligned}&\gamma(N,L)\textstyle\sum_n^S (E_n - E_N)\langle N0\,|p_i|\,n\rangle\langle n\,|p_i|\,N0\rangle \\ &= \textstyle\sum_n^S (E_n - E_N)\langle NL\,|p_i|\,n\rangle\langle n\,|p_i|\,NL\rangle \ln\frac{|E_n-E_N|}{\frac{1}{2}m(Z\alpha)^2}\end{aligned} \tag{296}$$

We use the dipole sum rule [150]

$$2\sum_n^S (E_n - E_N)\langle N\,|p_i|\,n\rangle\langle n\,|p_i|\,N\rangle = -\left\langle N\left|\nabla^2 V\right|N\right\rangle \tag{297}$$

and apply it for the Coulomb potential $\nabla^2 V(r) = 4\pi Z\alpha\delta(r)$. The use of the Bethe log allowed Bethe to take the logarithmic expression obtained from the frequency integration outside the summation over the states, and replace it with the average value. Only the S states contribute to the expectation value in Equation (297), giving, from Equation (282), an expression for the shift

$$\mathrm{Re}\,\Delta E_{NL} = \left[\frac{4m}{3\pi}\alpha(Z\alpha)^4\right]\frac{1}{N^3}\left\{\delta_{L0}\ln\frac{2}{(Z\alpha)^2} - \gamma(N,L)\right\}. \tag{298}$$

Comparing the shift in terms of M_{NL} Equation (294) to the shift in terms of $\gamma(N,L)$ we find that the Bethe log is

$$\gamma(N,L) = \int_0^{\phi_c} d\phi \sinh\phi\, e^{\phi}\int_0^{\infty} ds\, e^{\nu s}\frac{d}{ds}\left(\sinh^2\frac{s}{2}\,M_{NL}(s)\right) - \delta_{L0}ln\frac{1}{(Z\alpha)^2} \tag{299}$$

8.1. Generating Function for the Shifts

We can derive a generating function for the shifts for any eigenstate characterized by N and L if we multiply Equation (291) by $N^4e^{\beta N}$ and sum over all $N, N \geq L+1$. To simplify the right side of the resulting equation, we use the fact that the O(2,1) algebra of Γ_0, Γ_4, and S closes. We can compute the sum on the right hand side:

$$\sum_{N=L+1}^{\infty} e^{-\beta N}M_{NL} = \sum_{N=L+1}^{\infty}(NL|e^{-j\cdot\psi}|NL). \tag{300}$$

where

$$e^{-j\cdot\psi} \equiv e^{-\beta\Gamma_0}e^{-s\Gamma n(\xi)}. \tag{301}$$

We perform a $\boldsymbol{j}$ transformation, such that

$$e^{-\boldsymbol{j}\cdot\boldsymbol{\psi}} \to e^{-j_3\psi} = e^{-\Gamma_0\psi} \tag{302}$$

so

$$\sum_{N=L+1}^{\infty} e^{-\beta N} M_{NL} = \sum_{N=L+1}^{\infty} (NL|e^{-j_3\psi}|NL) = \sum_{N=L+1}^{\infty} e^{-N\psi} \tag{303}$$

$$= \frac{e^{-\psi(L+1)}}{1-e^{-\psi}}. \tag{304}$$

In order to find a particular M_{NL}, we must expand the right hand side of the equation in powers of $e^{-\beta}$ and equate the coefficients to those on the left hand side. First, we need an equation for $e^{-\psi}$. This can be obtained using the isomorphism between $\boldsymbol{j}$ and the Pauli σ matrices:

$$(\Gamma_4, S, \Gamma_0) \to (j_1, j_2, j_3) \to (\frac{i}{2}\sigma_1, \frac{i}{2}\sigma_2, \frac{1}{2}\sigma_3) \tag{305}$$

Using the formula

$$e^{\frac{i}{2}s\boldsymbol{n}\cdot\boldsymbol{\sigma}} = \cos\frac{s}{2} + i\boldsymbol{n}\cdot\boldsymbol{\sigma}\sin\frac{s}{2} \tag{306}$$

where $|\boldsymbol{n}| = 1$, we find

$$\cosh\frac{\psi}{2} = \cosh\frac{\beta}{2}\cosh\frac{s}{2} + \sinh\frac{\beta}{2}\sinh\frac{s}{2}\cosh\phi. \tag{307}$$

We can rewrite this equation in a form easier for expansion

$$e^{+\frac{1}{2}\psi} = de^{\frac{1}{2}\beta} + be^{-\frac{1}{2}\beta} - e^{-\frac{1}{2}\psi} \tag{308}$$

where

$$\begin{aligned} d &= \cosh\tfrac{s}{2} + \sinh\tfrac{s}{2}\cosh\phi \\ b &= \cosh\tfrac{s}{2} - \sinh\tfrac{s}{2}\cosh\phi \end{aligned}. \tag{309}$$

Let β become very large and iterate the equation for $e^{-\frac{1}{2}\psi}$ to obtain the result

$$e^{-\psi} = Ae^{-\beta}\left[1 + A_1e^{-\beta} + A_2e^{-2\beta} + \ldots\right] \tag{310}$$

where

$$\begin{aligned} A = A_0 &= \frac{1}{d^2} \\ A_1 &= -\left(\frac{2}{d}\right)\left(b - d^{-1}\right) \\ A_2 &= 3d^{-2}\left(b-d^{-1}\right)^2 - 2^{-2}\left(b-d^{-1}\right) \\ &\vdots \end{aligned} \tag{311}$$

Note $b - d^{-1} = -d^{-1}\sinh^2\frac{s}{2}\sinh^2\phi$.

8.2. The Shift between Degenerate Levels

Expressions for the energy shift between degenerate levels with the same value of N may be obtained directly from the generating function using Equations (294) and (304). We find

$$\sum_{N=L+1} e^{-\beta N} N^4 \operatorname{Re}\Delta E_{NL} - \sum_{N=L'+1}^{\infty} e^{-\beta N} N^4 \operatorname{Re}\Delta E_{NL'} =$$

$$\frac{4m\alpha(Z\alpha)^4}{3\pi}\int_0^{\phi_c} d\phi e^{\phi}\sinh\phi\int_0^{\infty} dse^{\nu s}\frac{d}{ds}\left(\sinh^2\frac{s}{2}\frac{e^{-\psi(L+1)}-e^{-\psi(L'+1)}}{1-e^{-\psi}}\right). \tag{312}$$

For an example, consider $L = 1, L' = 0$. For the shifts between levels we obtain

$$\sum_{N=2}^{\infty} e^{-\beta N}N^4\,\mathrm{Re}\,(\Delta E_{NO}-\Delta E_{N1})+\mathrm{Re}\,\Delta E_{10}\,e^{-\beta}=$$

$$\frac{4m\alpha(z\alpha)^4}{3\pi}\int_0^{\phi_c} d\phi\, e^{\phi}\sinh\phi\int_0^{\infty} dse^{\nu s}\frac{d}{ds}\left(\sinh^2\frac{s}{2}e^{-\psi}\right) \tag{313}$$

Substituting Equation (310) for $e^{-\psi}$, using the coefficient AA_{N-1} of $e^{-N\beta}$, gives

$$Re(\Delta E_{N0}-\Delta E_{N1})=\frac{4m\alpha(Z\alpha)^4}{3\pi N^4}\int_0^{\phi_c} d\phi e^{\phi}\sinh\phi\int_0^{\infty} dse^{\nu s}\frac{d}{ds}\left(\sinh^2\frac{s}{2}AA_{N-1}\right). \tag{314}$$

where A and A_{N-1} are given in Equation (311) in terms of the variables of integration s and ϕ.

General Expression for M_{NL}

Once we have a general expression for M_{NL}, we can use Equation (294) to calculate the shift for any level E_{NL}. We can obtain expressions for the values of M_{NL} by letting β become large, expanding the denominator in Equation (304) and equating coefficients of powers of $e^{-\beta}$. For large β, we have large ψ. We have

$$\frac{e^{-\psi(L+1)}}{1-e^{-\psi}}=\sum_{m=1}^{\infty} e^{-\psi(m+L)}$$

and for large β it follows from Equation (310) that

$$\sum_{N=L+1}^{\infty} e^{-\beta N}M_{NL}=\sum_{m=1}^{\infty}\left[e^{-\beta}A\left(1+A_1e^{-\beta}+\ldots\right)\right]^{m+L}. \tag{315}$$

Using the multinomial theorem [124], the right side of the equation becomes

$$\sum_{m=1}^{\infty} A^{m+L}\sum_{r,s,t,\ldots}\frac{(m+L)!}{r!s!t!\ldots}A_1{}^{s}A_2^{t}\ldots e^{-\beta(m+L+s+2t+\ldots)}. \tag{316}$$

where $r+s+t+\ldots=m+L$.

To obtain the expression for M_{NL}, we note N is the coefficient of β so $N=m+L+s+2t+\ldots=r+2s+3t+\ldots$ Accordingly we find

$$M_{NL}=\sum_{r,s,t,\ldots} A^{(r+s+t+\ldots)}\frac{(r+s+t+\ldots)!}{r!s!t!\ldots}A_1^{s}A_2^{t}\ldots \tag{317}$$

where $r+2s+3t+\ldots=N$ and $r+s+t+\ldots>$L.

By applying this formula, we obtain the results:

N = 1:

$$M_{10}=A \tag{318}$$

N = 2:

$$\begin{aligned} M_{20}&=A^2+AA_1\\ M_{21}&=A^2 \end{aligned} \tag{319}$$

Shifts for N = 1 and N = 2

To illustrate these results, we can calculate the shift for a given energy level using Equation (294). For $N=1$, we note from Equation (318) that $M_{10}=A$, and from Equation (311) that $A=1/d^2$. We find that the real part of the radiative shift for the $1S$ ground state is

$$Re\Delta E_{10} = \frac{4m\alpha(Z\alpha)^4}{3\pi} \int_0^{\phi_c} d\phi e^{\phi} \sinh\phi \int_0^{\infty} dse^{se^{-\phi}} \frac{d}{ds} \frac{1}{\left(\coth\frac{s}{2} + \cosh\phi\right)^2} \tag{320}$$

For the shift between two states Equation (314) can be used. For the N = 2 Lamb shift between 2S-2P states, the radiative shift to first order in α is

$$Re(\Delta E_{20} - \Delta E_{21}) = \frac{m\alpha(Z\alpha)^4}{6\pi} \int_0^{\phi_c} d\phi e^{\phi} \sinh^3\phi \int_0^{\infty} dse^{2se^{-\phi}} \frac{d}{ds} \frac{1}{\left(\coth\frac{s}{2} + \cosh\phi\right)^4} \tag{321}$$

The s integral can be computed in terms of a Jacobi function of the second kind [127].

As one check on our group theoretical methods, we can compare our matrix elements $(10|e^{iS\phi}|n0)$ with those of Huff [45]. To go from Equation (301) to Equation (302), we did a rotation $R(\phi) = e^{i\phi S}$ generated by S that transformed Γn into Γ_0. For $N, L = 1, 0$ we have

$$M_{10} = (10|e^{-\Gamma ns}|10) = (10|R(\phi)e^{-\Gamma_0 s}R^{-1}(\phi)|10) = \frac{1}{(\cosh\frac{s}{2} + \sinh\frac{s}{2}\cosh\phi)^2} \tag{322}$$

Expanding the hyperbolic functions, we get

$$\begin{aligned} M_{10} &= \frac{4e^{-s}}{(1+\cosh\phi)^2}\left[1 - e^{-s}\tanh^2\frac{s}{2}\right]^{-2} \\ &= \frac{4}{(1+\cosh\phi)^2}\sum_{n=1}^{\infty} ne^{-ns}\left(\tanh^2\frac{\phi}{2}\right)^{n-1}. \end{aligned} \tag{323}$$

We can also compute M_{10} by inserting a complete set of states and using $\Gamma_0|n0) = n|n0)$ in Equation (322). Because the generator S is a scalar, only states with $L = 0, m = 0$ can contribute:

$$M_{10} = \sum_{nlm} e^{-ns}|n(10|R(\phi)|n0)|^2. \tag{324}$$

Comparing this to Equation (323), we make the identification

$$|(10|R(\phi)|n0)|^2 = \frac{4n}{(1+\cosh\phi)^2}\left(\tanh^2\frac{\phi}{2}\right)^{n-1}. \tag{325}$$

Huff computes this matrix element by analytically continuing the known $O(3)$ matrix element of $e^{iJ_y\phi}$ obtaining

$$|\langle 10|R(\phi)|n0\rangle|^2 = \frac{4n}{\cosh^2\phi - 1}\left(\tanh^2\frac{\phi}{2}\right)^n \cdot [{}_2F_1(0,-1;n;\frac{1}{2}(1-\cosh\phi))]^2. \tag{326}$$

By algebraic manipulation and using ${}_2F_1 = 1$ for the arguments here, we see that this result agrees with our much more simply expressed result from group theory.

9. Conclusions and Future Research

Measuring and explaining the properties of the hydrogen atom has been central to the development of modern physics over the last century. One of the most useful and profound ways to understand its properties is through its symmetries, which we have explored in this paper, beginning with the symmetry of the Hamiltonian, which reflects the symmetry of the degenerate levels, then the larger non-invariance and spectrum-generating groups, which include all of the states. The successes in using symmetry to explore the hydrogen atom led to use of symmetry to understand and model other physical systems, particularly elementary particles.

The hydrogen atom will doubtless continue to be one of testing grounds for fundamental physics. Researchers are exploring the relationship between the hydrogen atom and quantum information [152], the effect of non-commuting canonical variables $[x_i, x_j] \neq 0$ on energy levels [153–155], muonic hydrogen spectra [156], and new physics using Rydberg states [157–162]. The ultra high precision of the measurement of the energy levels has led to new understanding of low Z two body systems, including muonium, positronium, and tritium [151]. As mentioned in the introduction, measurements of levels shifts are currently being used to measure the radius of the proton [2]. We can expect that further investigations of the hydrogen atom and hydrogenlike atoms will continue to reveal new vistas of physics and that symmetry considerations will play an important part.

Acknowledgments: I would especially like to thank Lowell S. Brown very much for the time and energy he has spent on my education as a physicist. I also thank Peter Milonni for his encouragement and his helpful comments and many insightful discussions and MDPI for the invitation to be the Guest Editor for this special issue of Symmetry on Symmetries in Quantum Mechanics.

References and Notes

1. Brown, L. Bound on Screening Corrections in Beta Decay. *Phys. Rev.* **1964**, *135*, B314. [CrossRef]
2. Beyer, A. The Rydberg constant and proton size from atomic hydrogen. *Science* **2017**, *358*, 79–85. [CrossRef]
3. Mohr, P.D.B.N.; Taylor, B.N. CODATA recommended values of the fundamental physical constants: 2014. *Rev. Mod. Phys.* **2016**, *88*, 035009. [CrossRef]
4. Rigden, J. *Hydrogen, The Essential Element*; Harvard University Press: Cambridge, MA, USA, 2002.
5. Lamb, W.; Retherford, R. Fine Structure of the Hydrogen Atom by a Microwave Method. *Phys. Rev.* **1947**, *72*, 241. [CrossRef]
6. Bethe, H. The Electromagnetic Shift of Energy Levels. *Phys. Rev.* **1947**, *72*, 339. [CrossRef]
7. Maclay, J. History and Some Aspects of the Lamb Shift. *Physics* **2020**, *2*, 8. [CrossRef]
8. Noether, E. Invariante Variationsprobleme. In *Nachrichten von der Gesellschaft der Wissenschaften zu Göttingen*; Akademie der Wissenschaften zu Göttingen: Göttingen, Germany, 1918; pp. 235–257.
9. Hamermesh, M. *Group Theory*; Adddison-Wesley Publishing Co.: Reading, MA, USA, 1962.
10. Weyl, H. *The Theory of Groups and Quantum Mechanics*, 2nd ed.; Dover Reprint; Dover Publications: New York, NY, USA, 1928.
11. Wigner, E. *Group Theory and Its Application to the Quantum Mechanics of Atomic Spectra*; Academic Press: New York, NY, USA, 1959.
12. Bargmann, V. Zur Theorie des Wasserstffatoms. *Z. Phys.* **1936**, *99*, 576. [CrossRef]
13. Laplace, P. *A Treatise of Celestial Mechanics*; Forgotten Books: Dublin, Ireland, 1827.
14. Pauli, W. Uber das Wasserstoffspektrum vom Standpunkt der neuen Quantummechanik. *Z. Phys.* **1926**, *36*, 336–363. [CrossRef]
15. McIntosh, H. On Accidental Degeneracy in Classical and Quantum Mechanics. *Am. J. Phys.* **1959**, *27*, 620–625. [CrossRef]
16. Hulthen, E. Über die quantenmechanische Herleitung der Balmerterme. *Z. Phys.* **1933**, *86*, 21–23. [CrossRef]
17. We employ natural Gaussian units so $\hbar = 1$, $c = 1$, and $\alpha = (e^2/\hbar c) \approx 1/137$. The notation for indices and vectors is $\mu, \nu, .. = 0, 1, 2, 3;\ \ i, j, . = 1, 2, 3;\ \ p_\mu p^\mu = -p_0^2 + \boldsymbol{p}^2,\ \ \boldsymbol{p} = (p_1, p_2, p_3),\ \ g_{\mu\nu} = (-1, 1, 1, 1)$.
18. Fock, V. Zur Theorie des Wasserstoffatoms. *Z. Phys.* **1935**, *98*, 145–154. [CrossRef]
19. Dirac, P. *Quantum Mechanics*, 1st ed.; Oxford University Press: Oxford, UK, 1930.
20. Gell-Mann, M. Symmetries of Baryons and Mesons. *Phys. Rev.* **1962**, *125*, 1067. [CrossRef]
21. Schwinger, J. Coulomb's Green's Function. *J. Math. Phys.* **1964**, *5*, 1606–1608. [CrossRef]
22. Ne'eman, Y. *Algebraic Theory of Particle Physics*; Benjamin: New York, NY, USA, 1967.
23. Ne'eman, Y. Derivation of strong interactions from a gauge invariance. *Nucl. Phys.* **1961**, *26*, 222–229. [CrossRef]
24. Gell-Mann, M. A schematic model of baryons and mesons. *Phys. Lett.* **1964**, *8*, 214–215. [CrossRef]
25. Gell-Mann, M.; Ne'eman, Y. *The Eightfold Way*; Benjamin: New York, NY, USA, 1964.

26. Dothan, Y.; Gell-Man, M.; Ne'eman, Y. Series of Hadron Energy Levels as Representations of Non-Compact Groups. *Phys. Letters* **1965**, *17*, 148. [CrossRef]
27. Nambu, Y. Infinite-Component Wave Equations with Hydrogenlike Mass Spectra. *Phys. Rev.* **1967**, *160*, 1171. [CrossRef]
28. Dyson, F. *Symmetry Groups in Nuclear and Particle Physics*; Benjamin: New York, NY, USA, 1966.
29. Thomas, L. On the Unitary Representations of the Group of de Sitter Space. *Ann. Math.* **1941**, *42*, 113–126. [CrossRef]
30. Harish-Chandra, Representations of Semisimple Lie Groups II. *Trans. Am. Math. Soc.* **1954**, *76*, 26. [CrossRef]
31. Barut, A.; Budini, P.; Fronsdal, C. Two examples of covariant theories with internal symmetries involving spin. *Proc. Roy. Soc.* **1966**, *A291*, 106–112.
32. Malkin, I.; Man'ko, V. Symmetry of the Hydrogen Atom. *Sov. Phys. Jetp Lett.* **1966**, *2*, 146.
33. Barut, A.; Kleinert, K. Transition Probabilities of the Hydrogen Atom from Noncompact Dynamical Groups. *Phy. Rev.* **1967**, *156*, 1541. [CrossRef]
34. Barut, A.; Kleinert, H. Transition Form Factors in the H Atom. *Phys. Rev.* **1967**, *160*, 1149. [CrossRef]
35. Bander, M.; Itzykson, C. Group Theory and the Hydrogen Atom (I). *Rev. Mod. Phys.* **1966**, *38*, 330. [CrossRef]
36. Bander, M.; Itzykson, C. Group Theory and the Hydrogen Atom (II). *Rev. Mod. Phys.* **1966**, *38*, 346. [CrossRef]
37. Fronsdal, C. Infinite Multiplets and Local Fields. *Phys. Rev.* **1967**, *156*, 1653. [CrossRef]
38. Fronsdal, C. Infinite Multiplets and the Hydrogen Atom. *Phys. Rev.* **1967**, *156*, 1665. [CrossRef]
39. Barut, A.; Fronsdal, C. On Non-Compact Groups. II Representations of the 2+1 Lorentz Group. *Proc. R. Soc.* **1965**, *A287*, 532–548.
40. Fronsdal, C. Relativistic Lagrangian Field Theory for Composite Systems. *Phys. Rev.* **1968**, *171*, 1811. [CrossRef]
41. Pratt, R.; Jordan, T. Coulomb Group Theory for and Spin. *Phys. Rev.* **1969**, *188*, 2534.
42. Fronsdal, C. Relativistic and Realistic Classical Mechanics of Two Interacting Point Particles. *Phys. Rev. D* **1971**, *4*, 1689. [CrossRef]
43. Kyriakopoulos, R. Dynamical Groups and the Bethe-Salpeter Equation. *Phys. Rev.* **1968**, *174*, 1846. [CrossRef]
44. Lieber, M. O(4) Symmetry of the Hydrogen Atom and the Lamb Shift. *Phys. Rev.* **1968**, *174*, 2037. [CrossRef]
45. Huff, R. Simplified Calculation of Lamb Shift Using Algebraic Techniques. *Phys. Rev.* **1969**, *186*, 1367. [CrossRef]
46. Musto, R. Generators of SO(4,1) for the Quantum Mechanical Hydrogen Atom. *Phys. Rev.* **1966**, *148*, 1274. [CrossRef]
47. Barut, A.; Bornzin, G. SO(4,2)-Formulation of the Symmetry Breaking in Relativistic Kepler Problems with of without Magnetic Charge. *J. Math. Phys.* **1971**, *12*, 841–846. [CrossRef]
48. Barut, A.; Kleinert, H. Current Operators and Majorana Equation for the Hydrogen Atom from Dynamical Groups. *Phys. Rev.* **1967**, *157*, 1180. [CrossRef]
49. Mack, G.; Todorov, I. Irreducibility of the Ladder representations when restricted to the Poincare Subgroup. *J. Math Phys.* **1969**, *10*, 2078–2085. [CrossRef]
50. Decoster, A. Realization of the Symmetry Groups of the Nonrelativistic Hydrogen Atom. *Nuovo Cimento* **1970**, *68A*, 105–117. [CrossRef]
51. Englefield, M. *Group Theory and the Coulomb Problem*; Wiley-Interscience: New York, NY, USA, 1972.
52. Barut, A. *Dynamical Groups*; University of Canterbury Press: Christchurch, New Zealand, 1972.
53. Bednar, M. Algebraic Treatment of Quantum-Mechanical Models with Modified Coulomb Potentials. *Ann. Phys.* **1973**, *75*, 305–331. [CrossRef]
54. Wulfman, C.; Takahata, Y. Noninvariance Groups in Molecular Quantum Mechanics. *J. Chem. Phys.* **1967**, *47*, 488–498. [CrossRef]
55. Wybourne, B. Symmetry Principles in Atomic Spectroscopy. *J. Phys.* **1970**, *31*, C4-33. [CrossRef]
56. Mariwalla, K. Dynamical Symmetries in Mechanics. *Phys. Rep.* **1975**, *20*, 287–362. [CrossRef]
57. Akyildiz, Y. On the dynamical symmetries of the Kepler problem. *J. Math. Phys.* **1980**, *21*, 665–670. [CrossRef]
58. Fronsdal, C.; Huff, R. Two-Body Problem in Quantum Field Theory. *Phys. Rev. D* **1971**, *3*, 933. [CrossRef]
59. Loebl, E. (Ed.) *Group Theory and Its Applications*; Academic Press: New York, NY, USA, 1971.
60. Barut, A.; Rasmussen, W. The hydrogen atom as a relativistic elementary particle I. The wave equation and mass formulae. *J. Phys.* **1973**, *B6*, 1695. [CrossRef]

61. Barut, A.; Rasmussen, W. The hydrogen atom as a relativistic elementary particle II. Relativistic scattering problems and photo-effect. *J. Phys.* **1973**, *B6*, 1713. [CrossRef]
62. Barut, A.; Bornzin, G. Unification of the external conformal symmetry group and the internal conformal dynamical group. *J. Math. Phys.* **1974**, *15*, 1000–1006. [CrossRef]
63. Barut, A.; Schneider, C.; Wilson, R. Quantum theory of infinite component fields. *J. Math. Phys.* **1979**, *20*, 2244–2256. [CrossRef]
64. Shibuya, T.; Wulfman, C. The Kepler Problem in Two-Dimensional Momentum Space. *Am. J. Phys.* **1965**, *33*, 570–574. [CrossRef]
65. Dahl, J. Physical Interpretation of the Runge-Lenz Vector. *Phys. Let.* **1968**, *27A*, 62–63. [CrossRef]
66. Collas, P. Algebraic Solution of the Kepler Problem Using the Runge-Lenz Vector. *Am. J. Phys.* **1970**, *38*, 253–255. [CrossRef]
67. Rodgers, H. Symmetry transformations of the classical Kepler problem. *J. Math. Phys.* **1973**, *14*, 1125–1129. [CrossRef]
68. Majumdar, S.; Basu, D. O(3,1) symmetry of the hydrogen atom. *J. Phys. Math. Nuc. Gen.* **1974**, *7*, 787. [CrossRef]
69. Stickforth, J. The classical Kepler problem in momentum space. *Am. J. Phys.* **1978**, *46*, 74–75. [CrossRef]
70. Ligon, T.; Schaaf, M. On the Global Symmetry of the Classical Kepler Problem. *Rep. Math. Phys.* **1976**, *9*, 281–300. [CrossRef]
71. Lakshmanan, M.; Hasegawa, H. On the canonical equivalence of the Kepler problem in coordinate and momentum space. *J. Phys. Math. Gen.* **1984**, *17*, L889. [CrossRef]
72. O'Connell, R.; Jagannathan, K. Illustrating dynamical symmetries in classical mechanics: The Laplace-Runge-Lenz vector revisited. *Am. J. Phys.* **2003**, *71*, 243–246. [CrossRef]
73. Valent, G. The hydrogen atom in electric and magnetic fields: Pauli's 1926 article. *Am. J. Phys.* **2003**, *71*, 171–175. [CrossRef]
74. Morehead, J. Visualizing the extra symmetry of the Kepler problem. *Am. J. Phys.* **2005**, *73*, 234–239. [CrossRef]
75. Huntington, L.; Nooijen, M. An SO(4) invariant Hamiltonian and the two-body bound state. I: Coulomb interaction between two spinless particles. *Int. J. Quant. Chem.* **2009**, *109*, 2885–2896. [CrossRef]
76. Barut, A.; Bohm, A.; Neeman, Y. *Dynamical Groups and Spectrum Generating Algebras*; World Scientific: Singapore, 1986.
77. Greiner, W.; Muller, B. *Quantum Mechanics, Symmetries*; Springer: Berlin, Germany, 1989.
78. Gilmore, R. *Lie Groups, Lie Algegras and Some of Their Applications*; Dover Books on Mathmatics; Dover: Mineola, NY, USA, 2005.
79. Kibler, M. On the use of the group SO(4,2) in atomic and molecular physics. *Mol. Phys.* **2004**, *102*, 1221–1229. [CrossRef]
80. Hammond, I.; Chu, S. Irregular wavefunction behavior in dimagnetic Rydberg atoms:a dynamical SO(4,2) group study. *Chem. Phys. Let.* **1991**, *182*, 63. [CrossRef]
81. Lev, F. Symmetries in Foundation of Quantum Theory and Mathematics. *Symmetry* **2020**, *12*, 409. [CrossRef]
82. Wulfman, C. *Dynamical Symmetry*; World Scientific Publishing: Singapore, 2011.
83. Johnson, M.; Lippmann, B.A. Relativistic Kepler problem. *Phys. Rev.* **1950**, *78*, 329.
84. Biedenharn, L. Remarks on the relativistic Kepler problem. *Phys. Rev.* **1962**, *126*, 845–851. [CrossRef]
85. Lanik, J. The Reformulations of the Klein-Gordon and Dirac Equations for the Hydrogen Atom to Algebraic Forms. *Czech. J. Phys.* **1969**, *B19*, 1540–1548. [CrossRef]
86. Stahlhofen, A. Algebraic solutions of relativistic Coulomb problems. *Helv. Phys. Acta* **1997**, *70*, 1141.
87. Chen, J.; Deng, D.; Hu, M. SO(4) symmetry in the relativistic hydrogen atom. *Phys. Rev. A* **2008**, *77*, 034102. [CrossRef]
88. Khachidze, T.; Khelashvili, A. The hidden symmetry of the Coulomb problem in relativistic quantum mechanics: From Pauli to Dirac. *Am. J. Phys.* **2006**, *74*, 628–632. [CrossRef]
89. Zhang, F.; Fu, B.; Chen, J. Dynamical symmetry of Dirac hydrogen atom with spin symmetry and its connection to Ginocchio's oscillator. *Phys. Rev. A* **2008**, *78*, 040101(R). [CrossRef]
90. Heine, V. *Group theory in Quantum Mechanics*; Dover Publications: New York NY, USA, 1993.
91. Noether, E.; Mort, T. Invariant Variation Problems. *Transp. Theory Stat. Phys.* **1971**, *1*, 186–207. [CrossRef]
92. Neuenschwander, D.E. *Emmy Noether's Wonderful Theorem*; Johns Hopkins University Press: Baltimore, MD, USA, 2010.

93. Hanca, J.; Tulejab, S.; Hancova, M. Symmetries and conservation laws: Consequences of Noether's theorem. *Am. J. Phys.* **2004**, *72*, 428–435. [CrossRef]
94. Byers, N.E. Noether's Discovery of the Deep Connection Between Symmetries and Conservation Laws. *arXiv* **1998**, arXiv:physics/9807044.
95. The daughter of a mathematician, she wanted to be a mathematician, but since women were not allowed to take classes at the University of Erlingen, she audited courses. She did so well in the exams, that she received a degree and was allowed to enroll in the university and received a PhD in 1907. She remained at the university, unpaid, in an unofficial status, for 8 years. Then she went to the University at Gottengen, where she worked for 8 years with no pay or status before being appointed as Lecturer with a modest salary. She was invited in 1915 by Felix Klein and David Hilbert, two of the most famous mathematicians in the world at the time, to work with them and address issues in Einstein's theory of General Relativity about energy conservation. She discovered Nother's First Theorem (and a second theorem also). She remained there until 1933 when she, as a Jew, lost her job. At Einstein's suggestion, she went to Bryn Mawr College in Pennsylvania. She died from ovarian cysts two years later.
96. A rotation in 4 dimension can be represented by an antisymmetric 4×4 matrix which has $3 + 2 + 1 = 6$ independent non-diagonal elements corresponding to 6 generators. Similarly a rotation in 5 dimensions has 10 independent elements or 10 generators.
97. Dothan, Y. Finite-Dimensional Spectrum-Generating Algebras. *Phys. Rev.* **1970**, *D2*, 2944. [CrossRef]
98. Mukanda, N.; Sudarshan, E. Characteristic Noninvariance Groups of Dynamical Systems. *Phys. Rev. Lett.* **1965**, *15*, 1041. [CrossRef]
99. Kyriakopoulos. E. Algebraic Equations for Bethe-Salpeter and Coulomb Green's Functions. *J. Math. Phys.* **1972**, *13*, 1729–1735. [CrossRef]
100. Lipkin, H. *Lie Groups for Pedestrians*; Dover Publications: New York, NY, USA, 2001.
101. Were it not for this displacement of the force center, the observation that a rotated circle projects onto a plane as an ellipse would manifest the four-dimensional symmetry of the hydrogenlike atom directly in configuration space. The elliptical orbits could be viewed as projections of a rotated hypercircle onto a three-dimensional hyperplane. These considerations can be applied with some modification to the three-dimensional harmonic oscillator for which the force center and the center of the ellipse coincide.
102. This equation and any other equation written in this specific coordinate system can be generalized to an arbitrary coordinate system by noting that the Cartesian unit vectors may be written in a manner that is independent of the coordinate system: $\boldsymbol{i} = \frac{\boldsymbol{A}}{A}, \quad \boldsymbol{j} = \frac{\boldsymbol{L}\times\boldsymbol{A}}{LA}, \quad \boldsymbol{k} = \frac{\boldsymbol{L}}{L}$.
103. Brown, L. University of Washington, Seattle, WA, Unpublished lecture notes, 1972. Unpublished lecture notes.
104. We define the angle between a three-dimensional hyper-plane and a line as $\pi/2$ minus the angle between the line and the normal to the hyperplane.
105. It is desirable to first show that $\boldsymbol{A}$ (and of course $\boldsymbol{L}$) generate rotations of the hypersphere or $\hat{U}$. However, as we prefer to do the necessary calculations in terms of commutators rather than Poisson brackets, we defer these considerations to Section 4. There we show that the generator L_i rotates $\hat{U}$ about the $i-4$ plane; the generator A_1 rotates $\hat{U}$ about the 2–3 plane, etc., thereby changing the orbit with respect to the 4-axis and changing the eccentricity.
106. Bois, G. *Tables of Indefinite Integrals*; Dover Pub1ications: New York, NY, USA, 1961; p. 123.
107. Using Equation (44) and [102], Equation (73) may be written as $cos^{-1}(\boldsymbol{U} \cdot \boldsymbol{A}/A) = \boldsymbol{p} \cdot \boldsymbol{r}/ar_c + \omega_{cl}t$. This agrees with the time dependent function $\phi = \boldsymbol{p} \cdot \boldsymbol{r}/ar_c - \omega_{cl}t$ Equation (70) defined in [97].
108. Brown, L. Forces giving no orbit precession. *Am. J. Phys.* **1978**, *46*, 930–931. [CrossRef]
109. Bacry, H. *Lectures in Theoretical Physics*; Brittin, W.E., Barut, A.O., Guenin, M., Eds.; Gordon and Breach: New York, NY, USA, 1967.
110. Barut, A. Dynamics of a Broken SU_N Symmetry for the Oscillator. *Phys. Rev.* **1965**, *139*, B1433. [CrossRef]
111. Boiteux, M. The Three-Dimensional Hydrogen Atom as a Restricted Four-Dimensional Harmonic Oscillator. *Physica* **1972**, *65*, 381–395. [CrossRef]
112. Hughes, J. The harmonic oscillator:values of the SU(3) invariants. *J. Phys. A Math. Gen.* **1973**, *6*, 453. [CrossRef]
113. Chen, A. Hydrogen atom as a four-dimensional oscillator. *Phys. Rev. A* **1980**, *22*, 333. [CrossRef]
114. Chen, A. Homomorphism between SO(4,2) and SU(2,2). *Phys. Rev. A* **1981**, *23*, 1653. [CrossRef]

115. Kibler, M.; Negadi, T. Connection between the hydrogen atom and the harmonic oscillator: The zero-energy case. *Phys. Rev. A* **1984**, *29*, 2891. [CrossRef]
116. Chen, A.; Kibler, M. Connection between the hydrogen atom and the four-dimensional oscillator. *Phys. Rev. A* **1985**, *31*, 3960. [CrossRef]
117. Gerry, C. Coherent states and the Kepler-Coulomb problem. *Phys. Rev. A* **1986**, *33*, 6. [CrossRef]
118. Chen, A. Coulomb–Kepler problem and the harmonic oscillator. *Am. J. Phys.* **1987**, *55*, 250–252. [CrossRef]
119. Van der Meer, J. The Kepler system as a reduced 4D oscillator. *J. Geom. Phys.* **2015**, *92*, 181–193. [CrossRef]
120. Bacry, H. The de Sitter Group $L_{4,1}$ and the Bound States of the Hydrogen Atom. *Nuovo Cimento* **1966**, *41*, 222–234. [CrossRef]
121. Biedenharn, L. Wigner Coefficients for the R_4 Group and Some Applications. *J. Math. Phys.* **1961**, *2*, 433–441. [CrossRef]
122. Shiff, L. *Quantum Mechanics*; McGraw Hill: New York, NY, USA, 1955.
123. Biedenharn, L.; Swamy, N. Remarks on the Relativistic Kepler Problem. II. Approximate Dirac-Coulomb Hamiltonian Possessing Two Vector Invariants. *Phys. Rev.* **1964**, *133*, B1353. [CrossRef]
124. Morse, P.; Feshbach, H. *Methods of Theoretical Physics, Vol. 1*; McGraw-Hill: New York, NY, USA, 1953.
125. The primes indicates eigenvalues of operators, and unprimed quantities indicate abstract operators. The quantity x' means the four-vector $(t', \vec{r'})$.
126. Morse, P.; Feshbach, H. *Methods of Theoretical Physics, Vol. 2*; McGraw-Hill: New York, NY, USA, 1953.
127. Erdeli, A. (Ed.) *Higher Transcendental Functions, Bateman Manuscript Project*; McGraw-Hill Book Co.: New York, NY, USA, 1953.
128. Makowski, A.; Pepłowski, P. Zero-energy wave packets that follow classical orbits. *Phys. Rev. A* **2012**, *86*, 042117. [CrossRef]
129. Bellomo, P.; Stroud, C., Jr. Classical evolution of quantum elliptical orbits. *Phys. Rev. A* **1999**, *59*, 2139. [CrossRef]
130. Berry, M.; Mount, K.E. Semiclassical approximations in wave mechanics. *Rep. Prog. Phys.* **1972**, *35*, 315. [CrossRef]
131. Eberly, J.; Stroud, C. Chapters 14 (Rydberg Atoms) and Chapter 73 (Coherent Transients). In *Springer Handbook of Atomic, Molecular, and Optical Physics*; Drake, G., Ed.; Springer Science and Business Media: New York, NY, USA, 2006.
132. Lakshmanan, M.; Ganesan, K. Rydberg atoms and molecules-Testing grounds for quantum manifestations of chaos. *Curr. Sci.* **1995**, *68*, 38–44.
133. Kay, K. Exact Wave Functions for the Coulomb Problem from Classical Orbits. *Phys. Rev.* **1999**, *25*, 5190. [CrossRef]
134. Lena, C.; Deland, D.; Gay, J. Wave functions of Atomic Elliptic States. *Europhys. Lett.* **1991**, *15*, 697. [CrossRef]
135. Bhaumik, D.; Dutta-Roy, B.; Ghosh, G. Classical limit of the hydrogen atom. *J. Phys. A Math. Gen.* **1986**, *19*, 1355. [CrossRef]
136. McAnally, D.; Bracken, A. Quasiclassical states of the Coulomb system and SO(4, 2). *J. Phys. A Math. Gen.* **1990**, *23*, 2027. [CrossRef]
137. Pitak, A.; Mostowski, J. Classical limit of position and matrix elements for Rydberg atoms. *Eur. J. Phys.* **2018**, *39*, 025402. [CrossRef]
138. Nauenberg, M. Quantum wavepackets on Kepler elliptical orbits. *Phys. Rev. A* **1989**, *40l*, 1133. [CrossRef] [PubMed]
139. Brown, L.S. Classical limit of the hydrogen atom. *Am. J. Phys.* **1973**, *41*, 525–530. [CrossRef]
140. Leonhardt, U. *Measuring the Quantum State of Light*; Cambridge University Press: Cambridge, UK, 1997.
141. Barut, A.; Girardello, L. New "coherent" states associated with non-compact groups. *Commun. Math. Phys.* **1971**, *21*, 41–55. [CrossRef]
142. Satyanarayana, M. Squeezed coherent states of the hydrogen atom. *J. Phys. A Math. Gen.* **1986**, *19*, 1973. [CrossRef]
143. Liu, Q.; Hu, B. The hydrogen atom's quantum-to-classical correspondence in Heisenberg's correspondence principle. *J. Phys. A Math. Gen.* **2001**, *34*, 5713. [CrossRef]
144. Zverev, V.; Rubinstein, B. Dynamical symmetries and well-localized hydrogenic wave packets. *Proc. Inst. Math. Nas Ukr.* **2004**, *50*, 1018.

145. The wave function in momentum space $\psi(p)$ is obtained by multiplying Y_{nlm} by the normalizing factor $\frac{(a_n)^{3/2}}{(1-U_4)^2}$, cf Equation (150).
146. Nandi, S.; Shastry, C. Classical limit of the two-dimensional and three-dimensional hydrogen atom. *J. Phys. A Math. Gen.* **1989**, *22*, 1005. [CrossRef]
147. Pauling, L.; Wilson, E.B. *Introduction to Quantum Mechanics*; McGraw-Hill: New York, NY, USA, 1935; p. 36.
148. Lamb, W.; Retherford, R. Fine Structure of the H Atom, Part I. *Phys. Rev.* **1950**, *79*, 549. [CrossRef]
149. Bethe, H.; Salpeter, E. *The Quantum Mechanics of One and Two Electron Atoms*; Springer: Berlin, Germany, 1957.
150. Milonni, P. *The Quantum Vacuum*; Academic Press: San Diego, CA, USA, 1994.
151. Eides, M.; Grotch, H.; Shelyuto, V. *Theory of Light Hydrogenic Bound States, Springer Tracts in Modern Physics 222*; Springer: Berlin, Germany, 2007.
152. Rau, A.; Alber, G. Shared symmetries of the hydrogen atom and the two-bit system. *J. Phys. B At. Mol. Opt.* **2017**, *50*, 242001. [CrossRef]
153. Castro, P.; Kullock, R. Physics of the $SO_p(4)$ Hydrogen Atom. *Theo. Math. Phys.* **2015**, *185*, 1678. [CrossRef]
154. Alavi, A.; Rezaei, N. Dirac equation, hydrogen atom spectrum and the Lamb shift in dynamical non-commutative spaces. *Pramana-J. Phys.* **2017**, *88*, 5. [CrossRef]
155. Gnatenko, K.P.; Krynytskyi, Y.S.; Tkachuk, V.M. Perturbation of the ns levels of the hydrogen atom in rotationally invariant noncommutative space. *Mod. Phys. Lett.* **2015**, *30*, 8. [CrossRef]
156. Haghighat, M.; Khorsandi, M. Hydrogen and muonic hydrogen atomic spectra in non-commutative space-time. *Eur. Phys. J.* **2015**, *75*, 1. [CrossRef]
157. Praxmeyer, L. Hydrogen atom in phase space: The Wigner representation. *J. Phys. A Math. Gen.* **2006**, *39*, 14143. [CrossRef]
158. Jones, M.; Potvliege, M.R.; Spannowsky, M. Probing new physics using Rydberg states of atomic hydrogen. *Phys. Rev. Res.* **2020**, *2*, 013244. [CrossRef]
159. Jentschura, U.; Mohr, P. Calculation of hydrogenic Bethe logarithms for Rydberg States. *Phys. Rev. A* **2002**, *72*, 012110. [CrossRef]
160. Jentschura, U.; LeBigot, E.; Evers, J.; Mohr, P.; Keitel, C. Relativistic and radiative shifts for Rydberg states. *J. Phys. B At. Mol. Opt. Phys.* **2005**, *38*, S97. [CrossRef]
161. Jentschura, U.; Mohr, P.; Tan, J. Fundamental constants and tests of theory in Rydberg states of one-electron ions. *J. Phys. B At. Mol. Opt. Phys.* **2010**, *43*, 074002. [CrossRef]
162. Cantu, S.H.; Venkatramani, A.V.; Xu, W. Repulsive photons in a quantum nonlinear medium. *Nat. Phys.* **2020**. [CrossRef]

5

Inflation with Scalar-Tensor Theory of Gravity

Dalia Saha *, Susmita Sanyal and Abhik Kumar Sanyal

Department of Physics, Jangipur College, Murshidabad 742213, West Bengal, India; susmitasanyal@yahoo.com (S.S.); sanyal_ak@yahoo.com (A.K.S.)
* Correspondence: daliasahamandal1983@gmail.com

Abstract: The latest released data from Planck in 2018 put up tighter constraints on inflationary parameters. In the present article, the in-built symmetry of the non-minimally coupled scalar-tensor theory of gravity is used to fix the coupling parameter, the functional Brans–Dicke parameter, and the potential of the theory. It is found that all the three different power-law potentials and one exponential pass these constraints comfortably, and also gracefully exit from inflation.

---

1. Introduction

The standard (FLRW) model of cosmology based on the basic assumption of homogeneity and isotropy, known as the 'cosmological principle', has successfully been able to explain several very important issues in connection with the evolution of the universe. First of all, it predicts the observed expansion of the universe being supported by the Hubble's law. It also postulates the existence of cosmic microwave background radiation (CMBR), formed since recombination when the electrons combined to form atoms, allowing photons to free stream, with extreme precession, being verified by Penzias and Wilson for the first time [1]. It further predicts with absolute precession the abundance of the light atomic nuclei ($\frac{^4\mathrm{He}}{\mathrm{H}} \sim 0.25$, $\frac{^2\mathrm{D}}{\mathrm{H}} \sim 10^{-3}$, $\frac{^3\mathrm{He}}{\mathrm{H}} \sim 10^{-4}$, $\frac{^7\mathrm{Li}}{\mathrm{H}} \sim 10^{-9}$, by mass and not by number) observed in the present universe [2–4]. Finally, assuming the presence of the seeds of perturbation in the early universe, it can explain the observed present structure of the universe. Despite such tremendous success, the model inevitably suffers from a plethora of pathologies. The problems at a glance are the following [5,6].

1. 'The singularity problem': Extrapolating the FLRW solutions back in time one encounters an unavoidable singularity, since all the physical parameters viz. the energy density (ρ), the thermodynamic pressure (p), the Ricci scalar (R), the Kretschmann scalar ($R_{\alpha\beta\gamma\delta}R^{\alpha\beta\gamma\delta}$) etc. diverge.

2a. 'The flatness problem': The model does not provide any explanation to the observed value of the density parameter $\Omega \approx 1$, which depicts that the universe is spatially flat.

2b. 'The horizon problem': It also can not provide any reason to the observed tremendous isotropy of the CMBR being split in 1.4×10^4 patches of the sky, that were never causally connected before emission of the CMBR.

2c. 'The structure formation problem': It does not also provide any clue to the seeds of perturbation responsible for the structure formation.

3. 'The dark energy problem': Finally, the standard FLRW model does not fit the redshift versus luminosity-distance curve plotted in view of the observed SN1a (Supernova type a) data.

In connection with the first problem, viz. the so called 'Big-Bang singularity', and also to understand the underlying physics of 'Black-Hole' being associated with Schwarzschild singularity, it has been realized long ago that 'General Theory of Relativity' (GTR) must have to be replaced by a quantum theory of gravity when and where gravity is strong enough. However, GTR is not renormalizable and a renormalized theory requires to include higher-order curvature invariant terms in the gravitational action [7]. Despite serious and intense research over several decades and formulation of new high energy physical theories like superstring [8,9] and supergravity [10,11] theories, a viable quantum theory of gravity is still far from being realized. In connection with last problem, a host of research is in progress over last two decades [12]. It has been realized that to fit the observed redshift versus luminosity-distance curve, it is either required to take into account some form of exotic matter in addition to the barotropic fluid (ordinary plus the cold dark matter) which violates the strong energy condition ($\rho + p \geq 0,\ \rho + 3p \geq 0$), and is dubbed as 'dark energy' (since it interacts none other than with the gravitational field) or to modify the theory of gravity by including additional curvature scalars in the Einstein–Hilbert action, known as 'the modified theory of gravity' [12]. It has been observed that both the possibilities lead to present accelerated expansion of the universe. The problem is thus rephrased as: why the universe undergoes an accelerated expansion at present? The pathology 2, in connection with the flatness, horizon and structure formation problems has however been solved under the hypothesis called 'Inflation' [5], which is our present concern.

Under the purview of cosmological principle, i.e., taking into account Robertson–Walker line element,

$$ds^2 = -dt^2 + a^2(t)\left[\frac{dr^2}{1-kr^2} + r^2(d\theta^2 + sin^2\theta d\phi^2)\right], \tag{1}$$

the co-moving distance (the present-day proper distance) traversed by light between cosmic time t_1 and t_2 in an expanding universe may be expressed as, $d_0(t_i, t_f) = a_0\int_{t_i}^{t_f}\frac{dt}{a(t)}$, where $a(t)$ is called the scale factor. The co-moving size of the particle horizon at the last-scattering surface of CMBR ($a_f = a_{lss}$) corresponds to $d_0 \sim 100$ Mpc, or approximately 1^0 (one degree) on the CMB sky today. In the decelerated radiation dominated era of the standard model of cosmology (FLRW model), for which $a \propto \sqrt{t}$ the integrand, $(\dot{a})^{-1} \sim 2\sqrt{t}$ decreases towards the past, and there exists a finite co-moving distance traversed by light since the Big Bang ($a_i \to 0$), called the particle horizon. The hypothesis of inflation [13,14] postulates a period of accelerated expansion, $\ddot{a} > 0$, in the very early universe ($t \sim 10^{-36}$ s) , prior to the hot Big-Bang era, administering certain initial conditions [15–21]. During a period of inflation e.g., a de-Sitter universe ($a \propto e^{\Lambda t}$) driven by a cosmological constant (say), $(\dot{a})^{-1} \sim (\Lambda e^{\Lambda t})^{-1}$ increases towards the past, and hence the integral diverges as ($a_i \to 0$). This allows an arbitrarily large causal horizon dependent only upon the duration of the accelerated expansion. Assuming that the universe inflates with a finite Hubble rate H_i, (instead of a constant exponent Λ) ending with $H_f < H_i$, we may have, $d_0(t_i, t_f) > (\frac{a_i}{a_f})H_i^{-1}(e^N - 1)$ where $N = \ln\left(\frac{a_f}{a_i}\right)$ is measured in terms of the logarithmic expansion (or 'e-folds'), and describes the duration of inflation. It has been found that a 40–60 e-folds of inflation can encompass our entire observable universe today, and thus solves the horizon and the flatness problem discussed earlier. In some situations, e-fold may range between $25 \leq N \leq 70$, depending on the model under consideration.

A false vacuum state can drive an exponential expansion, corresponding to a de-Sitter space-time with a constant Hubble rate on spatially-flat hypersurfaces. However, a graceful exit from such exponential expansion requires a phase transition to the true vacuum state. A second-order phase transition [22,23], under the slow roll condition of the scalar field (that can also drive the inflation instead of the cosmological constant), potentially leads to a smooth classical exit from the vacuum-dominated phase. Further, the quantum fluctuations of the scalar field, which essentially are the origin of the structures seen in the universe today, provides a source of almost scale-invariant density fluctuations [24–28], as detected in the CMBR. Accelerated expansion and primordial perturbations can also be produced in some modified theories of gravity (e.g., [13,29] and also in a host of models presently available in the literature), which introduce additional non-minimally coupled degrees of freedom. Such inflationary models are conveniently studied by transforming variables to the so-called 'Einstein frame', in which Einstein's equations apply with minimally coupled scalar fields [30,31], which we shall deal with, in the present manuscript.

Non-minimal coupling with the scalar field ϕ is unavoidable in a quantum theory, since such coupling is generated by quantum corrections, even if it is primarily absent in the classical action. Particularly, it is required by the renormalization properties of the theory in curved space-time background. Horndeski' theory [32] presents the most general scalar-tensor theory of gravity ensuring no more than second-order field equations to avoid ghost instability due to Ostrogradski theorem. Recently, in view of a general conserved current, obtained under suitable manipulation of the field Equations [33–36], a non-minimally coupled scalar-tensor theory of gravity, being a special case of Horndeski' model, has been studied extensively in connection with the cosmological evolution, starting from the very early stage (Inflationary regime) to the late-stage (presently accelerated matter-dominated era) via a radiation dominated era [37]. It has been found that such a theory admits a viable inflationary regime, since the inflationary parameters viz. the scalar-tensor ratio (r), and the spectral index n_s lie well within the limits of the constraints imposed by Planck's data, released in 2014 [38] and in 2016 [39]. Furthermore, the model passes through a Friedmann-like radiation era ($a \propto \sqrt{t}, q = 1$) and also an early stage of long Friedmann-like decelerating matter dominated era ($a \propto t^{\frac{2}{3}}$, $q = \frac{1}{2}$) till $z \approx 4$, where a, q and z denote the scale factor, the deceleration parameter and the red-shift respectively. The universe was also found to enter a recent accelerated expansion at a red-shift, $z \approx 0.75$, which is very much at par with recent observations. Further, the present numerical values of the cosmological parameters obtained in the process are also quite absorbing, since it revealed the age of the universe ($13.86 < t_0 < 14.26$) Gyr, the present value of the Hubble parameter ($69.24 < H_0 < 69.96$) Km.s^{-1}Mpc^{-1}, so that $0.991 < H_0t_0 < 1.01$ fit with the observation with appreciable precision. Numerical analysis also reveals that the state finder $\{r,s\} = \{1,0\}$, which establishes the correspondence of the present model with the standard ΛCDM universe. Last but not the least important outcome is: considering the CMBR temperature at decoupling ($z \sim 1080$) to be 3000 K, required for recombination, its present value is found to be 2.7255 K, which again fits the observation with extremely high precision. Thus, non-minimally coupled scalar-tensor theory of gravity appears to serve as a reasonably fair candidate for describing the evolution history of our observable universe, beyond quantum domain.

In the mean time, new Planck's data have been released [40,41], which imposed even tighter constraints on the inflationary parameters. In this manuscript, we therefore pose if the theory [37] admits these new constraints. However, earlier we considered a particular form of coupling parameter along with the potential in the form $V(\phi) = V_0\phi^4 - B\phi^2$, where, V_0 and B are constants [37]. Here instead, we choose different forms of the coupling parameters and also different potentials to study the inflationary regime. In Section 2, we describe the model, write down the field equations, find the parameters involved in the theory in view of a general conserved current. We also present the scalar-tensor equivalent form of the action in Einstein's frame to find the inflationary parameters. In Section 3, we choose different forms of the coupling parameters and associated potentials to test the viability of the model in view of the latest released data from Planck [40,41]. We conclude in Section 4.

2. The Model, Conserved Current, Scalar-Tensor Equivalence and Inflationary Parameters

As mentioned, here we concentrate upon pure non-minimally coupled scalar-tensor theory of gravity, for which the action is expressed in the form,

$$A = \int \left[f(\phi)R - \frac{\omega(\phi)}{\phi}\phi_{,\mu}\phi^{,\mu} - V(\phi) - \mathcal{L}_m\right]\sqrt{-g}d^4x, \tag{2}$$

where, $\mathcal{L}_m$ is the matter Lagrangian density, $f(\phi)$ is the coupling parameter, while, $\omega(\phi)$ is the variable Brans–Dicke parameter. As mentioned, action (2) is a special case of the generalized Horndeski' model [32] having an action S_H given by,

$$S_H = \int d^4x\sqrt{-g}\mathcal{L}_H = \int d^4x\sqrt{-g}\left(\mathcal{L}_2 + \mathcal{L}_3 + \mathcal{L}_4 + \mathcal{L}_5\right), \tag{3}$$

where, $\mathcal{L}_H$ is the Lagrangian density which is the sum of a quadratic, cubic, quartic and quintic terms:

$$\mathcal{L}_2 = K(\phi, X); \tag{4a}$$

$$\mathcal{L}_3 = -G_3(\phi, X)\Box\phi; \tag{4b}$$

$$\mathcal{L}_4 = G_4(\phi, X)R + G_{4X}[(\Box\phi)^2 - (\nabla_\mu\nabla_\nu\phi)^2]; \tag{4c}$$

$$\mathcal{L}_5 = G_5(\phi, X)G_{\mu\nu}\nabla^\mu\nabla^\nu\phi - \frac{G_{5X}}{6}[(\Box\phi)^3 - 3(\Box\phi)(\nabla_\mu\nabla_\nu\phi)^2 + 2(\nabla_\mu\nabla_\nu\phi)^3]. \tag{4d}$$

In the above, K and $G_{3,4,5}$ are generic functions of ϕ and X, while, the kinetic term $X := -\frac{\partial_\mu\phi\partial^\mu\phi}{2}$, R is the Ricci tensor, $G_{\mu\nu}$ is the Einstein tensor, $(\nabla_\mu\nabla_\nu\phi)^2 = \nabla_\mu\nabla_\nu\phi\nabla^\mu\nabla^\nu\phi$ and $G_{iX} = \frac{\partial G_i}{\partial X}$. It is important to mention that, gravity becomes dynamical only through mixing with a scalar field, a phenomenon dubbed as kinetic gravity braiding. General relativity is recovered by setting $K = G_3 = G_5 = 0$ and $G_4 = \frac{M_p^2}{2}$. Note that, when $G_4 = \frac{M_p^2}{2}$, $\mathcal{L}_4$ reduces to the Einstein–Hilbert term. We also obtain a non minimal coupling in the form $f(\phi)R$ from $\mathcal{L}_4$, by setting $G_4 = f(\phi)$. The Horndeski Theory and beyond, have been studied by numerous people from different perspectives [42–50]. In the context of inflation, the generalised (kinetically driven and potentially driven slow-roll) G-inflation has been studied to some details, although there had been no attempt to fit inflationary parameters with the observed data [42]. Additionally, Higgs G-inflation [43], inflation with Scalar-Tensor Horndeski Model [44] and K-Essence non-minimally coupled Gauss–Bonnet invariant for inflation [45], appear in the literature, as some Special cases of Horndeski model, since, it is extremely difficult to make a general study involving all the terms appearing in the Horndeski model. Likewise, present model (2) may be treated as a special case of the same, which may be obtained under the choice, $\mathcal{L}_2 = K(X,\phi) = K(\phi)X - V(\phi)$, where, $K(\phi) = 2\frac{\omega(\phi)}{\phi}$, $G_3 = 0$, $G_4 = f(\phi)$, $G_{4X} = 0$, and $G_5 = 0$, and may be dubbed as the potential-driven G-inflation.

The field equations corresponding to action (2) are,

$$\Big(R_{\mu\nu} - \frac{1}{2}g_{\mu\nu}R\Big)f(\phi) + g_{\mu\nu}\Box f(\phi) - f_{;\mu;\nu} - \frac{\omega(\phi)}{\phi}\phi_{,\mu}\phi_{,\nu} + \frac{1}{2}g_{\mu\nu}\Big(\phi_{,\alpha}\phi^{,\alpha} + V(\phi)\Big) = T_{\mu\nu}, \tag{5}$$

$$Rf' + 2\frac{\omega(\phi)}{\phi}\Box\phi + \Big(\frac{\omega'(\phi)}{\phi} - \frac{\omega(\phi)}{\phi^2}\Big)\phi_{,\mu}\phi^{,\mu} - V'(\phi) = 0, \tag{6}$$

where prime denotes derivative with respect to ϕ, and $\Box$ denotes D'Alembertian, such that, $\Box f(\phi) = f''\phi_{,\mu}\phi^{,\mu} - f'\Box\phi$. The model involves three parameters viz. the coupling parameter $f(\phi)$, the Brans–Dicke parameter $\omega(\phi)$ and the potential $V(\phi)$. It is customary to choose these parameters by hand in order to study the evolution of the universe. However, we have proposed a unique technique to relate the parameters in such a manner, that choosing one of these may fix the rest [33–37]. This follows in view of a general conserved current which is admissible by the above pair of field equations, briefly enunciated below.

The trace of the field equation (5) reads as,

$$Rf - 3\Box f - \frac{\omega(\phi)}{\phi}\phi_{,\mu}\phi^{,\mu} - 2V = T^\mu_\mu = T. \tag{7}$$

Now eliminating the scalar curvature between Equations (6) and (7), one obtains,

$$\Big(3f'^2 + \frac{2\omega f}{\phi}\Big)'\phi_{,\mu}\phi^{,\mu} + \Big(3f'^2 + \frac{2\omega f}{\phi}\Big)\Box\phi + 2f'V - fV = f'T, \tag{8}$$

which may then be expressed as,

$$\left[\left(3f'^2+\frac{2\omega f}{\phi}\right)^{\frac{1}{2}}\phi^{;\mu}\right]_{,\mu}-\frac{f^3}{2\left(3f'^2+\frac{2\omega f}{\phi}\right)^{\frac{1}{2}}}\left(\frac{V}{f^2}\right)'=\frac{f'}{2\left(3f'^2+\frac{2\omega f}{\phi}\right)^{\frac{1}{2}}}T, \tag{9}$$

and finally as,

$$\left(3f'^2+\frac{2\omega f}{\phi}\right)^{1/2}\left[\left(3f'^2+\frac{2\omega f}{\phi}\right)^{1/2}\phi^{;\mu}\right]_{;\mu}-f^3\left(\frac{V}{f^2}\right)'=\frac{f'}{2}T^{\mu}_{\mu}. \tag{10}$$

Thus there exists a conserved current J^{μ}, where,

$$J^{\mu}_{;\mu}=\left[\left(3f'^2+\frac{2\omega f}{\phi}\right)^{1/2}\phi^{;\mu}\right]_{;\mu}=0. \tag{11}$$

for trace-less matter field ($T^{\mu}_{\mu}=T=0$), provided

$$V(\phi)\propto f(\phi)^2. \tag{12}$$

To study cosmological consequence of such a conserved current, let us turn our attention to the minisuperspace Model (1), in which the conserved current (11), reads as

$$\sqrt{\left(3f'^2+\frac{2\omega f}{\phi}\right)}a^3\dot{\phi}=C_1, \tag{13}$$

in traceless vacuum dominated and also in radiation dominated eras. In the above, C_1 is the integration constant. Note that, fixing the form of the coupling parameter $f(\phi)$, the potential $V(\phi)$ is fixed in view of (12), once and forever. Further, we use a relation [37]

$$3f'^2+\frac{2\omega f}{\phi}=\omega_0^2, \tag{14}$$

where, ω_0 is a constant, to fix the Brans–Dicke parameter as well. As a result, we obtain the relation

$$a^3\dot{\phi}=\frac{C_1}{\omega_0}=C, \tag{15}$$

C being yet another non-vanishing constant. In the process, all the coupling parameters $f(\phi)$, $\omega(\phi)$ and the potential $V(\phi)$ are fixed a-priori. Note that one could have made other choice to fix the Brans–Dicke parameter, for example, an arbitrary functional form of the left hand side of (14). But then, additional functional parameter appears that would again require additional assumption. On the contrary, the above choice (14) finally leads to the conserved current associated with the canonical momenta conjugate to the scalar field ϕ, in the absence of a variable Brans–Dicke parameter, i.e., for $(\frac{\omega(\phi)}{\phi}=\frac{1}{2})$ with minimal coupling $\left(f(\phi)=(16\pi G)^{-1}=\frac{M_p^2}{2}\right)$, which is physically meaningful. We shall work, in the typical unit, $\frac{M_p^2}{2}=c=1$, and consider different forms of $f(\phi)$, that fixes the Brans–Dicke parameter $\omega(\phi)$ as well as the potential $V(\phi)$. In view of these known functional forms of the parameters of the theory, we focus our attention to study inflation, which must have occurred in the very early vacuum dominated universe. We relax the symmetry by adding an useful term in the potential $V(\phi)$, so that one of the terms act just as a constant in the effective potential. This ensures a constant value of the potential as the scalar field dies out, and this constant acts as an effective cosmological constant (Λ_e). In the process, the number of parameters increases to three (ω_0, V_0, V_1), which is essential to administer good fit with observation.

Scalar-Tensor Equivalence and Inflationary Parameters

As mentioned, it is convenient and hence customary to study inflationary evolution in the Einstein frame under suitable transformation of variables, where possible. Therefore, in order to study inflation, we consider very early vacuum dominated ($p = 0 = \rho$, for which trace of the matter field identically vanishes and symmetry holds) era, and express the action (2) in the form,

$$A = \int \left[f(\phi)R - \frac{K(\phi)}{2}\phi_{,\mu}\phi^{,\mu} - V(\phi)\right]\sqrt{-g}\, d^4x, \tag{16}$$

where, $K(\phi) = 2\frac{\omega(\phi)}{\phi}$. The above action (16) may be translated to the Einstein frame under the conformal transformation ($g_{E\mu\nu} = f(\phi)g_{\mu\nu}$) to take the form [51],

$$A = \int \left[R_E - \frac{1}{2}\sigma_{E,\mu}\sigma_E{}^{,\mu} - V_E(\sigma(\phi))\right]\sqrt{-g_E}\, d^4x, \tag{17}$$

where, the subscript 'E' stands for Einstein's frame. The effective potential (V_E) and the field (σ) in the Einstein frame may be found from the following expressions,

$$V_E = \frac{V(\phi)}{f^2(\phi)}; \qquad \text{and,} \qquad \left(\frac{d\sigma}{d\phi}\right)^2 = \frac{K(\phi)}{f(\phi)} + 3\frac{f'^2(\phi)}{f^2(\phi)} = \frac{2\omega(\phi)}{\phi f(\phi)} + 3\frac{f'^2(\phi)}{f^2(\phi)}. \tag{18}$$

In view of the action (17), it is also possible to cast the field equations, viz. the Klein–Gordon and the $\binom{0}{0}$ equations of Einstein as,

$$\ddot{\sigma} + 3H\dot{\sigma} + V_E' = 0; \qquad 3H^2 = \frac{1}{2}\dot{\sigma}^2 + V_E, \tag{19}$$

where, $H = \frac{\dot{a}_E}{a_E}$ denotes the expansion rate, commonly known as the Hubble parameter. The slow-roll parameters and the number of e-foldings, then admit the following forms,

$$\epsilon = \left(\frac{V_E'}{V_E}\right)^2\left(\frac{d\sigma}{d\phi}\right)^{-2}; \quad \eta = 2\left[\left(\frac{V_E''}{V_E}\right)\left(\frac{d\sigma}{d\phi}\right)^{-2} - \left(\frac{V_E'}{V_E}\right)\left(\frac{d\sigma}{d\phi}\right)^{-3}\frac{d^2\sigma}{d\phi^2}\right]; \quad N = \int_{t_i}^{t_f} H dt = \frac{1}{2\sqrt{2}}\int_{\phi_e}^{\phi_b}\frac{d\phi}{\sqrt{\epsilon}}\frac{d\sigma}{d\phi}, \tag{20}$$

where, t_i, t_f stand for the initiation time and the end time, while ϕ_b, ϕ_e stand for the values of the scalar field at the beginning and at the end of inflation respectively. Comparing expression for the primordial curvature perturbation on super-Hubble scales produced by single-field inflation ($P_\zeta(k)$) with the primordial gravitational wave power spectrum ($P_t(k)$), one obtains the tensor-to-scalar ratio for single-field slow-roll inflation $r = \frac{P_t(k)}{P_\zeta(k)} = 16\epsilon$, while, the scalar tilt, conventionally defined as $n_s - 1$ may be expressed as $n_s - 1 = -6\epsilon + 2\eta$, or equivalently $n_s = 1 - 6\epsilon + 2\eta$, dubbed as scalar spectral index. According to the latest released results, the scalar to tensor ratio $r \leq 0.16$ (TT,TE,EE+lowEB+lensing), while $r \leq 0.07$ (TT,TE,EE+lowE+lensing+BK14+BAO) [40,41]. Further, combination of all the data (TT+lowE, EE+lowE, TE+lowE, TT,TE,EE+lowE, TT,TE,EE+lowE+lensing) constrain the scalar spectral index to $0.9569 \leq n_s \leq 0.9815$ [40,41]. It is useful to emphasize that under the present choice of unit $\frac{M_P^2}{2} = c = 1$ (which although appears to be a bit unusual but does not cause any harm), ϕ controls the cosmological evolution in the manner $\phi > 1$ corresponds to the inflationary stage, $\phi \sim 1$ describes the end of inflation while $\phi < 1$ is the low energy regime which triggers matter dominated era. We also point out that the above scalar-tensor equivalent form appearing in (17) may be achieved starting from $F(R)$ theory of gravity. For example, in view of $F(R) \propto R + R^n + R^m$, theory of gravity, a recent study of inflationary regime has been performed under conformal transformation to Einstein's frame and a good fit is obtained with the Planck's data by Bhattacharyya et al [52].

3. Inflation with Power Law and Exponential Potentials

In the non-minimal theory, the flat section of the potential $V(\phi)$ responsible for slow-roll is usually distorted. However, flat potential is still obtainable if the Einstein frame potential V_E is asymptotically constant [53,54]. Note that, the symmetry explored in Section 2, makes the potential (V_E) in the Einstein frame to be constant, once and forever. Thus, we need to relax the symmetry as required by the condition (12), by taking into account additional term in the potential, viz. a constant term V_0, or even a functional form, to ensure that the effective potential in the Einstein frame (V_E) is asymptotically constant. At the end of inflation, the universe becomes cool due to sudden large (exponential, in the present case) expansion. Therefore, in order that the structure we live in are formed, the universe must be reheated and take the state of a hot thick soup of plasma (the so called hot Big-Bang). This phenomena is possible if at the end of inflation, the scalar field starts oscillating rapidly on the Hubble time scale, about the minimum of the potential. In the process, particles are created under standard quantum field theoretic (in curved space-time) approach, which results in the re-heating of the universe. The universe then eventually transits to the radiation dominated era. At that epoch, the additional term may be absorbed in the potential if it is a constant term (V_0), or may even be neglected in case it is a function (since ϕ goes below the Planck's mass), without any loss of generality, to reassure symmetry. The symmetry leads to the first integral of certain combination of the field equations, which helps in solving the field equations leading to a Friedmann-like radiation dominated era, as shown earlier [37]. However, in the present manuscript, we only concentrate upon inflationary regime and of-course study the possibility of graceful exit from inflation. In the following subsection, we shall study different power law potentials, while in the next we shall deal with exponential potential. We consider de-Sitter solution in the form $a \propto e^{Ht}$, where the Hubble parameter H is slowly varying during inflation. We repeat that according to our current choice of units $(\frac{M_P^2}{2} = 1)$, which although is uncommon, but does not create any problem whatsoever, the value of $f(\phi)$ at the end of inflation must be a little greater than 1.

3.1. *Power Law Potential $f(\phi) = \phi^n$:*

Under the choice, $f(\phi) = \phi^n$, the potential is $V(\phi) \propto \phi^{2n}$. We shall take into account three different values of n, viz. $n = 1, \frac{3}{2}$ and 2, in the following three sub-subsections. For each value of n we shall study different cases taking into account different additive terms. In the first place however, we shall consider an additive constant V_0 in all the three cases, viz.

$$\text{Case-1: } V(\phi) = V_1\phi^{2n} + V_0, \tag{21}$$

corresponding to which, one can now find the expression for the Brans–Dicke parameter $\omega(\phi)$, the potential $V_E(\sigma)$ in the Einstein frame, the expression for $\frac{d\sigma}{d\phi}$, and the slow-roll parameters ϵ, η along with the number of e-foldings N, in view of the Equations (14), (18) and (20) respectively as,

$$\text{Case-1:}\begin{cases} \omega(\phi) = \dfrac{\omega_0^2 - 3n^2\phi^{2(n-1)}}{2\phi^{(n-1)}}, \quad V_E = V_1 + V_0\phi^{-2n}, \quad \left(\dfrac{d\sigma}{d\phi}\right)^2 = \dfrac{\omega_0^2}{\phi^{2n}}, \\ \epsilon = \dfrac{4n^2V_0^2\phi^{2(n-1)}}{\omega_0^2(V_0 + V_1\phi^{2n})^2}, \quad \eta = \dfrac{4n(n+1)V_0\phi^{2(n-1)}}{\omega_0^2(V_0 + V_1\phi^{2n})}, \quad N = \dfrac{\omega_0^2}{4\sqrt{2}nV_0}\displaystyle\int_{\phi_e}^{\phi_b} \dfrac{V_0 + V_1\phi^{2n}}{\phi^{2n-1}}d\phi. \end{cases} \tag{22}$$

The effect of the constant term V_0 is now clearly noticeable, since when ϕ is large, second term in the Einstein frame potential (V_E) becomes insignificantly small, for $n \geq 1$, and it almost becomes (non-zero) constant, assuring slow-roll. On the contrary, if V_0 is set to vanish from the very beginning, the Einstein frame potential $V_E = V_1$ would remain flat always, and the universe would have been ever-inflating.

We shall also consider a functional additive term in the potential for all the cases under consideration, such the the potential reads as

$$\text{Case-2: } V(\phi) = V_1\phi^{2n} + V_0\phi^m. \tag{23}$$

In view of the above potential (23) it is possible to find the expression for the Brans–Dicke parameter $\omega(\phi)$, the potential $V_E(\sigma)$ in the Einstein frame, the expression for $\frac{d\sigma}{d\phi}$, and the slow-roll parameters ϵ, η along with the number of e-foldings N, in view of the Equations (14), (18) and (20) respectively as,

$$\text{Case-2:}\begin{cases} \omega(\phi) = \dfrac{\omega_0^2 - 3n^2\phi^{2(n-1)}}{2\phi^{(n-1)}}, \quad V_E = V_1 + V_0\phi^{m-2n}, \quad \left(\dfrac{d\sigma}{d\phi}\right)^2 = \dfrac{\omega_0^2}{\phi^{2n}}, \quad \epsilon = \dfrac{(m-2n)^2V_0^2\phi^{2(m-n-1)}}{\omega_0^2[V_1 + V_0\phi^{(m-2n)}]^2}, \\ \eta = \dfrac{2V_0(m-2n)(m-n-1)\phi^{(m-2)}}{\omega_0^2[V_1 + V_0\phi^{(m-2n)}]}, \quad N = \dfrac{\omega_0^2}{2\sqrt{2}(m-2n)V_0}\displaystyle\int_{\phi_e}^{\phi_b} \dfrac{V_1 + V_0\phi^{m-2n}}{\phi^{(m-1)}}d\phi. \end{cases} \tag{24}$$

In the following sub-subsections, we shall take three different values of n, as already mentioned, and present the data set in tabular form along with appropriate plots, to demonstrate the behaviour of the slow-roll parameters in comparison with the latest data set released by Planck [40,41]. Different additive terms, as indicated, will be considered in each subcase separately. In the subsection (3.1.3), we shall consider an additional case with a pair of additive terms in the form of a whole square.

3.1.1. $n = 1$, $f(\phi) = \phi$

Case-1: Under the choice $n = 1$, the potential (21) takes the form $V(\phi) = V_0 + V_1\phi^2$, and thus the parameters of the theory under consideration (22) read as,

$$\begin{aligned} &\omega(\phi) = \frac{\omega_0^2 - 3}{2}, \quad \frac{d\sigma}{d\phi} = \frac{\omega_0}{\phi}, \quad V_E = V_1 + V_0\phi^{-2}, \quad \epsilon = \frac{4V_0^2}{\omega_0^2(V_0 + V_1\phi^2)^2}, \\ &\eta = \frac{8V_0}{\omega_0^2(V_0 + V_1\phi^2)}, \quad N = \frac{\omega_0^2}{4\sqrt{2}V_0}\left[V_1\left(\frac{\phi_b^2}{2} - \frac{\phi_e^2}{2}\right) + V_0(\ln\phi_b - \ln\phi_e)\right]. \end{aligned} \tag{25}$$

In view of the above forms of the slow roll parameters (25), we present Tables 1 and 2, underneath, corresponding to two different values of the parameter $V_1 > 0$. The wonderful fit with the latest data sets released by Planck [40,41] is appreciable particularly because $0.968 \approx n_s < 0.982$, while $r < 0.0278$. Further, the number of e-fold ($36 \leq N \leq 62$) is sufficient to alleviate the horizon and flatness problems. Figures 1 and 2 are the two plots r versus n_s and r versus ω_0 respectively, presented for visualization. For example, the figures clearly depict that the plot which represents data sets corresponding to Table 2 appears to be even better.

One very interesting feature is that the above data sets remain unaltered even if the sign of V_0 and V_1 are interchanged. Note that, second derivative of the potential has to be positive, since it represents effective mass of the scalar field. In view of the forms of the potentials $V(\phi)$ and $V_E(\sigma)$ presented in (21) and (22) the effective mass of the scalar fields ϕ and σ respectively are,

$$\frac{d^2V}{d\phi^2} = 2V_1; \quad \frac{d^2V_E}{d\sigma^2} = 6\frac{V_0}{\phi^4}. \tag{26}$$

In our data set, we keep $V_1 > 0$, since ϕ is the scalar field under consideration, while translation to σ only amounts to handling the situation with considerable ease. However, as a matter of taste if one favours Einstein's frame over Jordan's frame, it is possible to revert the sign and keep $V_0 > 0$, without changing the data set.

Table 1. $f(\phi) = \phi$, (case-1): $\phi_b = 2.0$. $V_0 = -0.9 \times 10^{-13}\mathrm{T}^{-2}$, $V_1 = 0.9 \times 10^{-13}\mathrm{T}^{-2}$.

ω_0	$\lvert\eta\rvert$	$r = 16\epsilon$	ϕ_e	n_s	N
16.0	0.010418	0.0278	1.060	0.9688	36
16.5	0.009795	0.0261	1.059	0.9706	38
17.0	0.009227	0.0246	1.057	0.9723	41
17.5	0.008707	0.0232	1.056	0.9739	43
18.0	0.008230	0.0219	1.054	0.9753	46
18.5	0.007792	0.0208	1.053	0.9766	48
19.0	0.007387	0.0197	1.051	0.9778	51
19.5	0.007013	0.0187	1.050	0.9790	54
20.0	0.006667	0.0178	1.049	0.9800	56
20.5	0.006345	0.0169	1.048	0.9810	59
21.0	0.006047	0.0161	1.047	0.9819	62

Table 2. $f(\phi) = \phi$, (case-1): $\phi_b = 2.0$. $V_0 = -0.9 \times 10^{-13}\mathrm{T}^{-2}$, $V_1 = 1.0 \times 10^{-13}\mathrm{T}^{-2}$.

ω_0	$\lvert\eta\rvert$	$r = 16\epsilon$	ϕ_e	n_s	N
14.5	0.011047	0.0257	1.012	0.9682	36
15.0	0.01032	0.0239	1.009	0.9703	38
15.5	0.009667	0.0225	1.008	0.9722	41
16.0	0.009072	0.0210	1.006	0.9739	43
16.5	0.008531	0.0198	1.005	0.9755	46
17.0	0.008037	0.0187	1.003	0.9769	49
17.5	0.007584	0.0176	1.001	0.9782	52
18.0	0.007168	0.0166	1.000	0.9794	55
18.5	0.006786	0.0158	0.9986	0.9805	59
19.0	0.006433	0.0149	0.9973	0.9815	62

As mentioned, at the end of inflation, the scalar field must oscillate rapidly so that particles are produced and the universe turns to the phase of: a hot thick soup of plasma, commonly called the 'hot big-bang'. This phenomena is dubbed as graceful exit, which is required for the structure formation together with the formation of CMB. We therefore proceed to check if the present model admits graceful exit from inflation. Here, $V_E = V_1 + \frac{V_0}{\phi^2}$, and so one can express (19) as,

$$\frac{3H^2}{V_1} = \frac{\dot{\sigma}^2}{2V_1} + \left(1 + \frac{V_0}{V_1\phi^2}\right). \tag{27}$$

At large value of the scalar field, which in the present unit $\phi > 1$, we obtained slow-roll. However, as the scalar field falls below the Planck's mass M_p, then the Hubble rate H also decreases, and once it falls below the effective mass V_1, i.e., $H \ll V_1$, then the above equation may be approximated to,

$$\dot{\sigma}^2 = 2i^2\left(V_1 + \frac{V_0}{\phi^2}\right) \quad \longrightarrow \quad \dot{\phi} = i\frac{\phi}{\omega_0}\sqrt{2\left(V_1 + \frac{V_0}{\phi^2}\right)}, \tag{28}$$

where, $\dot{\sigma} = \frac{\omega_0}{\phi}\dot{\phi}$, in view of (25). Thus, finally we get,

$$\phi = \frac{1}{2V_1}\left[(1 - V_0V_1)\cos\left(\frac{\sqrt{2V_1}}{\omega_0}t\right) + i(1 + V_0V_1)\sin\left(\frac{\sqrt{2V_1}}{\omega_0}t\right)\right], \tag{29}$$

which is an oscillatory solution, and the field then oscillates many times over a Hubble time. This coherent oscillating field corresponds to a condensate of non-relativistic massive (inflaton) particles, which ensures graceful exit from the inflationary regime, driving a matter-dominated era at the end of inflation. There is a long standing debate regarding the physical frame. It appears that most of the people favour Einstein's frame over Jordan's frame (we have briefly discussed the issue in conclusion). In this regard, it is important to mention that since in view of (25) $\sigma = \omega_0 \ln \phi$, therefore σ executes oscillatory behaviour as well.

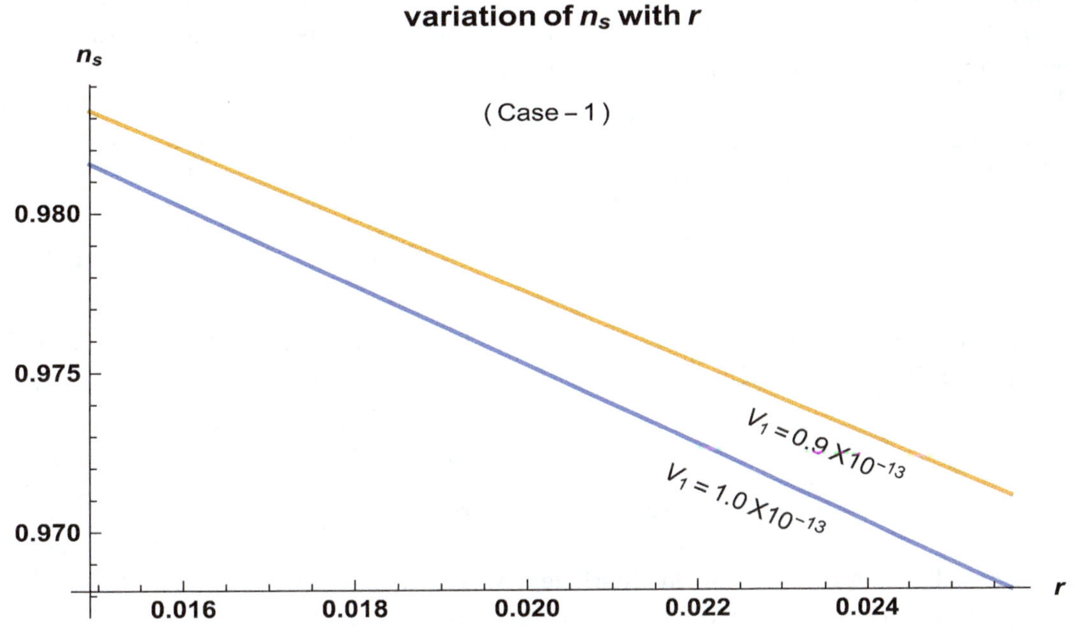

Figure 1. $f = \phi$, (case-1): Yellow ochre and blue colours represent Table 1 and 2 data, respectively.

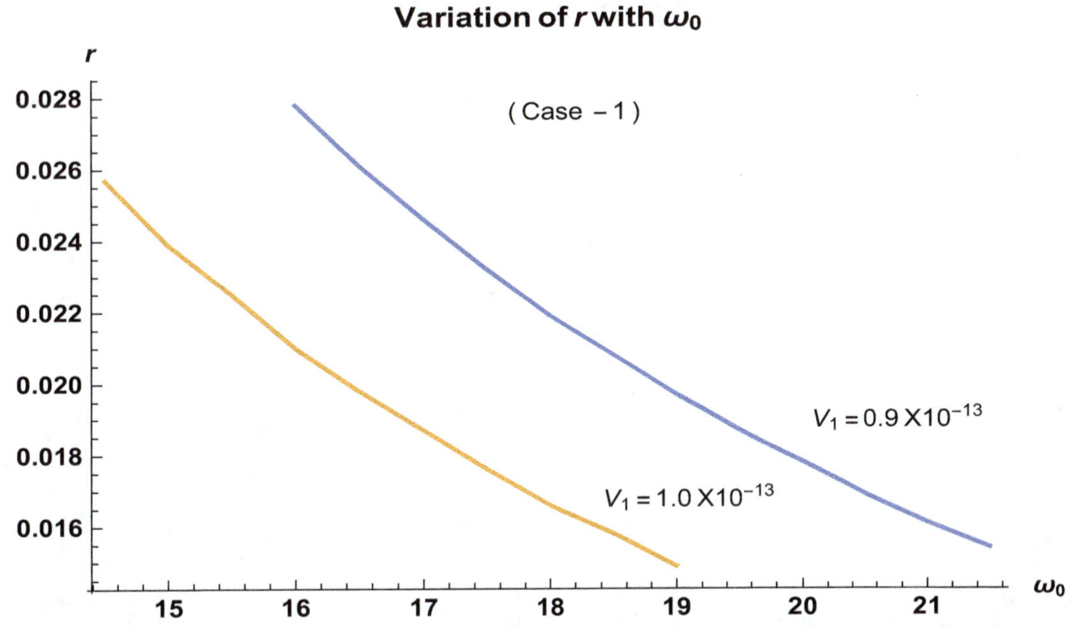

Figure 2. $f = \phi$, (case-1): Blue and yellow ochre colours represent Tables 1 and Table 2 data, respectively.

Case-2: Under the same situation $f(\phi) = \phi$, let us now consider, $V(\phi) = V_1\phi^2 + V_0\phi^4$, where instead of a constant term, we have added a quartic term in the potential. The expression for the Brans–Dicke parameter ($\omega(\phi)$), the potential (V_E) in the Einstein frame, $\frac{d\sigma}{d\phi}$, the slow-roll parameters ϵ, η and the number of e-foldings N, may then be found in view of the Equation (24), respectively as,

$$\omega(\phi) = \frac{\omega_0^2 - 3}{2\phi}, \quad \frac{d\sigma}{d\phi} = \frac{\omega_0}{\phi}, \quad V_E = V_1 + V_0\phi^2, \quad \epsilon = \frac{4V_0^2\phi^4}{\omega_0^2(V_1 + V_0\phi^2)^2},$$
$$\eta = \frac{8V_0\phi^2}{\omega_0^2(V_1 + V_0\phi^2)}, \quad N = \int_{\phi_e}^{\phi_b} \frac{\omega_0^2(V_1 + V_0\phi^2)}{4\sqrt{2}V_0\phi^3} d\phi. \tag{30}$$

Although, the potential V_E does not appear to attend a flat section, the smallness of the value of η confirms that there indeed exists a flat section, admitting slow-roll. In fact, in the Einstein frame (17), this is just the case of a standard inflation field theory with quadratic potential. Following Tables 3 and 4, for $V_1 > 0$, together with the associated plots n_s versus r in Figure 3 and r versus ω_0 in Figure 4 here again depict appreciably good fit with the recent released Planck's data set, particularly because $0.97 \leq n_s \leq 0.98$ while $r \leq 0.098$. Figure 3 depicts that the data of Table 3 are somewhat better.

As before here again we test if the model associated with a different potential admits graceful exit. Here $V_E = (V_1 + V_0\phi^2)$, and so from (19) one obtains,

$$\frac{3H^2}{V_1} = \frac{\dot{\sigma}^2}{2V_1} + \left(1 + \frac{V_0}{V_1}\phi^2\right) \tag{31}$$

As the Hubble rate $H \ll V_1$, the above equation can be approximated to, $\dot{\sigma}^2 = 2i^2(V_1 + V_0\phi^2)$, yielding $\dot{\phi} = i\frac{\phi}{\omega_0}\sqrt{2(V_1 + V_0\phi^2)}$, where, $\dot{\sigma} = \dot{\phi}\frac{\omega_0}{\phi}$. Finally we get,

$$\phi = \frac{2V_1\left[(1 - V_0V_1)\cos\left(\frac{\sqrt{2V_1}}{\omega_0}t\right) + i(1 + V_0V_1)\sin\left(\frac{\sqrt{2V_1}}{\omega_0}t\right)\right]}{1 + V_0^2V_1^2 - 2V_0V_1\cos\left(\frac{2\sqrt{2V_1}}{\omega_0}t\right)}. \tag{32}$$

The oscillatory behaviour of the scalar field clearly ensures graceful exit from inflationary regime, as already discussed, and in view of (30) σ also executes oscillatory behaviour.

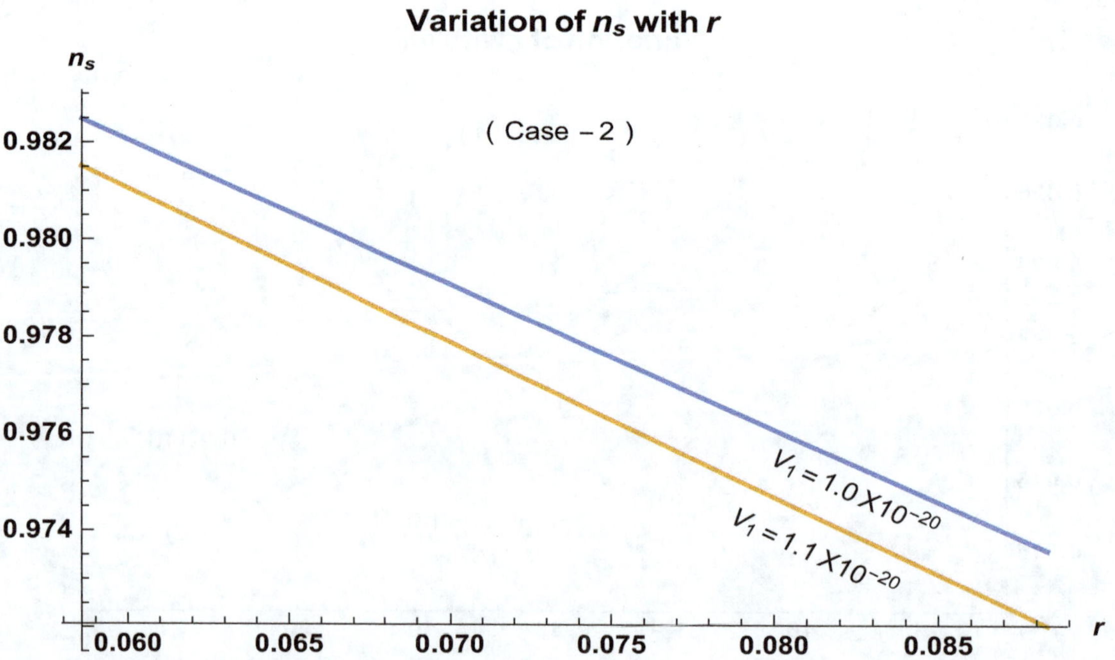

Figure 3. $f = \phi$. Yellow ochre and blue colours represent Tables 3 and 4 respectively.

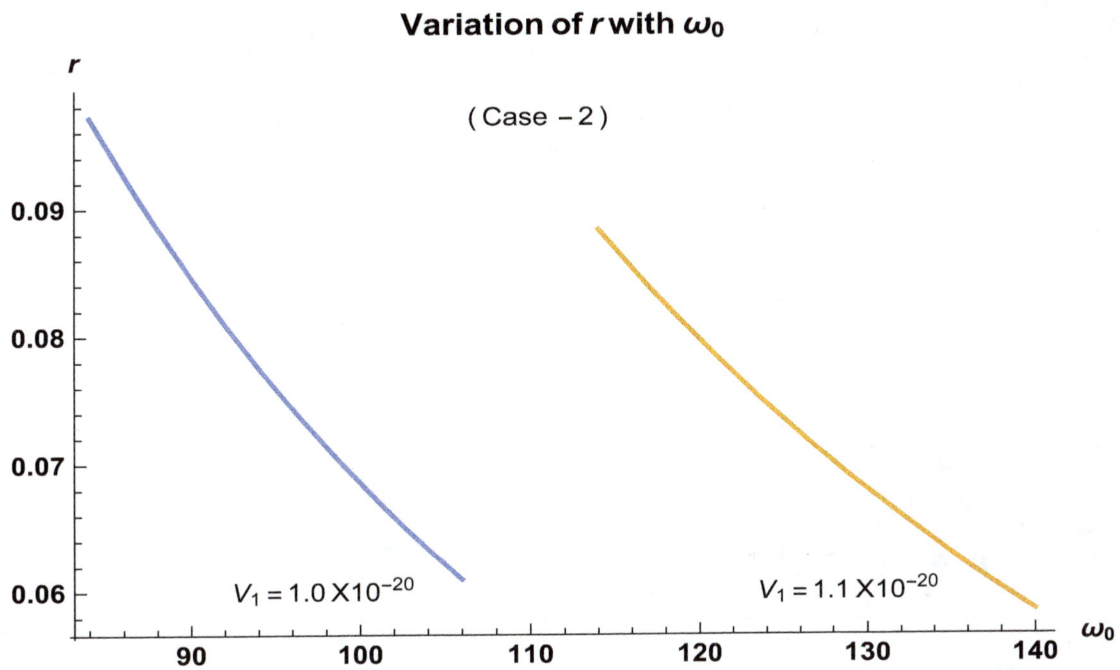

Figure 4. $f = \phi$. Yellow ochre and blue colours represent Tables 3 and 4 respectively, unlike previous cases.

Table 3. $f(\phi) = \phi$, (case-2): $\phi_b = 1.2$, $V_0 = -1.0 \times 10^{-20}\mathrm{T}^{-2}; V_1 = 1.1 \times 10^{-20}\mathrm{T}^{-2}$.

ω_0	$\|\eta\|$	$r = 16\epsilon$	ϕ_e	n_s	N
114	0.002607	0.08834	1.0581	0.9721	38
116	0.002518	0.08532	1.0580	0.9730	39
118	0.002433	0.08245	1.0578	0.9739	41
120	0.002353	0.07972	1.0577	0.9748	42
122	0.002276	0.07713	1.0575	0.9756	44
124	0.002204	0.07466	1.0574	0.9764	45
126	0.002134	0.07231	1.0572	0.9772	46
128	0.002068	0.07007	1.0571	0.9779	48
130	0.002005	0.06793	1.0570	0.9785	49
132	0.001945	0.06589	1.0569	0.9792	51
134	0.001887	0.06393	1.0567	0.9798	53
136	0.001832	0.06207	1.0566	0.9804	54
138	0.001780	0.06028	1.0565	0.9810	56
140	0.001729	0.05857	1.0564	0.9815	57

Table 4. $f(\phi) = \phi$, (case-2): $\phi_b = 1.2$, $V_0 = -1.0 \times 10^{-20}\mathrm{T}^{-2}$; $V_1 = 1.0 \times 10^{-20}\mathrm{T}^{-2}$.

ω_0	$\|\eta\|$	$r = 16\epsilon$	ϕ_e	n_s	N
84	0.003711	0.09715	1.0121	0.9710	36
86	0.003540	0.09268	1.0118	0.9723	38
88	0.003381	0.08852	1.0115	0.9736	40
90	0.003232	0.08463	1.0113	0.9747	42
92	0.003093	0.08099	1.0110	0.9758	44
94	0.002963	0.07758	1.0108	0.9768	46
96	0.002841	0.07438	1.0105	0.9778	47
98	0.002726	0.07138	1.0103	0.9787	49
100	0.002618	0.06855	1.0101	0.9795	51
102	0.002517	0.06589	1.0099	0.9803	54
104	0.002421	0.0634	1.0097	0.9810	56
106	0.002330	0.06101	1.0096	0.9818	58

3.1.2. $n = \frac{3}{2}$, $f(\phi) = \phi^{\frac{3}{2}}$

Case-1: Under the choice $n = \frac{3}{2}$, $f(\phi) = \phi^{\frac{3}{2}}$, and the potential (21) takes the cubic form, $V(\phi) = V_0 + V_1\phi^3$. Thus the expressions for the parameters of the theory under consideration along with the slow-roll parameters (22) are,

$$\omega(\phi) = \frac{4\omega_0^2 - 27\phi}{8\sqrt{\phi}}, \quad \frac{d\sigma}{d\phi} = \frac{\omega_0}{\phi^{\frac{3}{2}}}; \qquad V_E = V_1 + V_0\phi^{-3}, \qquad \epsilon = \frac{9V_0^2\phi}{\omega_0^2(V_0 + V_1\phi^3)^2},$$
$$\eta = \frac{15V_0\phi}{\omega_0^2(V_0 + V_1\phi^3)}, \qquad N = \frac{\omega_0^2}{6\sqrt{2}V_0}\left[V_0\left(\frac{1}{\phi_e} - \frac{1}{\phi_b}\right) + \frac{V_1}{2}(\phi_b^2 - \phi_e^2)\right]. \tag{33}$$

As before, we present two sets of data in Tables 5 and 6, for two different values of $V_1 > 0$. Figures 5 and 6 depict the variations of the spectral index n_s with the scalar-tensor ratio r and the scalar-tensor ratio r with the Brans–Dicke parameter ω_0 respectively. Here again we observe that $r \leq 0.0278$, and $0.96 < n_s < 0.9815$, which are very much within the stipulated observational range [40,41], while number of e-folding N is sufficient to remove the flatness and the horizon problems.

Table 5. $f(\phi) = \phi^{\frac{3}{2}}$, (case-1): $\phi_b = 1.7$, $V_0 = -0.9 \times 10^{-13}\mathrm{T}^{-2}$, $V_1 = 0.9 \times 10^{-13}\mathrm{T}^{-2}$.

ω_0	$\|\eta\|$	$r = 16\epsilon$	ϕ_e	n_s	N
24	0.011314	0.0278	1.0408	0.9670	36
25	0.010427	0.0256	1.0392	0.9696	39
26	0.009640	0.0237	1.0377	0.9719	42
27	0.008939	0.0219	1.0364	0.9739	46
28	0.008312	0.0204	1.0351	0.9757	49
29	0.007749	0.0190	1.0339	0.9773	52
30	0.007241	0.0178	1.0328	0.9788	56
31	0.006781	0.0166	1.0318	0.9802	60
32	0.006364	0.0156	1.0308	0.9814	64

Table 6. $f(\phi) = \phi^{\frac{3}{2}}$, (case-1): $\phi_b = 1.7$, $V_0 = -0.9 \times 10^{-13}\mathrm{T}^{-2}$, $V_1 = 1.0 \times 10^{-13}\mathrm{T}^{-2}$.

ω_0	$\lvert\eta\rvert$	$r = 16\epsilon$	ϕ_e	n_s	N
22.0	0.011816	0.0254	1.0076	0.9668	36
22.5	0.011217	0.0243	1.0067	0.9683	38
23.0	0.010811	0.0233	1.0059	0.9697	39
23.5	0.010356	0.0223	1.0050	0.9709	41
24.0	0.009928	0.0214	1.0042	0.9721	43
24.5	0.009528	0.0205	1.0034	0.9733	45
25.0	0.009150	0.0197	1.0027	0.9743	47
25.5	0.008795	0.0189	1.0019	0.9753	48
26.0	0.008460	0.0182	1.0013	0.9762	50
26.5	0.008143	0.0175	1.0006	0.9771	52
27.0	0.007845	0.0169	1.000	0.9780	54

Variation of n_s with r

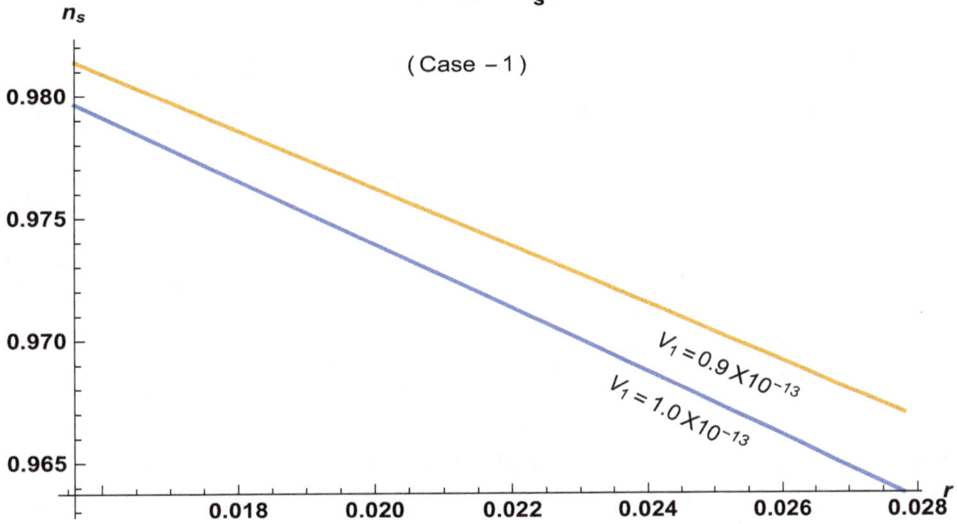

Figure 5. $f(\phi) = \phi^{\frac{3}{2}}$. Yellow ochre and blue colours represent Tables 5 and 6 respectively.

Variation of r with ω_0

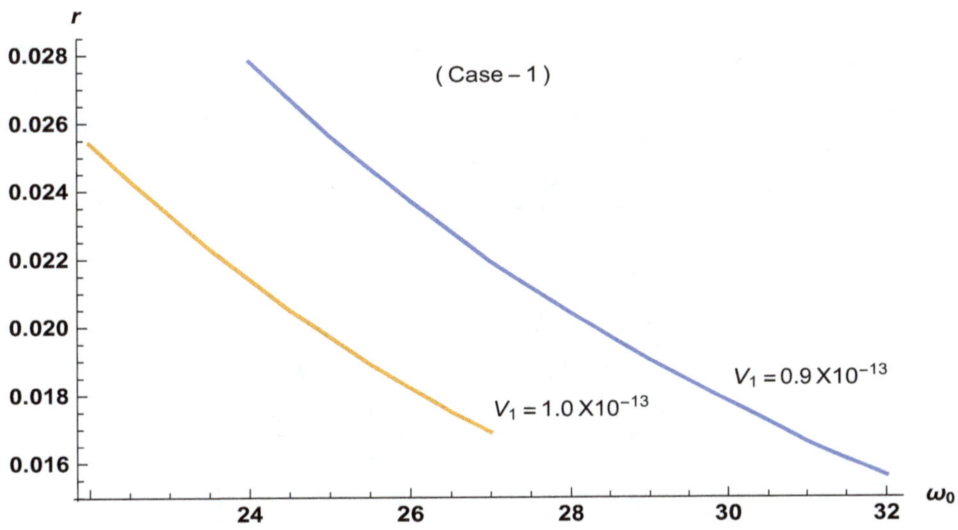

Figure 6. $f(\phi) = \phi^{\frac{3}{2}}$. Blue and yellow ochre colours represent Tables 5 and 6 respectively.

To check the behaviour of the scalar field, we proceed as before, and find,

$$\frac{3H^2}{V_1} = \frac{\omega_0^2\dot{\phi}^2}{2V_1\phi^3} + \frac{V_0}{V_1\phi^3} + 1 \Longrightarrow \dot{\phi}^2 = -2\frac{(V_0+V_1\phi^3)}{\omega_0^2}, \tag{34}$$

under suitable approximation, as the Hubble rate $H \ll V_1$, and using the relation $\dot{\sigma} = \omega_0\frac{\dot{\phi}}{\phi^{\frac{3}{2}}}$, in view of (33). The solution reads as,

$$\begin{aligned}\phi(t) =&\text{InverseFunction}\left[\frac{2i}{\sqrt[4]{3}\sqrt[3]{V_1}\sqrt{V_0-\#1^3V_1}}\text{EllipticF}\left[\text{Sin}^{-1}\left(\frac{1}{\sqrt[4]{3}}\sqrt{-(-1)^{\frac{5}{6}}-i\#1\sqrt[3]{\frac{V_1}{V_0}}}\right),\sqrt[3]{-1}\right]\right.\\ &\left.\sqrt[3]{V_0}\sqrt{(-1)^{5/6}\left(\#1\sqrt[3]{\frac{V_1}{V_0}}-1\right)}\sqrt{\#1^2\sqrt[3]{\frac{V_1^2}{V_0^2}}+\#1\sqrt[3]{\frac{V_1}{V_0}}+1\&}\right][c_1-t].\end{aligned} \tag{35}$$

In the above the hash tag (#n) denotes nth argument of a pure function, and c_1 is a constant. Although, the solution is not obtainable in closed form, rather is a complicated inverse elliptic function, nevertheless its oscillatory behaviour is quite apparent, and $\sigma = -2\frac{\omega_0}{\sqrt{\phi}}$ also oscillates as well.

Case-2: Under the choice $n = \frac{3}{2}$, $f(\phi) = \phi^{\frac{3}{2}}$, and taking the potential in the form, $V(\phi) = V_1\phi^3 + V_0\phi$, the expressions for the parameters of the theory under consideration along with the slow-roll parameters (24) are,

$$\begin{aligned}&\omega(\phi) = \frac{4\omega_0^2-27\phi}{8\sqrt{\phi}}, \quad \frac{d\sigma}{d\phi} = \frac{\omega_0}{\phi^{\frac{3}{2}}}; \qquad V_E = V_1 + \frac{V_0}{\phi^2}, \qquad \epsilon = \frac{4V_0^2\phi}{\omega_0^2(V_0+V_1\phi^2)^2},\\ &\eta = \frac{6V_0\phi}{\omega_0^2(V_0+V_1\phi^2)}, \qquad N = \frac{\omega_0^2}{4\sqrt{2}V_0}\left[V_0\left(\frac{1}{\phi_e}-\frac{1}{\phi_b}\right)+V_1(\phi_b-\phi_e)\right].\end{aligned} \tag{36}$$

We present two sets of data in Tables 7 and 8 underneath, for two different values of $V_1 > 0$. Figures 7 and 8 depict the variations of the spectral index n_s with the scalar-tensor ratio r, and the scalar-tensor ratio r with the Brans–Dicke parameter ω_0, respectively. Here also we observe that $r \leq 0.039$, and $0.971 < n_s < 0.983$, which are again in excellent agreement of Planck's data [40,41], while the number of e-folding N is also sufficient to remove the flatness and the horizon problems.

Table 7. $f(\phi) = \phi^{\frac{3}{2}}$, (case-2): $\phi_b = 1.7$, $V_0 = -0.9\times10^{-13}\mathrm{T}^{-2}$, $V_1 = 0.90\times10^{-13}\mathrm{T}^{-2}$.

ω_0	$\lvert\eta\rvert$	$r = 16\epsilon$	ϕ_e	n_s	N
28	0.006884	0.03885	1.0357	0.9717	40
29	0.006417	0.03623	1.0345	0.9736	43
30	0.005996	0.03384	1.0333	0.9753	46
31	0.005616	0.03169	1.0323	0.9769	49
32	0.005270	0.02974	1.0313	0.9783	52
33	0.004956	0.02797	1.0303	0.9796	55
34	0.004669	0.02635	1.0294	0.9808	59
35	0.004406	0.02486	1.0286	0.9819	62
36	0.004164	0.02350	1.0278	0.9829	66

Table 8. $f(\phi) = \phi^{\frac{3}{2}}$, (case-2): $\phi_b = 1.7$, $V_0 = -0.9 \times 10^{-13}\mathrm{T}^{-2}$, $V_1 = 0.95 \times 10^{-13}\mathrm{T}^{-2}$.

ω_0	$\|\eta\|$	$r = 16\epsilon$	ϕ_e	n_s	N
26	0.007358	0.03828	1.0103	0.9709	39
27	0.006823	0.03549	1.0089	0.9730	42
28	0.006345	0.03300	1.0076	0.9749	45
29	0.005915	0.03077	1.0064	0.9766	48
30	0.005527	0.02875	1.0053	0.9782	52
31	0.005176	0.02693	1.0043	0.9796	55
32	0.004858	0.02527	1.0033	0.9808	59
33	0.004568	0.02376	1.0024	0.9820	63
34	0.004303	0.02238	1.0016	0.9830	67

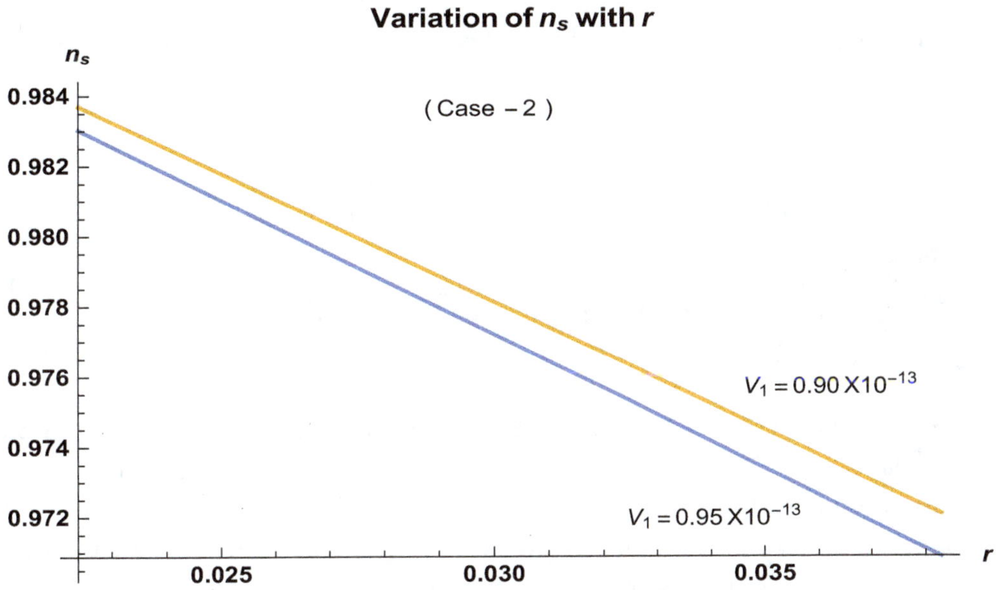

Figure 7. $f(\phi) = \phi^{\frac{3}{2}}$. Yellow ochre and blue colours represent Tables 7 and 8 respectively.

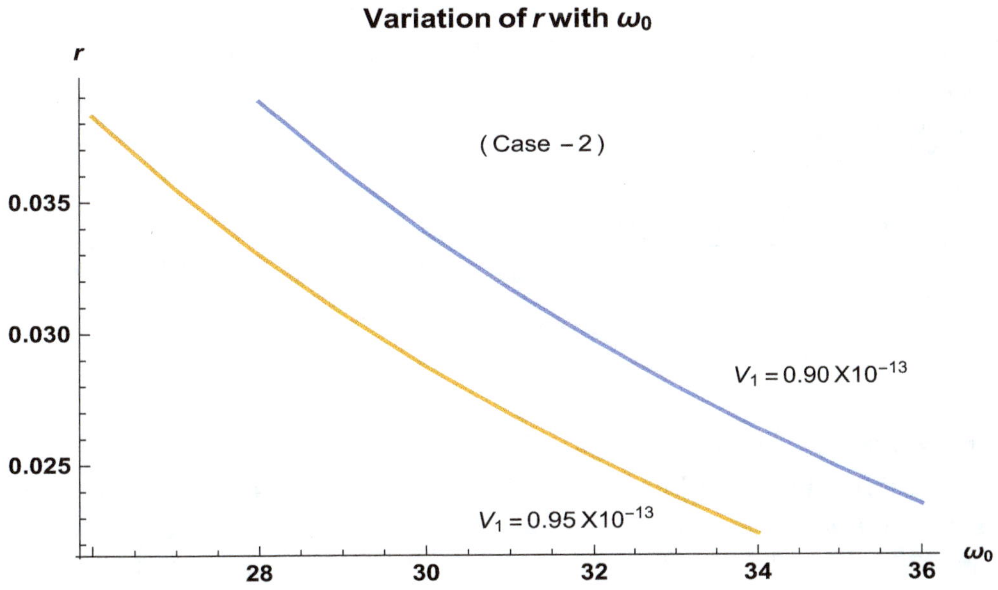

Figure 8. $f(\phi) = \phi^{\frac{3}{2}}$. Blue and yellow ochre colours represent Tables 7 and 8 respectively.

In order to study the behaviour of the scalar field at the end of inflation, we start with the Einstein frame potential as before, $V_E = (V_1 + \frac{V_0}{\phi^2})$, and express the field Equation (19) as,

$$\frac{3H^2}{V_1} = \frac{\dot{\sigma}^2}{2V_1} + \left(1 + \frac{V_0}{V_1\phi^2}\right) \tag{37}$$

As the Hubble rate falls, and $H \ll V_1$, the above equation may be approximated to, $\dot{\sigma}^2 = 2i^2(V_1 + \frac{V_0}{\phi^2})$, which in terms of the scalar field ϕ reads as,

$$\dot{\phi}^2 + \frac{2}{\omega_0^2}(V_0 + V_1\phi^2)\phi = 0. \tag{38}$$

The above equation when solved, is found to execute oscillatory behaviour as before, along with $\sigma = -2\frac{\omega_0}{\sqrt{\phi}}$, as well.

$$\phi(t) = \text{InverseFunction}\left[\frac{2i\#1^{3/2}\sqrt{\frac{V_0}{\#1^2 V_1} + 1}\,\text{EllipticF}\left(i\sinh^{-1}\left(\frac{\sqrt{-\frac{i\sqrt{V_0}}{\sqrt{V_1}}}}{\sqrt{\#1}}\right)\Big| -1\right)}{\sqrt{-\frac{i\sqrt{V_0}}{\sqrt{V_1}}}\sqrt{-\#1\left(\#1^2 V_1 + V_0\right)}}\&\right]\left[c_1 + \frac{\sqrt{2}t}{\omega_0}\right], \tag{39}$$

Case-3: Cubic potentials with additive term have important consequence. For example, a potential in the form $V = \frac{1}{2}m\omega^2x^2 - \frac{1}{3}bx^3$ can be used to model decay of metastable states [55], and it also describes the global flow [56]. Further, the tunnelling rate in real time in the semiclassical limit may be found for arbitrary energy levels, while its ground state agrees well with the result found by the instanton method [57]. It is therefore worth to continue the present study in view of such an additive form in the cubic potential.

Under the choice $n = \frac{3}{2}$, $f(\phi) = \phi^{\frac{3}{2}}$, and taking the potential as cubic form added with a quadratic term, i.e., $V(\phi) = V_1\phi^3 + V_0\phi^2$, the expressions for the parameters of the theory under consideration along with the slow-roll parameters (24) are,

$$\begin{gathered}\omega(\phi) = \frac{4\omega_0^2 - 27\phi}{8\sqrt{\phi}}, \quad \frac{d\sigma}{d\phi} = \frac{\omega_0}{\phi^{\frac{3}{2}}}; \qquad V_E = V_1 + \frac{V_0}{\phi}, \qquad \epsilon = \frac{V_0^2\phi}{\omega_0^2(V_0 + V_1\phi)^2}, \\ \eta = \frac{V_0\phi}{\omega_0^2(V_0 + V_1\phi)}, \qquad N = \frac{\omega_0^2}{2\sqrt{2}V_0}\left[V_0\left(\frac{1}{\phi_e} - \frac{1}{\phi_b}\right) + V_1\ln(\phi_b - \phi_e)\right].\end{gathered} \tag{40}$$

We present two sets of data in Tables 9 and 10 underneath, for two different values of $V_1 > 0$. Figures 9 and 10 depict the variations of the spectral index n_s with the scalar-tensor ratio r and the scalar-tensor ratio r with the Brans–Dicke parameter ω_0 respectively. Here also we observe that $r \le 0.062$, and $0.973 < n_s < 0.983$, which are again in excellent agreement of Planck's data [40,41], while the number of e-folding N is also sufficient to remove the flatness and the horizon problems. It is interesting to note that the variation n_s with r for the two sets of data almost overlap in Figure 9.

Table 9. $f(\phi) = \phi^{\frac{3}{2}}$, (case-3): $\phi_b = 1.7$, $V_0 = -0.9 \times 10^{-13}\mathrm{T}^{-2}$, $V_1 = 0.9 \times 10^{-13}\mathrm{T}^{-2}$.

ω_0	$\|\eta\|$	$r = 16\epsilon$	ϕ_e	n_s	N
30	0.00270	0.0617	1.0339	0.9715	38
31	0.00253	0.0578	1.0328	0.9733	40
32	0.00237	0.0542	1.0317	0.9749	43
33	0.00223	0.0509	1.0308	0.9764	46
34	0.00210	0.0480	1.0299	0.9778	48
35	0.00198	0.0453	1.0290	0.9790	51
36	0.00187	0.0428	1.0282	0.9802	54
37	0.00177	0.0405	1.0273	0.9812	57
38	0.00168	0.0384	1.0267	0.9822	61

Table 10. $f(\phi) = \phi^{\frac{3}{2}}$, (case-3): $\phi_b = 1.7$, $V_0 = -0.9 \times 10^{-13}\mathrm{T}^{-2}$, $V_1 = 0.92 \times 10^{-13}\mathrm{T}^{-2}$.

ω_0	$\|\eta\|$	$r = 16\epsilon$	ϕ_e	n_s	N
29	0.00274	0.0594	1.0122	0.9723	39
30	0.00256	0.0555	1.0110	0.9741	41
31	0.00240	0.0520	1.0097	0.9757	44
32	0.00225	0.0488	1.0090	0.9772	47
33	0.00212	0.0459	1.0080	0.9786	50
34	0.00199	0.0432	1.0071	0.9798	53
35	0.00188	0.0408	1.0063	0.9809	56
36	0.00178	0.0386	1.0055	0.9820	60
37	0.00168	0.0365	1.0048	0.9829	63

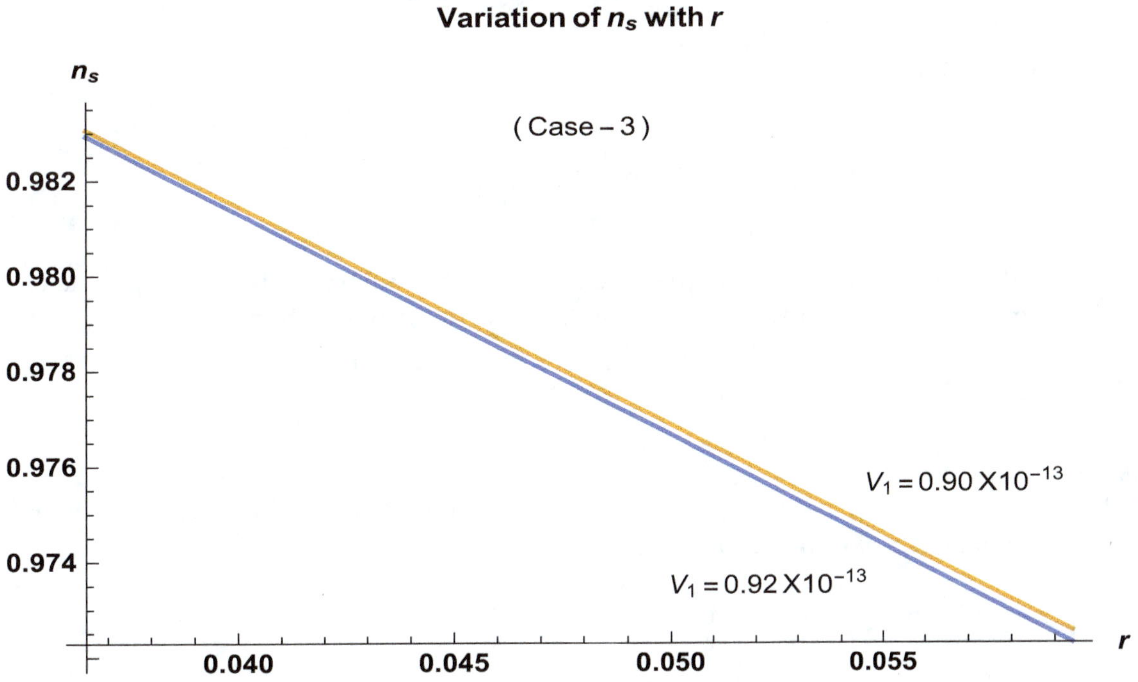

Figure 9. $f(\phi) = \phi^{\frac{3}{2}}$. Yellow ochre and blue colours represent Tables 9 and 10 respectively.

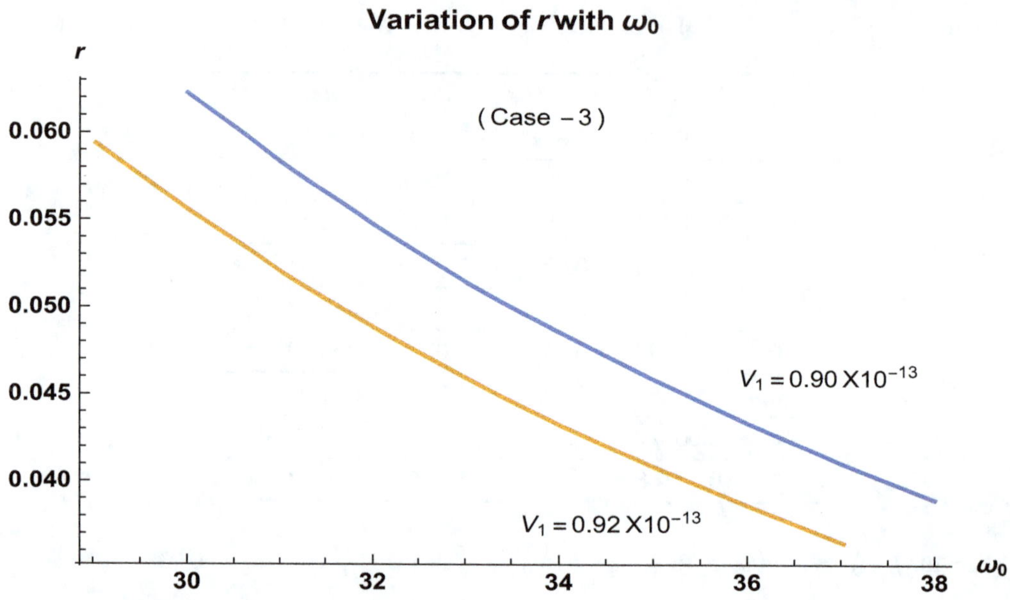

Figure 10. $f(\phi) = \phi^{\frac{3}{2}}$. Blue and yellow ochre colours represent Tables 9 and 10 respectively.

In order to study the behaviour of the scalar field at the end of inflation, we start with the Einstein frame potential as before, $V_E = (V_1 + \frac{V_0}{\phi})$, and express the field Equation (19) as,

$$\frac{3H^2}{V_1} = \frac{\dot{\sigma}^2}{2V_1} + \left(1 + \frac{V_0}{V_1\phi}\right) \tag{41}$$

As the Hubble rate falls, and $H \ll V_1$, the above equation may be approximated to, $\dot{\sigma}^2 = 2i^2(V_1 + \frac{V_0}{\phi})$, which in terms of the scalar field ϕ reads as,

$$\dot{\phi}^2 + \frac{2}{\omega_0^2}(V_0 + V_1\phi)\phi^2 = 0. \tag{42}$$

The above equation may be solved to find

$$\phi = \frac{V_0}{V_1}\left[-1 - \tan\left\{\frac{\sqrt{V_0}}{2}\left(c_1 + \frac{\sqrt{2}}{\omega_0}t\right)\right\}^2\right], \tag{43}$$

which unfortunately is not oscillatory. Perhaps, due to the asymmetry of the potential, the oscillatory behaviour of the scalar field with an additive quadratic term is not exhibited.

3.1.3. $n = 2,\ f(\phi) = \phi^2$

Case-1: Under the choice $n = 2$, $f(\phi) = \phi^2$, and the potential (21) is now chosen as $V(\phi) = V_0 + V_1\phi^4$. Therefore the Brans–Dicke parameter and the Einstein frame potential, together with the slow-roll parameters (22), take the following forms,

$$\omega(\phi) = \frac{\omega_0^2 - 12\phi^2}{2\phi}, \quad \frac{d\sigma}{d\phi} = \frac{\omega_0}{\phi^2}; \qquad V_E = V_1 + V_0\phi^{-4}, \qquad \epsilon = \frac{16V_0^2\phi^2}{\omega_0^2(V_0 + V_1\phi^4)^2},$$
$$\eta = \frac{24V_0\phi^2}{\omega_0^2(V_0 + V_1\phi^4)}, \qquad N = \frac{\omega_0^2}{8\sqrt{2}V_0}\left[\frac{V_0}{2}\left(\frac{1}{\phi_e^2} - \frac{1}{\phi_b^2}\right) + \frac{V_1}{2}(\phi_b^2 - \phi_e^2)\right]. \tag{44}$$

It is quite transparent that for large value of the scalar field ϕ, a flat Einstein's frame potential is realizable here too. As before, we take two sets of data corresponding to two different values of $V_1 > 0$, and tabulate the parametric values in Tables 11 and 12. One can see that the scalar tensor ratio $r \leq 0.0202$, and the spectral index lies between $0.9645 \leq n_s \leq 0.9809$, which are in excellent agreement with Planck's data [40,41]. Further number of e-folding N is also sufficient to alleviate the flatness and the horizon problems. The n_s versus r and r versus ω_0 plots are presented in Figures 11 and 12, as well.

Equation (19) now reads as,

$$3H^2 = \frac{1}{2}\dot{\sigma}^2 + V_1 + \frac{V_0}{\phi^4}, \tag{45}$$

which, as H falls below V_1, i.e., $H \ll V_1$, may be approximated to, $\frac{1}{2}\dot{\sigma}^2 + \frac{V_0}{\phi^4} + V_1 = 0$. In terms of the scalar field ϕ it is expressed as $\dot{\phi}^2 + \frac{2}{\omega_0^2}\left(V_1\phi^4 + V_0\right) = 0$, since, $\dot{\sigma} = \omega_0\frac{\dot{\phi}}{\phi^2}$, in view of (44). The solution is exhibited either as,

$$\phi(t) = -\frac{(-1)^{3/4}\sqrt[4]{V_0}\text{JacobiSN}\left(c_1\sqrt[4]{-V_0V_1}\left(1 + i\frac{\sqrt{2}}{\omega_0}t\right)\Big| -1\right)}{\sqrt[4]{V_1}}, \tag{46}$$

(where JacobiSN is a meromorphic function in both arguments, which for certain special arguments may automatically be evaluated to exact values. In any case, under numerical simulation the above solution is found to exhibit oscillatory behaviour of the scalar field ϕ), or as inverse elliptic function as before,

$$\phi(t) = \text{InverseFunction}\left[-\frac{i\sqrt{\frac{\#1^4V_1}{V_0} + 1}\text{EllipticF}\left(i\sinh^{-1}\left(\#1\sqrt{-\frac{i\sqrt{V_1}}{\sqrt{V_0}}}\right)\Big| -1\right)}{\sqrt{-\frac{i\sqrt{V_1}}{\sqrt{V_0}}}\sqrt{\#1^4(-V_1) - V_0}}\&\right]\left[c_1 + \frac{\sqrt{2}t}{\omega_0}\right], \tag{47}$$

It is also clear that $\sigma = -\frac{\omega_0}{\phi}$ oscillates as well, and the universe transits from inflationary regime to the matter dominated era.

Table 11. $f(\phi) = \phi^2$, (case-1): $\phi_b = 1.7$, $V_0 = -0.9 \times 10^{-13}\text{T}^{-2}$, $V_1 = 0.9 \times 10^{-13}\text{T}^{-2}$.

ω_0	$\lvert\eta\rvert$	$r = 16\epsilon$	ϕ_e	n_s	N
26	0.013956	0.0202	1.0377	0.9645	37
27	0.012941	0.0188	1.0364	0.9671	39
28	0.012033	0.0175	1.0351	0.9694	42
29	0.01122	0.0163	1.0339	0.9715	45
30	0.010482	0.0152	1.0328	0.9733	49
31	0.009817	0.0142	1.0317	0.9750	52
32	0.009213	0.0134	1.0308	0.9766	56
33	0.008663	0.0126	1.0298	0.9780	59
34	0.008161	0.0118	1.0289	0.9792	63
35	0.007701	0.0112	1.0282	0.9804	67

Table 12. $f(\phi) = \phi^2$, (case-1): $\phi_b = 1.7$. $V_0 = -0.9 \times 10^{-13}\mathrm{T}^{-2}$, $V_1 = 1.0 \times 10^{-13}\mathrm{T}^{-2}$.

ω_0	$\|\eta\|$	$r = 16\epsilon$	ϕ_e	n_s	N
24	0.014543	0.0187	1.0127	0.9639	37
25	0.013403	0.0173	1.0112	0.9667	40
26	0.012392	0.0160	1.0098	0.9692	43
27	0.011491	0.0148	1.0085	0.9714	46
28	0.010685	0.0138	1.0073	0.9735	50
29	0.00996	0.0128	1.0062	0.9753	54
30	0.00931	0.0120	1.0051	0.9769	57
31	0.008717	0.0112	1.0041	0.9784	61
32	0.008180	0.0105	1.0032	0.9797	65
33	0.007692	0.0099	1.0023	0.9809	69

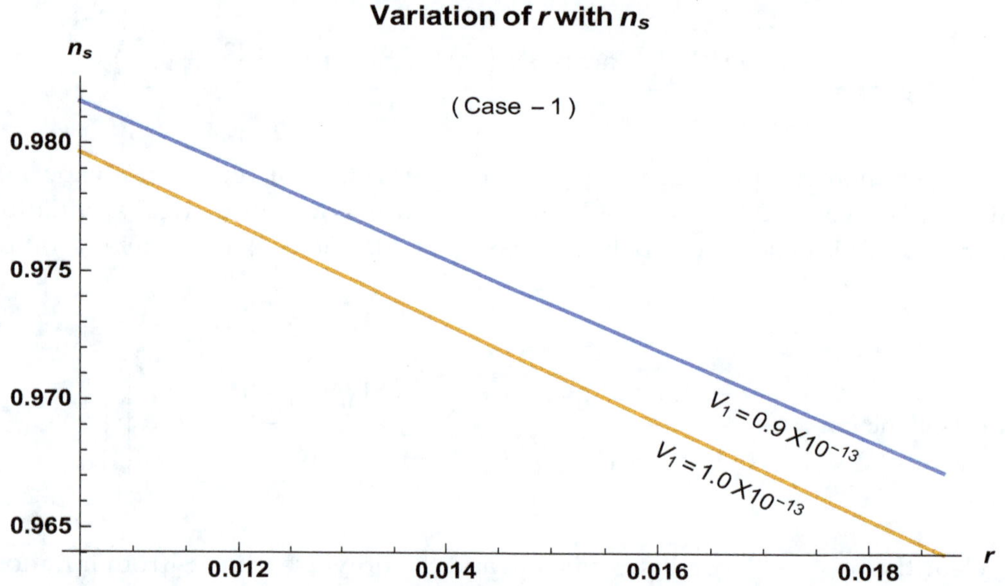

Figure 11. $f(\phi) = \phi^2$. Blue and yellow ochre colours represent Tables 11 and 12 respectively.

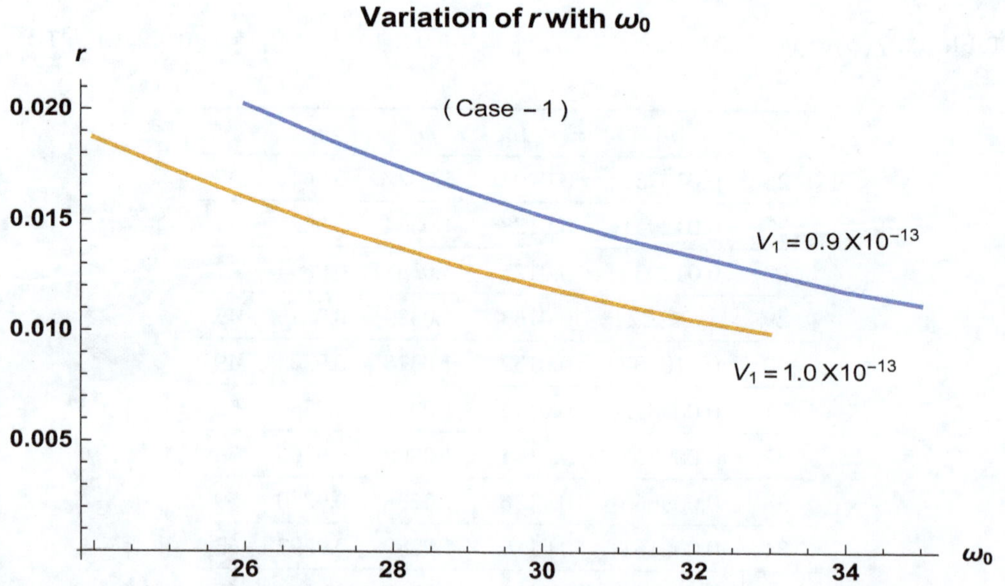

Figure 12. $f(\phi) = \phi^2$. Blue and yellow ochre colours represent Tables 11 and 12 respectively.

Case-2: Here, for $f(\phi) = \phi^2$, we consider the potential in the form, $V = V_1\phi^4 + V_0\phi^2$, i.e., instead of a constant additive term, we consider $V_0\phi^2$ in addition. This case was earlier studied in [37]. However, as already mentioned, over the years, Planck's data put up tighter constraints on inflationary parameters, and so it is quite reasonable to check if this form of potential passes the said constraints [40,41]. One can now find the expression for the Brans–Dicke parameter $\omega(\phi)$, the potential V_E in the Einstein frame, $\frac{d\sigma}{d\phi}$, the slow-roll parameters ϵ, η and the number of e-folding N, in view of the Equations (24), respectively as,

$$\begin{aligned} \omega(\phi) &= \frac{\omega_0^2 - 12\phi^2}{2\phi}, \quad \frac{d\sigma}{d\phi} = \frac{\omega_0}{\phi^2}; \qquad V_E = V_1 + V_0\phi^{-2}, \qquad \epsilon = \frac{4V_0^2\phi^2}{\omega_0^2(V_1\phi^2 + V_0)^2}, \\ \eta &= \frac{4V_0\phi^2}{\omega_0^2(V_1\phi^2 + V_0)}, \qquad N = \frac{\omega_0^2}{4\sqrt{2}V_0}\left[V_1 \ln(\phi_b - \phi_e) - \frac{V_0}{2}\left(\frac{1}{\phi_b^2} - \frac{1}{\phi_e^2}\right)\right]. \end{aligned} \tag{48}$$

Note that the Einstein frame potential now takes the same form as in case-1 for $n = 1$, and a flat section of the potential is still realizable at large value of the scalar field ϕ. We present two Tables 13 and 14, as before for different values of $V_1 > 0$. The scalar to tensor ratio $r \leq 0.0636$ and the spectral index $0.9715 \leq n_s \leq 0.9831$ lie very much within the Planck's data, while the number of e-folding N is again sufficient to alleviate the horizon and flatness problems. The Figures 13 and 14 represent n_s versus r and r versus ω_0 respectively. In view of the plots, the data for Table 14, here appears to be even better.

To check if the scalar field executes oscillatory behaviour at the end of inflation, we note that here $V_E = V_1 + \frac{V_0}{\phi^2}$. So in view of Equation (19) one obtains,

$$\frac{3H^2}{V_1} = \frac{\dot{\sigma}^2}{2V_1} + \left(1 + \frac{V_0}{V_1\phi^2}\right) \tag{49}$$

As, H falls below V_1, and $H \ll V_1$, the above equation can be approximated to, $\dot{\sigma}^2 = 2i^2(V_1 + \frac{V_0}{\phi^2})$, yielding, $\dot{\phi} = i\frac{\phi}{\omega_0}\sqrt{2(V_0 + V_1\phi^2)}$, where, $\dot{\sigma} = \omega_0\frac{\dot{\phi}}{\phi^2}$. Thus, we obtain,

$$\frac{\phi}{\left(\sqrt{V_0} + \sqrt{(V_0 + V_1\phi^2)}\right)} = \sqrt{V_0}e^{i\frac{\sqrt{2V_0}}{\omega_0}t}. \tag{50}$$

It is also possible to solve for ϕ and express it in the following form,

$$\phi(t) = \frac{\sqrt{-\beta(t) + \sqrt{\beta(t)^2 - 4\alpha(t)\gamma(t)}}}{\sqrt{2\alpha(t)}}, \tag{51}$$

$$\text{where,} \quad \alpha(t) = 1 - V_0V_1e^{i2\frac{\sqrt{2V_0}}{\omega_0}t}, \quad \beta(t) = -4V_0^3V_1e^{i4\frac{\sqrt{2V_0}}{\omega_0}t}, \quad \gamma(t) = -4V_0^4e^{i4\frac{\sqrt{2V_0}}{\omega_0}t}.$$

It is now quite apparent that the scalar field $\phi(t)$ executes oscillatory behaviour and therefore graceful exit from inflation is realizable. Since in view of (48) $\sigma = -\frac{\omega_0}{\phi}$, therefore σ also executes oscillatory behaviour.

Table 13. $f(\phi) = \phi^2$, (case-2): $\phi_b = 1.26$. $V_0 = -1.0 \times 10^{-20}\mathrm{T}^{-2}$; $V_1 = 1.0 \times 10^{-20}\mathrm{T}^{-2}$.

ω_0	$\lvert\eta\rvert$	$r = 16\epsilon$	ϕ_e	n_s	N
68	0.002337	0.0636	1.0148	0.9715	37
70	0.002206	0.0601	1.0144	0.9731	40
72	0.002085	0.0568	1.0140	0.9745	42
74	0.001974	0.0537	1.0136	0.9759	44
76	0.001871	0.0509	1.0132	0.9772	47
78	0.001776	0.0484	1.0129	0.9783	49
80	0.001689	0.0460	1.0126	0.9793	52
82	0.001607	0.0438	1.0122	0.9804	55
84	0.001532	0.0417	1.0119	0.9813	57
86	0.001461	0.0398	1.0117	0.9821	60
88	0.001396	0.0380	1.0114	0.9830	63

Table 14. $f(\phi) = \phi^2$, (case-2.): $\phi_b = 1.26$. $V_0 = -1.0 \times 10^{-20}\mathrm{T}^{-2}$; $V_1 = 1.1 \times 10^{-20}\mathrm{T}^{-2}$.

ω_0	$\lvert\eta\rvert$	$r = 16\epsilon$	ϕ_e	n_s	N
60	0.002363	0.0507	1.0656	0.9762	38
61	0.002287	0.0490	1.0653	0.9770	39
62	0.002213	0.0475	1.0650	0.9778	40
63	0.002144	0.0460	1.0648	0.9785	41
64	0.002078	0.0445	1.0646	0.9792	43
65	0.002014	0.0432	1.0643	0.9798	44
66	0.001953	0.0419	1.0641	0.9804	46
67	0.001895	0.0406	1.0638	0.9810	47
68	0.001840	0.0394	1.0636	0.9815	49
69	0.001787	0.0383	1.0634	0.9821	50
70	0.001737	0.0372	1.0632	0.9826	51
71	0.001688	0.0362	1.0630	0.9831	53

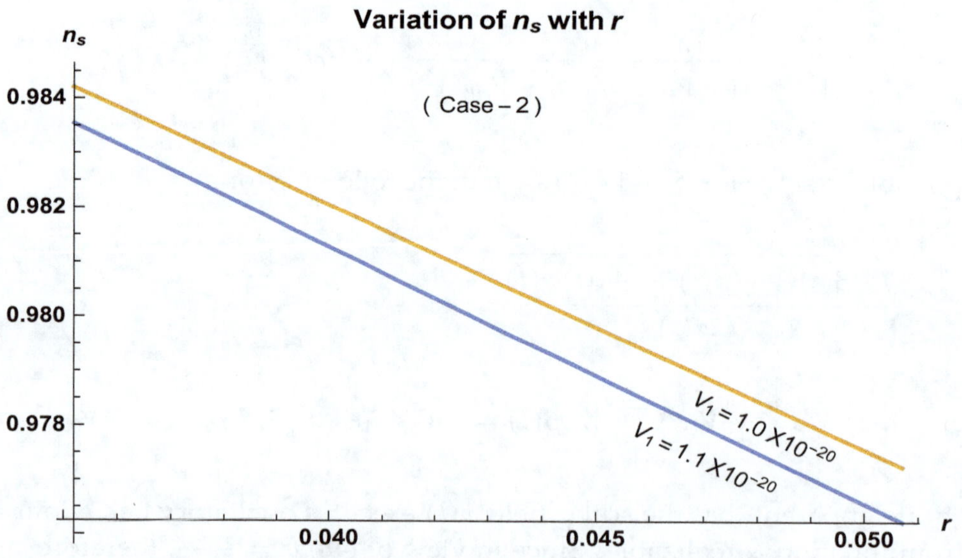

Figure 13. $f(\phi) = \phi^2$. Yellow ochre and blue colours represent Tables 13 and 14 respectively.

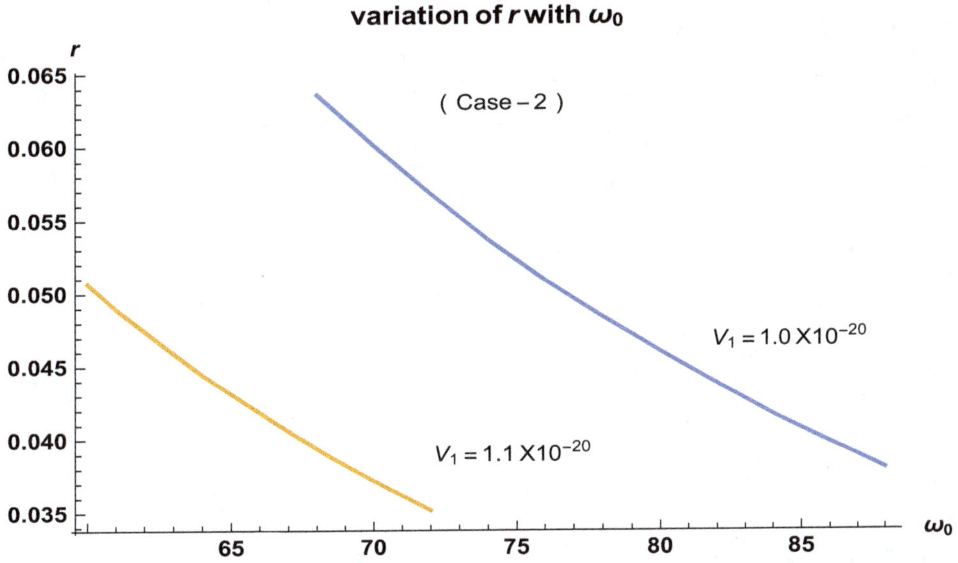

Figure 14. $f(\phi) = \phi^2$. Blue and yellow ochre colours represent Tables 13 and 14 respectively.

Case-3: We consider yet another case for $n = 2$, i.e., taking $f(\phi) = \phi^2$, with the potential $V(\phi)$ being represented by two additional terms apart from ϕ^4, as it should be, to make it a perfect square: $V(\phi) = (\sqrt{V_1}\phi^2 - \sqrt{V_0}\phi)^2 = V_1\phi^4 + V_0\phi^2 - 2\sqrt{V_0V_1}\phi^3$. As before, one can now find the expression for the Brans–Dicke parameter $\omega(\phi)$, the potential V_E in the Einstein frame, $\frac{d\sigma}{d\phi}$, the slow-roll parameters ϵ, η and the number of e-foldings N, in view of the Equations (14), (18) and (20) respectively as,

$$\omega(\phi) = \frac{\omega_0^2 - 12\phi^2}{2\phi}, \quad \frac{d\sigma}{d\phi} = \frac{\omega_0}{\phi^2}; \qquad V_E = V_1 - 2\frac{\sqrt{V_0V_1}}{\phi} + \frac{V_0}{\phi^2}, \qquad \epsilon = \frac{4V_0^2\phi^2}{\omega_0^2(\sqrt{V_0V_1}\phi - V_0)^2},$$
$$\eta = \frac{4V_0^2\phi^2}{\omega_0^2(\sqrt{V_0V_1}\phi - V_0)^2}, \qquad N = \int_{\phi_e}^{\phi_b} \frac{\omega_0^2}{4\sqrt{2}V_0}\frac{(\sqrt{V_0V_1}\phi - V_0)}{\phi^3}d\phi. \tag{52}$$

One can clearly see that the flat section of the potential is still attainable for large value of the scalar field ϕ. Tables 15 and 16 depict that the scalar to tensor ratio $r < 0.1$), is quite reasonable, while the spectral index $0.9752 \leq n_s \leq 0.981$ fits perfectly with Planck's data [40,41]. Figures 15 and 16 represent n_s versus r and r versus ω_0 plots respectively. Interestingly, two r versus n_s plots (Figure 15) corresponding to the two sets of data (Tables 15 and 16) merge almost perfectly.

The scalar field executes oscillatory behaviour here too, as we demonstrate below. Here, $V_E = V_1 - 2\frac{\sqrt{V_0V_1}}{\phi} + \frac{V_0}{\phi^2}$, and so from (19) we find,

$$\frac{3H^2}{V_1} = \frac{\dot{\sigma}^2}{2V_1} + \left(1 - \frac{2}{\phi}\sqrt{\frac{V_0}{V_1}}\right) + \frac{V_0}{V_1\phi^2}. \tag{53}$$

As H falls below V_1, and $H \ll V_1$, the above equation can be approximated as, $\dot{\sigma}^2 = 2i^2(\frac{\sqrt{V_0}}{\phi} - \sqrt{V_1})^2$, which yields $\dot{\phi} = i\frac{\phi}{\omega_0}\sqrt{2}(\sqrt{V_0} - \sqrt{V_1}\phi)$ where, $\dot{\sigma} = \omega_0\frac{\dot{\phi}}{\phi^2}$. Therefore finally we obtain,

$$\phi(t) = \frac{\sqrt{V_0}e^{i\frac{\sqrt{2V_0}}{\omega_0}t}}{1 + \sqrt{V_1}e^{i\frac{\sqrt{2V_0}}{\omega_0}t}}. \tag{54}$$

Clearly, ϕ executes oscillatory behaviour, and graceful exit from inflation may be realized here too. Here again since in view of (52) $\sigma = -\frac{\omega_0}{\phi}$, therefore σ executes oscillatory behaviour, as well.

Table 15. $f(\phi) = \phi^2$, (case-3): $\phi_b = 1.3$. $V_1 = 0.9 \times 10^{-20}\mathrm{T}^{-2}$; $V_0 = 0.9 \times 10^{-20}\mathrm{T}^{-2}$.

ω_0	η	$r = 16\epsilon$	ϕ_e	n_s	N
110	0.006208	0.0993	1.0185	0.9752	57
112	0.005988	0.0958	1.0182	0.9760	59
114	0.005780	0.0925	1.0178	0.9769	61
116	0.005582	0.0893	1.0175	0.9777	63
118	0.005394	0.0863	1.0172	0.9784	65
120	0.005216	0.0835	1.0170	0.9791	67
122	0.005046	0.0807	1.0167	0.9798	70
124	0.004885	0.0782	1.0164	0.9804	72
126	0.004731	0.0757	1.0161	0.9810	74

Table 16. $f(\phi) = \phi^2$, (case-3): $\phi_b = 1.3$. $V_1 = 1.0 \times 10^{-20}\mathrm{T}^{-2}$; $V_0 = 0.9 \times 10^{-20}\mathrm{T}^{-2}$.

ω_0	η	$r = 16\epsilon$	ϕ_e	n_s	N
142	0.006160	0.0986	1.0386	0.9754	57
144	0.005990	0.0958	1.0388	0.9760	59
146	0.005827	0.0932	1.0391	0.9767	60
148	0.005671	0.0907	1.0393	0.9773	62
150	0.005521	0.0883	1.0395	0.9779	64
152	0.005376	0.0860	1.0397	0.9785	65
154	0.005237	0.0838	1.0399	0.9791	67
156	0.005104	0.0817	1.0400	0.9796	69
158	0.004976	0.0796	1.0402	0.9801	71
160	0.004852	0.0776	1.0404	0.9806	72

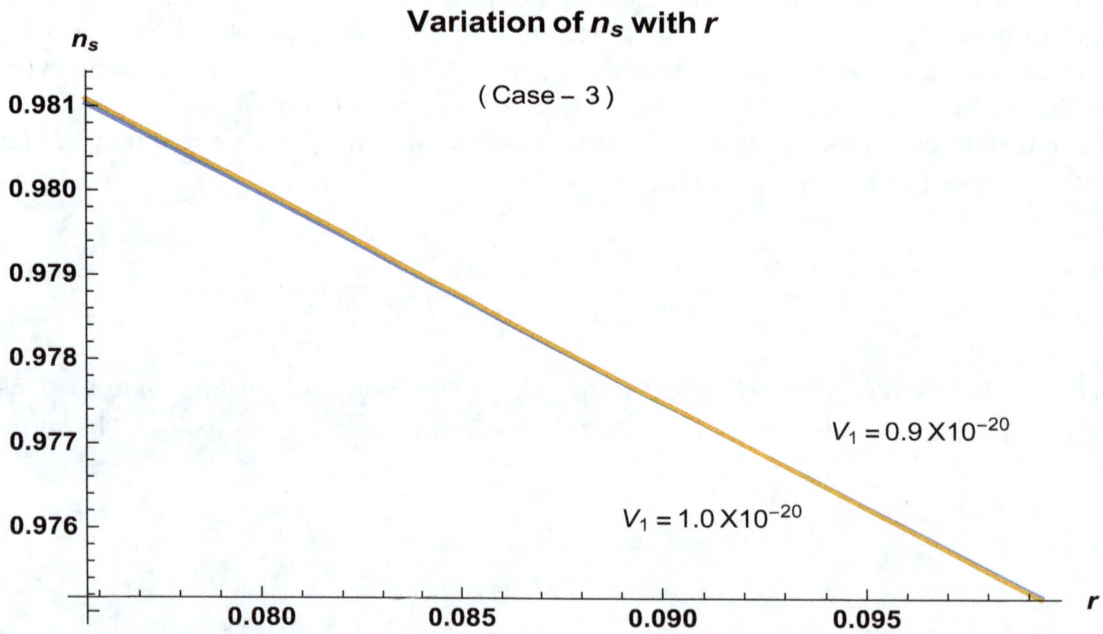

Figure 15. $f(\phi) = \phi^2$. Blue and yellow ochre colours represent Tables 15 and 16 respectively.

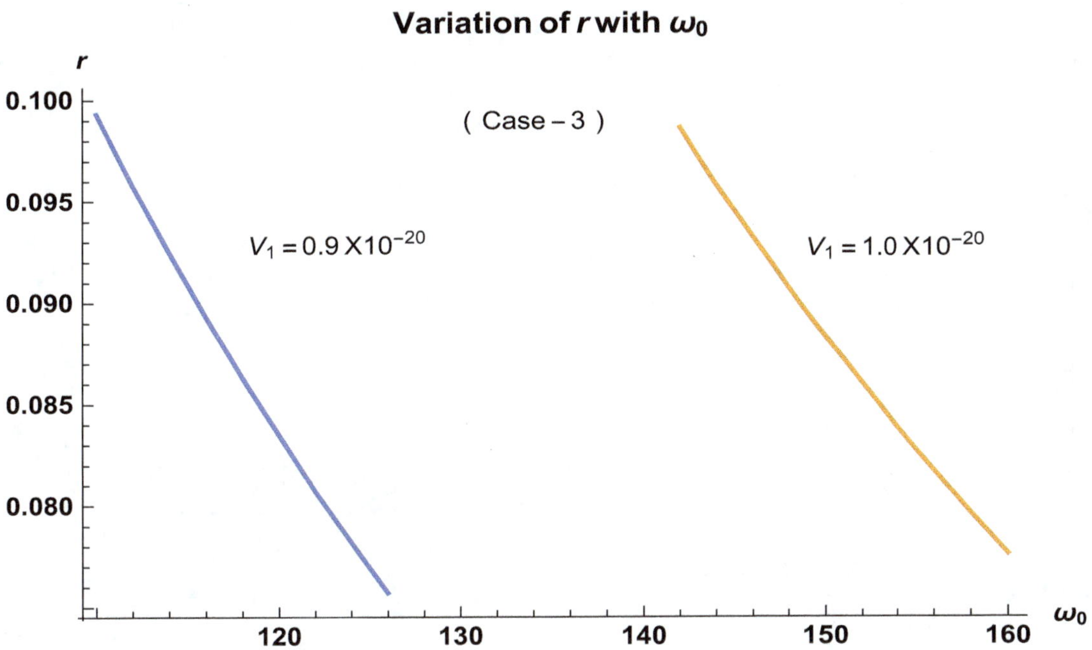

Figure 16. $f(\phi) = \phi^2$. Blue and yellow ochre colours represent Tables 15 and 16 respectively.

3.2. Exponential Potential

Finally, we consider an exponential form of the potential with: $f(\phi) = e^{\frac{\lambda\phi}{2}}$, with $V(\phi) = V_1 e^{\lambda\phi} + V_0$. It is possible to find the expression for the Brans–Dicke parameter $\omega(\phi)$, the potential V_E in the Einstein frame, $\frac{d\sigma}{d\phi}$, the slow-roll parameters ϵ, η and the number of e-folding N, in view of the Equations (14), (18) and (20) respectively as,

$$\omega(\phi) = \frac{(\omega_0^2 - \frac{3}{4}\lambda^2 e^{\lambda\phi})\phi}{2e^{\frac{\lambda\phi}{2}}}; \quad \frac{d\sigma}{d\phi} = \frac{\omega_0}{e^{\frac{\lambda\phi}{2}}}; \quad V_E = V_1 + V_0 e^{-\lambda\phi}, \quad \epsilon = \frac{\lambda^2 V_0^2 e^{\lambda\phi}}{\omega_0^2 (V_1 e^{\lambda\phi} + V_0)^2},$$
$$\eta = \frac{\lambda^2 V_0 e^{\lambda\phi}}{\omega_0^2 (V_1 e^{\lambda\phi} + V_0)}, \quad N = \int_{\phi_e}^{\phi_b} \frac{\omega_0^2 (V_0 + V_1 e^{\lambda\phi})}{2\sqrt{2} V_0 \lambda e^{\lambda\phi}} d\phi. \tag{55}$$

We present our results in the following Table 17, under the only choice of the parameter $\lambda = -1$. The data shows good agreement $r < 0.06$, and $0.9612 \leq n_s \leq 0.9836$ with Planck's data [40,41]. Figure 17, represents a plot for n_s versus r.

Table 17. $|\phi_b| = 1.2$, $V_0 = -1.0 \times 10^{-20}\mathrm{T}^{-2}$, $V_1 = 1.0 \times 10^{-20}\mathrm{T}^{-2}$.

ω_0	η	$r = 16\epsilon$	$\lvert\phi_e\rvert$	n_s	N
13	0.008468	0.0584	0.07690	0.9612	30
14	0.007301	0.0503	0.07141	0.9665	34
15	0.00636	0.0439	0.06665	0.9708	40
16	0.005589	0.0385	0.06249	0.9745	45
17	0.004952	0.0341	0.05882	0.9773	51
18	0.004417	0.0305	0.05555	0.9797	57
19	0.003964	0.02734	0.05263	0.9818	64
20	0.003578	0.02467	0.04999	0.9836	71

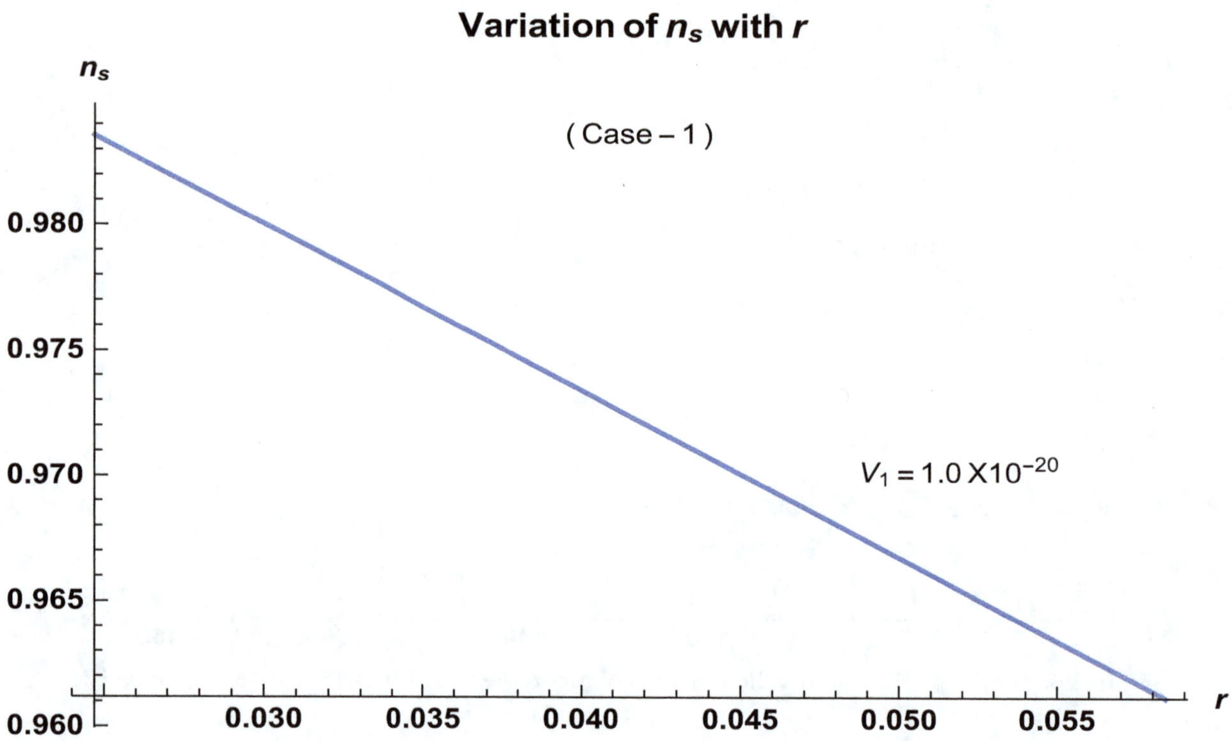

Figure 17. $f(\phi) = e^{-\frac{\phi}{2}}$.

The scalar field executes oscillatory behaviour here too, as we demonstrate below. Here $V_E = V_1 + V_0 e^{\phi}$, and so in view of Equation (19), one can calculate,

$$\frac{3H^2}{V_1} = \frac{\dot{\sigma}^2}{2V_1} + \left(1 + \frac{V_0}{V_1}e^{\phi}\right) \tag{56}$$

Again as H falls below V_1, and $H \ll V_1$, the above equation can be approximated to, $\dot{\sigma}^2 = 2i^2(V_1 + V_0 e^{\phi})$, which yields $\dot{\phi} = i\frac{e^{-\frac{\phi}{2}}}{\omega_0}\sqrt{2(V_1 + V_0 e^{\phi})}$, where,$\dot{\sigma} = \omega_0 \dot{\phi} e^{\frac{\phi}{2}}$. Finally, therefore

$$\phi = \ln\left[\frac{1}{4V_0^2}\left(e^{i\frac{\sqrt{2V_0}}{\omega_0}t} - 2V_0V_1 + V_0^2V_1^2 e^{-i\frac{\sqrt{2V_0}}{\omega_0}t}\right)\right]. \tag{57}$$

The oscillatory behaviour of the scalar field here again assures graceful exit from inflationary regime. In view of (55), $\sigma = 2\omega_0 e^{\frac{\phi}{2}}$, for $\lambda = -1$, which we have considered, hence σ also executes oscillatory behaviour.

4. Concluding Remarks

Scalar-tensor theories of gravity are generalizations of the Brans–Dicke theory, in which the coupling parameter is a function of the scalar field, i.e., $\omega_{BD} = \omega(\phi)$, and therefore is a variable. The requirement for such generalization of Brans–Dicke theory generated from the tight constraints on ω_{BD} established by the solar system experiments [58]. There exists various classification of scalar-tensor theory of gravity [59]. In the present manuscript we have considered standard non-minimal coupling, where the coupling parameter $f(\phi)$ is arbitrary. It has been noticed earlier that such a theory has an in-built symmetry being associated with a conserved current for trace-free fields, such as vacuum and radiation dominated eras for barotropic fluids. In view of such a symmetry, it is possible to fix all the variables of the theory, including the potential function, fixing the form of one of those. In this manuscript, we have chosen different forms of the coupling parameter $f(\phi)$, which fixed $\omega(\phi)$

and $V(\phi)$, to study the cosmological evolution of the very early universe in the context of inflation. Inflation is a quantum mechanical phenomenon, and has occurred around Planck's era. However, it has been argued that since the radiative corrections to the potential are negligible, hence the inflationary parameters can be computed using the classical Lagrangian [60]. This argument leads in general, to calculate inflationary parameters in view of the classical Lagrangian, which we have done in the present manuscript. The so called unification programmes, which essentially claim to unify early inflationary regime with late-time cosmic acceleration have no credentials, since none of the models passes through a well behaved radiation and early matter dominated era. However, a history of cosmic evolution starting from inflationary regime, followed by a Friedmann-like radiation ($a \propto \sqrt{t}$) and early matter dominated eras ($a \propto t^{\frac{2}{3}}$), that finally ends up to a late-time accelerated universe ($z = 0.75$), has already been explored in view of the present model [37]. In this connection, the present model makes a reasonably viable attempt to unify early inflation with late-time cosmic acceleration. Nevertheless earlier, only a single form of the coupling parameter $f(\phi)$ together with a particular form of $V(\phi)$ had been treated. Here, we have extended our work considering combinations of a host of power potentials ($V(\phi) = V_0 + V_1\phi^2, V_0 + V_1\phi^3, V_0 + V_1\phi^4, V_0\phi + V_1\phi^3, V_0\phi^2 + V_1\phi^3, V_0\phi^2 + V_1\phi^4, V_0\phi^2 + V_1\phi^4 - 2\sqrt{V_0V_1}\phi^3$), together with an exponential potential ($V(\phi) = V_0 + V_1e^{\lambda\phi}$). We find that all the quadratic, cubic, quartic and exponential potentials pass the tighter constraints on inflationary parameters released by latest Planck's data [40,41] quite comfortably. Further, all these potentials admit graceful exit from inflation, except one case of cubic potential associated with a square potential, i.e., ($V(\phi) = V_0\phi^2 + V_1\phi^3$), for which unfortunately the scalar field does not show up oscillatory behaviour at the end of inflation.

For the purpose of the present analysis, we have translated the non-minimally coupled Jordan's frame action to the Einstein frame, under conformal transformation. It is therefore worth to make certain comments in this regard. There is an age old debate regarding physical equivalence between the two: Jordan's and Einstein's frames, which are related under conformal transformation. Now, indeed if the two formulations are not equivalent, the problem arises in selecting the physically preferred frame. It emerges from the work of several authors, in different contexts on Kaluza–Klein and Brans–Dicke theories, that the formulations of a scalar-tensor theory in the two conformal frames are physically inequivalent [61–68]. Furthermore, the Jordan frame formulation of a scalar-tensor theory is not viable because the energy density of the gravitational scalar field present in the theory is not bounded from below, which amounts to the violation of the weak energy condition [69]). The system therefore is unstable and decays toward a lower and lower energy state ad infinitum [66–68]. Although, a quantum system may have states with negative energy density [69–71], such feature is not acceptable for a viable classical theory of gravity. In fact, a classical theory must have a ground state that is stable against small perturbations. The violation of the weak energy condition by scalar-tensor theories formulated in the Jordan conformal frame makes them unviable descriptions of classical gravity, while the Einstein frame formulation of scalar-tensor theories is free from such problem. However, in the Einstein frame also there is a violation of the equivalence principle due to the anomalous coupling of the scalar field to ordinary matter. Nevertheless, this violation is small and compatible with the available tests of the equivalence principle [72]. Further, Einstein's frame is indeed regarded as an important low energy manifestation of compactified theories [72–77]. However, in search of Noether symmetries of $F(R)$ theory of gravity, the two frames have been found to be physically equivalent [78]. So although the debate persists, but somehow it is quite relevant to consider Einstein's frame to be the physical frame. Therefore, in view of the above discussions, it is justified to study the physics associated with non-minimally coupled scalar-tensor theory of gravity, after translating it to the Einstein frame, as we have done in the present article.

Author Contributions: D.S., S.S. and A.K.S. have contributed to the extent they could. All authors have read and agreed to the published version of the manuscript.

References

1. Partridge, R.B. *3K: The Cosmic Microwave Background Radiation*; Cambridge Univ. Press: Cambridge, UK, 1995.
2. Turner, M.S.; Kolb, E.W. *The Early Universe*; Addison-Wesley Publishing Company: Boston, MA, USA, 1990.
3. Peebles, P.J.E. *The Large Scale Structure of the Universe*; Princeton Univ. Press: Princeton, NJ, USA, 1980.
4. Peebles, P.J.E. *Principles of Physical Cosmology*; Princeton Univ. Press: Princeton, NJ, USA, 1993.
5. Baumann, D. The Physics of Inflation—A Course for Graduate Students in Particle Physics and Cosmology. *arXiv* **2015**, arXiv:0907.5424.
6. Sanyal, A.K. Enlightening the dark universe. *Indian J. Theor. Phys.* **2014**, *62*, 211–262.
7. Stelle, K.S. Renormalization of higher-derivative quantum gravity. *Phys. Rev.* **1977**, *953*, D16.
8. Sen, A. Recent Developments in Superstring Theory. *Nucl. Phys. Proc. Suppl.* **2001**, *94*, 35.
9. Mukhi, S. String theory: A perspective over the last 25 years. *Class. Quant. Grav.* **2011**, *28*, 153001.
10. Lahanas, A.B.; Nanopoulos, D.V. The road to no-scale supergravity. *Phys. Rep.* **1987**, *145*, 1.
11. Ferrara1, S.; Sagnotti, A. Supergravity at 40: Reflections and Perspectives. *J. Phys. Conf. Ser.* **2016**, 873.
12. Kenath, A.; Gudennavar, S.B.; Sivaram, C. Dark matter, dark energy, and alternate models: A review. *Adv. Space Res.* **2017**, *60*, 166.
13. Starobinsky, A.A. A new type of isotropic cosmological models without singularity. *Phys. Lett.* **1980**, *99*, B91.
14. Guth, A.H. Inflationary universe: A possible solution to the horizon and flatness problems. *Phys. Rev. D* **1981**, *23*, 347.
15. Olive, K.A. Inflation. *Phys. Rept.* **1990**, *190*, 307.
16. Lyth, D.H.; Riotto, A. Particle physics models of inflation and the cosmological density perturbation. *Phys. Rep.* **1999**, *314*, 1–146.
17. Liddle, A.R.; Lyth, D.H. *Cosmological Inflation and Large-Scale Structure*; Cambridge University Press: Cambridge, UK, 2000.
18. Baumann, D. TASI lectures on inflation. *arXiv* **2009**, arXiv:0907.5424.
19. Martin, J.; Ringeval, C.; Vennin, V. Encyclopaedia inflationaris. *Phys. Dark Univ.* **2014**, *5–6*, 75–235.
20. Martin, J.; Ringeval, C.; Trotta, R.; Vennin, V. The best inflationary models after Planck. *JCAP* **2014**, *1403*, 39.
21. Martin, J. The observational status of cosmic inflation after Planck. In *The Cosmic Microwave Background*; Fabris, J., Piattella, O., Rodrigues, D., Velten, H., Zimdahl, W., Eds.; Springer: Cham, Switzerland, 2015.
22. Linde, A.D. A new inflationary universe scenario: A possible solution of the horizon, flatness, homogeneity, isotropy and primordial monopole problems. *Phys. Lett. B* **1982**, *108*, 389.
23. Albrecht, A.; Steinhardt, P.J. Cosmology for grand unified theories with radiatively induced symmetry breaking. *Phys. Rev. Lett.* **1982**, *48*, 1220.
24. Press, W.H. Spontaneous production of the Zel'dovich spectrum of cosmological fluctuations. *Phys. Scr.* **1980**, *21*, 702.
25. Hawking, S.W. The development of irregularities in a single bubble inflationary universe. *Phys. Lett. B* **1982**, *115*, 295.
26. Starobinsky, A.A. Dynamics of phase transition in the new inflationary universe scenario and generation of perturbations. *Phys. Lett. B* **1982**, *117*, 175.
27. Guth, A.H.; Pi, S.Y. Fluctuations in the new inflationary universe. *Phys. Rev. Lett.* **1982**, *49*, 1110.
28. Bardeen, J.M.; Steinhardt, P.J.; Turner, M.S. Spontaneous creation of almost scale-free density perturbations in an inflationary universe. *Phys. Rev. D* **1983**, *28*, 679.
29. Mukhanov, V.F.; Chibisov, G.V. Quantum fluctuations and a nonsingular universe. *JETP Lett.* **1981**, *33*, 532.
30. Whitt, B. Fourth-order gravity as general relativity plus matter. *Phys. Lett. B* **1984**, *145*, 176.
31. Wands, D. Extended gravity theories and the Einstein–Hilbert action, Class. *Quant. Grav.* **1994**, *11*, 269.
32. Horndeski, G.W. Second-order Scalar-tensor field equations in a four-dimentional Space. *Int. J. Theor. Phys.* **1974**, *10*, 363.
33. Sanyal, A.K. Scalar–tensor theory of gravity carrying a conserved current. *Phys. Lett. B* **2005**, *624*, 81.
34. Sanyal, A.K. Study of symmetry in F(R) theory of gravity. *Mod. Phys. Lett. A* **2010**, *25*, 2667.
35. Sk, N.; Sanyal, A.K. Field independent cosmic evolution. *J. Astrophys.* **2013**, *2013*, 590171.
36. Sarkar, K.; Sk, N.; Debnath, S.; Sanyal, A.K. Viability of Noether symmetry of F(R) theory of gravity. *Int. J. Theor. Phys.* **2013**, *52*, 1194.
37. Tajahmad, B.; Sanyal, A.K. Unified cosmology with scalar–tensor theory of gravity. *Eur. Phys. J.* **2017**, *77*, 217.

38. Ade, P.A.; Aghanim, N.; Armitage-Caplan, C.; Arnaud, M.; Ashdown, M.; Atrio-Barandela, F.; Aumont, J.; Baccigalupi, C.; Banday, A.J.; Barreiro, R.B.; et al. Planck 2013 results. XVI. Cosmological parameters. *Astron. Astrophys.* **2014**, *571*, A16.
39. Ade, P.A.; Aghanim, N.; Arnaud, M.; Ashdown, M.; Aumont, J.; Baccigalupi, C.; Banday, A.J.; Barreiro, R.B.; Bartlett, J.G.; Bartolo, N.; et al. Planck 2015 resultss. XII. Cosmological parameters. *Astron. Astrophys.* **2016**, *594*, A13.
40. Aghanim, N.; Akrami, Y.; Ashdown, M.; Aumont, J.; Baccigalupi, C.; Ballardini, M.; Banday, A.J.; Barreiro, R.B.; Bartolo, N.; Basak, S.; et al. Planck 2018 Results. VI. Cosmological Parameters, (Planck Collaboration). *arXiv* **2018**, arXiv:1807.06209.
41. Akrami, Y.; Arroja, F.; Ashdown, M.; Aumont, J.; Baccigalupi, C.; Ballardini, M.; Banday, A.J.; Barreiro, R.B.; Bartolo, N.; Basak, S.; et al. Planck 2018 results. X. Constraints on inflation, (Planck Collaboration). *arXiv* **2018**, arXiv:1807.06211.
42. Kobayashi, T.; Yamaguchi, M.; Yokoyama, J. Generalised G-inflation: Inflation with most general second order field equation. *Prog. Theor. Phys.* **2011**, *126*, 511.
43. Kamada, K.; Kobayashi, T.; Yamaguchi, M.; Yokoyama, J. Higgs G-inflation. *Phys. Rev. D* **2011**, *83*, 083515.
44. Myrzakulov, R.; Sebastiani, L. Scalar tensor Horndeski Models: Simple cosmological applications. *Astrophys. Space Sci.* **2016**, *361*, 62.
45. Myrzakulov, R.; Sebastiani, L. K-Essence Non-Minimally Coupled with Gauss-Bonnet Invariant for Inflation. *Symmetry* **2016**, *8*, 57.
46. Gleyzes, J.; Langlois, D.; Piazza, F.; Vernizzi, F. Healthy theories beyond Horndeski. *Phys. Rev. Lett.* **2015**, *114*, 211101.
47. Gleyzes, J.; Langlois, D.; Piazza, F.; Vernizzi, F. Exploring gravitational theories beyond Horndeski. *JCAP* **2015**, *2*, 018.
48. Gao, X. Unifying framework for Scalar-tensor theories of gravity. *Phys. Rev. D* **2014**, *90*, 081501.
49. Zumalacárregui, M.; García-Bellido, J. Trasforming gravity: From derivative couplings to matter to second-order scalar-tensor theories beyond the Horndeski Lagrangian. *Phys. Rev. D* **2014**, *89*, 064046.
50. Crisostomi, M.; Hull, M.; Koyama, K.; Tasinato, G. Horndeski: Beyond, or not beyond? *JCAP* **2016**, *3*, 38.
51. Simone, A.D.; Hertzberg, M.P.; Wilczek, F. Running inflation in the standard model. *Phys. Lett. B* **2009**, *678*, 1.
52. Bhattcharyya, S.; Das, K.; Dutta, K. Attractor models in Scalar-Tensor Theories of Inflation. *Int. J. Mod. Phys.* **2018**, *27*, 1850079.
53. Park, S.C. Inflation in the nonminimal theory with 'K(phi)R' term. *AIP Conf. Proc.* **2009**, *1078*, 524.
54. Park, S.C.; Yamaguchi, S. Inflation by non-minimal coupling. *arXiv* **2008**, arXiv:0801.1722.
55. Kleinert, H.; Mustapic, I. Decay rates of metastable states in cubic potential by variational perturbation theory. *Int. J. Mod. Phys. A* **1996**, *11*, 4383.
56. Falconi, M.; Lacomba, E.A.; Vidal, C. The flow of classical mechanical cubic potential systems. *Discont. Cont. Dyn. Syst.* **2004**, *11*, 827.
57. Wartak, M.S.; Krzeminski, S. On tunnelling in the cubic potential. *J. Phys. A: Math. Gen.* **1989**, *22*, L1005.
58. Will, C.M. The confrontation between general relativity and experiment. *Living Rev. Rel.* **2014**, *17*, 4.
59. Quiros, I.; De Arcia, R.; García-Salcedo, R.; Gonzalez, T.; Horta-Rangel, F.A. An issue with the classification of the scalar-tensor theories of gravity. *Int. J. Mod. Phys. D* **2020**, *29*, 7.
60. Bezrukov, F.L.; Shaposhnikov, M. The standard model Higgs boson as the inflaton. *Phys. Lett. B* **2008**, *659*, 703.
61. Bombelli, L.; Koul, R.K.; Kunstatter, G.; Lee, J.; Sorkin, R.D. On energy in 5-dimensional gravity and the mass of the Kaluza-Klein monopole. *Nucl. Phys. B* **1987**, *289*, 735.
62. Sokolowski, L.M.; Golda, Z.A. Instability of Kaluza-Klein cosmology. *Phys. Lett. B* **1987**, *195*, 349.
63. Sokolowski, L.M. Uniqueness of the metric line element in dimensionally reduced theories. *Class. Quant. Grav.* **1989**, *6*, 59.
64. Cho, Y.M. Unified cosmology. *Phys. Rev. D* **1990**, *41*, 2462.
65. Cho, Y.M. Violation of equivalence principle in Brans–Dicke theory. *Class. Quant. Grav.* **1997**, *14*, 2963.
66. Magnano, G.; Sokolowski, L.M. On Physical Equivalence between Nonlinear Gravity Theories. *Phys. Rev. D* **1994**, *50*, 5039 .
67. Faraoni, V.; Gunzig, E.; Nardone, P. Conformal transformations in classical gravitational theories and in cosmology. *Fund. Cosmic Phys.* **1998**, *20*, 121.

68. Faraoni, V.; Gunzig, E.; Nardone, P. Einstein Frame or Jordan Frame? *Int. J. Theory Phys.* **1999**, *38*, 217.
69. Witten, E. Instability of the Kaluza-Klein vacuum. *Nucl. Phys. B* **1982**, *195*, 481.
70. Ford, L.H.; Roman, T.A. Cosmic flashing in four dimensions. *Phys. Rev. D* **1992**, *46*, 1328.
71. Ford, L.H.; Roman, T.A. Averaged energy conditions and quantum inequalities. *Phys. Rev. D* **1995**, *51*, 4277.
72. Cho, Y.M. Reinterpretation of Jordan-Brans–Dicke theory and Kaluza-Klein cosmology. *Phys. Rev. Lett.* **1992**, *68*, 3133.
73. Taylor, T.R.; Veneziano, G. Dilaton couplings at large distances. *Phys. Lett. B* **1988**, *213*, 450.
74. Cvetic, M. Low energy signals from moduli. *Phys. Lett. B* **1989**, *229*, 41.
75. Ellis, J.; Kalara, S.; Olive, K.A.; Wetterich, C. Density-dependent couplings and astrophysical bounds on light scalar particles. *Phys. Lett. B* **1989**, *228*, 264.
76. Damour, T.; Polyakov, A.M. The String Dilaton and a Least Coupling Principle. *Nucl. Phys. B* **1994**, *423*, 532.
77. Damour, T.; Polyakov, A.M. String theory and gravity. *Gen. Rel. Gravit.* **1994**, *26*, 1171.
78. Sk, N.; Sanyal, A.K. On the equivalence between different canonical forms of F(R) theory of gravity. *Int. J. Mod. Phys. D* **2018**, *27*, 1850085.

6

Deformed Shape Invariant Superpotentials in Quantum Mechanics and Expansions in Powers of $\hbar$

Christiane Quesne

Physique Nucléaire Théorique et Physique Mathématique, Université Libre de Bruxelles, Campus de la Plaine CP229, Boulevard du Triomphe, B-1050 Brussels, Belgium; cquesne@ulb.ac.be

Abstract: We show that the method developed by Gangopadhyaya, Mallow, and their coworkers to deal with (translational) shape invariant potentials in supersymmetric quantum mechanics and consisting in replacing the shape invariance condition, which is a difference-differential equation, which, by an infinite set of partial differential equations, can be generalized to deformed shape invariant potentials in deformed supersymmetric quantum mechanics. The extended method is illustrated by several examples, corresponding both to $\hbar$-independent superpotentials and to a superpotential explicitly depending on $\hbar$.

Keywords: quantum mechanics; supersymmetry; shape invariance; curved space; position-dependent mass

PACS: 03.65.Fd; 03.65.Ge

---

1. Introduction

Exactly solvable (ES) Schrödinger equations (SE) allow us to understand some physical phenomena and to test some approximation schemes. Supersymmetric quantum mechanics (SUSYQM) [1–4] is known to be a very powerful method for generating such ES models, especially whenever the corresponding potential is (translationally) shape invariant (SI) [5]. SUSYQM may be considered as a modern version of the old Darboux transformation [6] and of the factorization method used by Schrödinger [7–9] and by Infeld and Hull [10].

Some ten years ago, the list of (translational) SI potentials, whose bound-state wavefunctions can be expressed in terms of classical orthogonal polynomials (COP) [1] has been completed (see [11] and references quoted therein) by introducing [12–15] some rational extensions of these potentials, connected with the novel field of exceptional orthogonal polynomials (EOP) [16]. The latter are polynomial sets which are orthogonal and complete, but, in contrast with COP, admit a finite number of gaps in the sequence of their degrees.

ES models for some unconventional SE are also very interesting. These unconventional equations may be of three different kinds. They may occur whenever the standard commutation relations are replaced by deformed ones, associated with nonzero minimal uncertainties in position and/or momentum [17–19], as suggested by several investigations in string theory and quantum gravity [20]. They may also appear whenever the constant mass of the conventional SE is replaced by a position-dependent mass (PDM). The latter is an essential ingredient in the study of electronic properties of semiconductor heterostructures [21,22], quantum wells and quantum dots [23,24], helium clusters [25], graded crystals [26], quantum liquids [27], metal clusters [28], nuclei [29,30], nanowire structures [31], and neutron stars [32]. A third possibility corresponds to the replacement of the Euclidean space by a curved one. The study of the Kepler–Coulomb problem on the sphere dates back to the work of Schrödinger [7], Infeld [33], and Stevenson [34], and was generalized to a hyperbolic space by Infeld and Schild [35]. Since then, many studies have been devoted to this topic (see [36,37]).

As shown elsewhere [38], there are some intimate connections between these three types of unconventional SE, occurring whenever a specific relation exists between the deforming function $f(x)$, the PDM $m(x)$, and the (diagonal) metric tensor $g(x)$. Such unconventional SE may then be discussed in the framework of deformed SUSYQM (DSUSYQM), where the standard SI condition is replaced by a deformed one (DSI) [39,40]. On starting from the known superpotentials of SI potentials [1], a procedure has been devised in [39] to maintain the solvability of the DSI condition, thereby resulting in a list of deformed superpotentials and deforming functions giving rise to bound-state spectra. In such a deformed case, physically acceptable wavefunctions have not only to be square integrable on the defining interval of the potential, but also must ensure the Hermiticity of the Hamiltonian. More recently [41], this list of deformed superpotentials and deforming functions has been completed by considering the case of some rationally-extended potentials, connected with one-indexed families of EOP.

In the case of conventional SUSYQM, Gangopadhyaya, Mallow, and their coworkers proposed an interesting approach to SI potentials, consisting in replacing the SI condition, which is a difference-differential equation, by an infinite set of partial differential equations. The latter is obtained by expanding the superpotential in powers of $\hbar$ and expressing that the coefficient of each power must separately vanish [42]. This procedure enabled them to prove that the SI superpotentials connected with COP are those with no explicit dependence on $\hbar$, while the new ones related to EOP have such an explicit dependence [43]. They also showed that the list of the former given in [1] is complete [44] and constructed a novel example of SI superpotential with an explicit $\hbar$-dependence [45]. Furthermore, they encountered a pathway for going from those superpotentials of [1] corresponding to SE that can be reduced to the confluent hypergeometric equation to those related to SE connected with the hypergeometric equation [46].

It is the purpose of the present paper to propose an extension of the approach of Gangopadhyaya, Mallow, and their coworkers to the case of DSI potentials in DSUSYQM, both without and with explicit dependence of the superpotential on $\hbar$. We plan to illustrate this method by re-examining the known pairs of deformed superpotentials and deforming functions of [39–41] along these lines. In the present work, we restrict ourselves to unbroken DSUSYQM and only consider the discrete part of the spectrum.

After reviewing the general formalism of DSUSYQM and obtaining the DSI condition in Section 2, we will show in Section 3 how to convert such a condition into a set of partial differential equations in the case where the superpotential does not contain any dependence on $\hbar$. The case where the superpotential has such an explicit dependence is then treated in Section 4. Finally, Section 5 contains the conclusion.

2. Deformed Shape Invariance in Deformed Supersymmetric Quantum Mechanics

In DSUSYQM [39–41], a general Hamiltonian H_- is written in terms of linear operators

$$A^{\pm} = A^{\pm}(a) = \mp\hbar\sqrt{f(x)}\frac{d}{dx}\sqrt{f(x)} + W(x,a), \tag{1}$$

where $f(x)$ is some positive-definite function of x, known as the deforming function, and $W(x,a)$ is a real function of x and a parameter a, called the superpotential. Both $f(x)$ and $W(x,a)$ in general depend on some extra parameters. The Hamiltonian H_- is given by

$$H_- = A^+A^- = -\hbar^2\sqrt{f(x)}\frac{d}{dx}f(x)\frac{d}{dx}\sqrt{f(x)} + V_-(x,a), \tag{2}$$

where

$$V_-(x,a) = W^2(x,a) - \hbar f(x)\frac{dW(x,a)}{dx}. \tag{3}$$

It may be interpreted [38] as a Hamiltonian describing a PDM system with $m(x) = 1/f^2(x)$, the ordering of the latter and the differential operator d/dx being that proposed by Mustafa and Mazharimousavi [47], or as a Hamiltonian in a curved space with a diagonal metric tensor $g(x) = 1/f^2(x)$.

The product of operators $A^- A^+$ generates the so-called partner of H_-,

$$H_+ = A^- A^+ = -\hbar^2\sqrt{f(x)}\frac{d}{dx}f(x)\frac{d}{dx}\sqrt{f(x)} + V_+(x,a), \tag{4}$$

with

$$V_+(x,a) = W^2(x,a) + \hbar f(x)\frac{dW(x,a)}{dx}. \tag{5}$$

The pair of Hamiltonians intertwine with A^+ and A^- as

$$A^- H_- = H_+ A^-, \qquad A^+ H_+ = H_- A^+. \tag{6}$$

The Hamiltonian H_- is assumed to have a ground-state wavefunction $\psi_0^{(-)}(x,a)$, such that

$$A^- \psi_0^{(-)}(x,a) = 0. \tag{7}$$

From (1) and (2), the latter is therefore such that $E_0^{(-)} = 0$ and

$$\psi_0^{(-)}(x,a) = \frac{N_0^{(-)}}{\sqrt{f(x)}}\exp\left(-\int^x \frac{W(x',a)}{\hbar f(x')}dx'\right), \tag{8}$$

where $N_0^{(-)}$ is the normalization coefficient.

The intertwining relations (6) then imply the following isospectrality relationships among the eigenvalues and eigenfunctions of the two partners,

$$E_{n+1}^{(-)} = E_n^{(+)}, \tag{9}$$

$$\psi_n^{(+)}(x,a) = \frac{A^-}{\sqrt{E_n^{(+)}}}\psi_{n+1}^{(-)}(x,a), \qquad \psi_{n+1}^{(-)}(x,a) = \frac{A^+}{\sqrt{E_n^{(+)}}}\psi_n^{(+)}(x,a), \tag{10}$$

for all $n \geq 0$ such that physically acceptable wavefunctions exist. In the deformed case considered here, this imposes that they satisfy two conditions [39]:
(i) As for conventional SE, they should be square integrable on the (finite or infinite) interval of definition (x_1, x_2) of the potentials $V_\pm(x,a)$—i.e.,

$$\int_{x_1}^{x_2} dx\, |\psi_n^{(\pm)}(x,a)|^2 < \infty. \tag{11}$$

(ii) They should ensure the Hermiticity of $H_\pm$. This amounts to the condition

$$|\psi_n^{(\pm)}(x,a))|^2 f(x) \to 0 \qquad \text{for } x \to x_1 \text{ and } x \to x_2, \tag{12}$$

which implies an additional restriction whenever $f(x) \to \infty$ for $x \to x_1$ and/or $x \to x_2$. Equations (11) and (12) ensure the self-adjointness of $H_\pm$ and guarantee the relation $(A^\pm)^\dagger = A^\mp$.

The knowledge of the eigenvalues and eigenfunctions of H_- automatically implies the same for its partner H_+ (or vice versa). However, whenever the partner potentials V_- and V_+ are similar in shape and differ only in the parameters that appear in them—i.e.,

$$V_+(x, a_0) + g(a_0) = V_-(x, a_1) + g(a_1), \tag{13}$$

where a_1 is some function of a_0 and $g(a_0)$, $g(a_1)$ do not depend on x, then the spectrum of either Hamiltonian can be derived without reference to its partner. Here we restrict ourselves to the case of translational (or additive) shape invariance—i.e., a_1 and a_0 only differ by some additive constant. Considering then a set of parameters a_i, $i = 0, 1, 2, \ldots$, and extending condition (13) to

$$V_+(x, a_i) + g(a_i) = V_-(x, a_{i+1}) + g(a_{i+1}), \qquad i = 0, 1, 2, \ldots, \tag{14}$$

we get from Equations (3) and (5) the so-called DSI condition

$$W^2(x, a_i) + \hbar f(x) \frac{dW(x, a_i)}{dx} + g(a_i) = W^2(x, a_{i+1}) - \hbar f(x) \frac{dW(x, a_{i+1})}{dx} + g(a_{i+1}). \tag{15}$$

The eigenvalues and eigenfunctions of H_- turn out to be given by

$$E_n^{(-)}(a_0) = g(a_n) - g(a_0), \qquad n = 0, 1, 2, \ldots, \tag{16}$$

$$\psi_n^{(-)}(x, a_0) \propto A^+(a_0) A^+(a_1) \ldots A^+(a_{n-1}) \psi_0^{(-)}(x, a_n), \qquad n = 0, 1, 2, \ldots, \tag{17}$$

with $\psi_0^{(-)}(x, a_n)$ as expressed in (8).

3. Deformed Shape Invariance for Superpotentials with no Explicit Dependence on $\hbar$

As in [42–46], let us assume that the additive constant that allows us to get a_{i+1} from a_i is just $\hbar$—i.e., $a_{i+1} = a_i + \hbar$. Note that, with respect to conventions used elsewhere where the system of units is such that $\hbar = 1$, this implies some parameter re-normalization. In Appendix A, we summarize the transformations that have to be carried out on the parameters and possibly the variable used in [39–41] in order to arrive at the conventions employed here.

In the present section, we will also suppose that the dependence of $W(x, a_i)$ on $\hbar$ is entirely contained in a_i, thus leaving the case of an explicit dependence of W on $\hbar$ to Section 4.

Since Equation (15) must hold for an arbitrary value of $\hbar$, we can expand it in powers of $\hbar$ and require that the coefficient of each power vanishes. It is straightforward to show that the coefficient of $\hbar$ leads to the condition

$$W \frac{\partial W}{\partial a} - f(x) \frac{\partial W}{\partial x} + \frac{1}{2} \frac{dg}{da} = 0. \tag{18}$$

Then, the coefficient of $\hbar^2$ yields

$$\frac{\partial}{\partial a} \left[W \frac{\partial W}{\partial a} - f(x) \frac{\partial W}{\partial x} + \frac{1}{2} \frac{dg}{da} \right] = 0, \tag{19}$$

which is automatically satisfied if Equation (18) is so. Finally, the coefficient of $\hbar^n$ for $n \geq 3$ gives the condition

$$\frac{(2-n) f(x)}{n!} \frac{\partial^n W}{\partial a^{n-1} \partial x} = 0, \qquad n = 3, 4, \ldots. \tag{20}$$

This set of equations is satisfied, provided

$$\frac{\partial^3 W}{\partial a^2 \partial x} = 0. \tag{21}$$

We are therefore left with two independent conditions (18) and (21). This is similar to what happens in SUSYQM [43,44], the only difference being the appearance of the deforming function $f(x)$ in the first equation.

Before giving the set of results, we shall discuss in detail two examples, a simple one and a more involved one.

3.1. Example of the Pöschl-Teller Potential

Let us consider a deforming function $f(x) = 1 + \alpha \sin^2 x$ with $-1 < \alpha \neq 0$ and $-\frac{\pi}{2} < x < \frac{\pi}{2}$, as well as a superpotential

$$W(x,a) = (1+\alpha)a \tan x, \qquad -\frac{\pi}{2} < x < \frac{\pi}{2}, \tag{22}$$

where

$$a = \frac{1}{2(1+\alpha)}[(1+\alpha)\hbar + \Delta], \qquad \Delta = \sqrt{(1+\alpha)^2\hbar^2 + 4A(A-\hbar)}, \qquad A > \hbar. \tag{23}$$

We note that this W automatically satisfies Equation (21) and that, on inserting it in Equation (18), we obtain

$$\frac{dg}{da} = 2(1+\alpha)a, \tag{24}$$

from which $g(a) = (1+\alpha)a^2$, up to some additive constant.

From Equation (3), the starting potential can be written as

$$\begin{aligned} V_-(x,a) &= (1+\alpha)^2 a(a-\hbar)\sec^2 x - (1+\alpha)a[(1+\alpha)a - \hbar\alpha] \\ &= A(A-\hbar)\sec^2 x - \left\{A(A-\hbar) + \frac{1}{2}\hbar[(1+\alpha)\hbar + \Delta]\right\}, \end{aligned} \tag{25}$$

and therefore corresponds to the Pöschl–Teller (PT) potential $V = A(A-\hbar)\sec^2 x$ with ground-state energy $E_0 = A(A-\hbar) + \frac{1}{2}\hbar[(1+\alpha)\hbar + \Delta]$. On the other hand, from (16), we get

$$E_n^{(-)} = g(a+n\hbar) - g(a) = \hbar^2(1+\alpha)n(n+1) + \hbar\Delta n. \tag{26}$$

The results obtained here may be compared with those derived in [40] for $\bar{V} = \bar{A}(\bar{A}-1)\sec^2 \bar{x}$, with bound-state energies $\bar{E}_n = (E_n^{(-)} + E_0)/\hbar^2 = (\bar{\lambda}+n)^2 - \bar{\alpha}(\bar{\lambda} - n^2)$, where $\bar{\lambda} = (1+\bar{\alpha})a/\hbar$ is changed into $\bar{\lambda} + 1 + \bar{\alpha}$ when going to the partner.

3.2. Example of the Radial Harmonic Oscillator Potential

Let us now consider a deforming function $f(x) = 1 + \alpha x^2$ with $\alpha > 0$ and $0 < x < \infty$, as well as a superpotential

$$W(x,a) = a\left(-\frac{1}{x} + \alpha x\right) - b\left(\frac{1}{x} + \alpha x\right), \qquad 0 < x < \infty, \tag{27}$$

where

$$\begin{aligned} &a = \frac{1}{2}\left(l + 1 + \frac{1}{2}\hbar + \frac{\Delta}{2\alpha}\right), \qquad b = \frac{1}{2}\left(l + 1 - \frac{1}{2}\hbar - \frac{\Delta}{2\alpha}\right), \\ &\Delta = \sqrt{\omega^2 + \hbar^2\alpha^2}, \qquad \omega > 0, \qquad l = 0, 1, 2, \ldots. \end{aligned} \tag{28}$$

Here, when going to the partner, a is assumed to change into $a + \hbar$, while b remains constant. This W automatically satisfies Equation (21) again and Equation (18) leads to

$$\frac{dg}{da} = 8\alpha a, \tag{29}$$

from which $g(a) = 4\alpha a^2$, up to some additive constant.

Equation (3) shows that the starting potential is given by

$$\begin{aligned} V_-(x,a) &= \alpha^2(a-b)(a-b-\hbar)x^2 + \frac{(a+b)(a+b-\hbar)}{x^2} - 2\alpha(a^2-b^2+\hbar a) \\ &= \frac{1}{4}\omega^2x^2 + \frac{(l+1)(l+1-\hbar)}{x^2} - \left[\left(l+1+\frac{\hbar}{2}\right)\Delta + \hbar\alpha\left(2l+2+\frac{1}{2}\hbar\right)\right] \end{aligned} \tag{30}$$

and therefore corresponds to the radial harmonic oscillator (RHO) potential $V = \frac{1}{4}\omega^2x^2 + \frac{(l+1)(l+1-\hbar)}{x^2}$ with ground-state energy $E_0 = \left(l+1+\frac{\hbar}{2}\right)\Delta + \hbar\alpha\left(2l+2+\frac{1}{2}\hbar\right)$. Furthermore, Equation (16) leads to

$$E_n^{(-)} = g(a+n\hbar) - g(a) = 4\hbar\alpha n\left[\left(n+\frac{1}{2}\right)\hbar + l + 1\right] + 2\hbar n\Delta. \tag{31}$$

These results are comparable with those obtained in [40] for $\bar{V} = \frac{1}{4}\bar{\omega}^2\bar{x}^2 + \bar{l}(\bar{l}+1)/\bar{x}^2$ with bound-state energies $\bar{E}_n = E_n^{(-)} + E_0 = 2\bar{\lambda}\bar{\mu} - \bar{\alpha}\bar{\lambda} + \bar{\mu} - 4(\bar{\alpha}\bar{\lambda} - \bar{\mu})n + 4\bar{\alpha}n^2$, where $\bar{\lambda} = -(a+b)/\hbar$ and $\bar{\mu} = (a-b)\hbar\alpha$ are changed into $\bar{\lambda} - 1$ and $\bar{\mu} + \bar{\alpha}$ when going to the partner, respectively.

3.3. Lists of Results

On proceeding, as in Sections 3.1 and 3.2, we analyzed the other sets of potentials and deforming functions considered in [39–41]. They include the Scarf I (S), radial Coulomb (C), Morse (M), Eckart (E), Rosen-Morse I (RM), shifted harmonic oscillator (SHO), deformed radial harmonic oscillator (DRHO), and deformed radial Coulomb (DC) potentials. The list of them is given in Table 1 in the notations used in this paper. In all the cases, except for the PT and DC potentials, the deformed superpotential is written in terms of two combinations of parameters, the first one a being changed into $a + \hbar$ and the second one b remaining constant when going to the partner. The corresponding results are listed in Table 2. In all cases, it turns out that Equation (21) is automatically satisfied, while Equation (18) leads to the expressions of $g(a)$ listed in Table 3, together with the resulting energies $E_n^{(-)}$.

Table 1. Potentials and deforming functions.

Name	V	f
PT	$A(A-\hbar)\sec^2 x$	$1+\alpha\sin^2 x$
	$-\frac{\pi}{2} < x < \frac{\pi}{2}, A > \hbar$	$-1 < \alpha \neq 0$
RHO	$\frac{1}{4}\omega^2x^2 + \frac{(l+1)(l+1-\hbar)}{x^2}$	$1+\alpha x^2$
	$0 < x < +\infty$	$\alpha > 0$
S	$[A(A-\hbar)+B^2]\sec^2 x - B(2A-\hbar)\sec x\tan x$	$1+\alpha\sin x$
	$-\frac{\pi}{2} < x < \frac{\pi}{2}, A-\hbar > B > 0$	$0 < \|\alpha\| < 1$
C	$-\frac{e^2}{x} + \frac{(l+1)(l+1-\hbar)}{x^2}$	$1+\alpha x$
	$0 < x < +\infty$	$\alpha > 0$
M	$B^2e^{-2x} - B(2A+\hbar)e^{-x}$	$1+\alpha e^{-x}$
	$-\infty < x < +\infty, A, B > 0$	$\alpha > 0$

Table 1. *Cont.*

Name	V	f
E	$A(A-\hbar)\,\mathrm{csch}^2 x - 2B\coth x$	$1+\alpha e^{-x}\sinh x$
	$0<x<+\infty$, $A\geq\frac{3}{2}\hbar$, $B>A^2$	$-2<\alpha\neq 0$
RM	$A(A-\hbar)\csc^2 x + 2B\cot x$	$1+\sin x(\alpha\cos x+\beta\sin x)$
	$0<x<\pi$, $A\geq\frac{3}{2}\hbar$	$\frac{\lvert\alpha\rvert}{2}<\sqrt{1+\beta}$, $\beta>-1$
SHO	$\frac{1}{4}\omega^2\left(x-\frac{2d}{\omega}\right)^2$	$1+\alpha x^2+2\beta x$
	$-\infty<x<+\infty$	$\alpha>\beta^2\geq 0$
DRHO	$\frac{\omega(\omega+2\hbar\lambda)x^2}{4(1+\lambda x^2)}+\frac{(l+1)(l+1-\hbar)}{x^2}$	$\sqrt{1+\lambda x^2}$
	$0<x<+\infty$ if $\lambda>0$	
	$0<x<1/\sqrt{\lvert\lambda\rvert}$ if $\lambda<0$	
DC	$-\frac{e^2}{x}\sqrt{1+\lambda x^2}+\frac{(l+1)(l+1-\hbar)}{x^2}$	$\sqrt{1+\lambda x^2}$
	$0<x<+\infty$ if $\lambda>0$	
	$0<x<1/\sqrt{\lvert\lambda\rvert}$ if $\lambda<0$	

Table 2. Superpotentials and combinations of parameters.

Name	W	Parameters
PT	$(1+\alpha)a\tan x$	$a=\frac{1}{2(1+\alpha)}[(1+\alpha)\hbar+\Delta]$
		$\Delta=\sqrt{(1+\alpha)^2\hbar^2+4A(A-\hbar)}$
RHO	$a\left(-\frac{1}{x}+\alpha x\right)-b\left(\frac{1}{x}+\alpha x\right)$	$a=\frac{1}{2}\left(l+1+\frac{1}{2}\hbar+\frac{\Delta}{2\alpha}\right)$
		$b=\frac{1}{2}\left(l+1-\frac{1}{2}\hbar-\frac{\Delta}{2\alpha}\right)$
		$\Delta=\sqrt{\omega^2+\hbar^2\alpha^2}$
S	$a(\tan x+\alpha\sec x)$	$a=\frac{1}{2}\left(\hbar+\frac{\alpha-1}{2\alpha}\Delta_+ +\frac{\alpha+1}{2\alpha}\Delta_-\right)$
	$+b(\tan x-\alpha\sec x)$	$b=\frac{1}{4\alpha}[(\alpha+1)\Delta_+ +(\alpha-1)\Delta_-]$
		$\Delta_\pm=\sqrt{\frac{1}{4}\hbar^2(1\mp\alpha)^2+(A\pm B)(A\pm B-\hbar)}$
C	$-\frac{a+b}{x}+\frac{2b}{a+b}-\frac{\alpha}{2}(a+b)$	$a=-\frac{1}{4}\{e^2+(l+1)[\alpha(l+1-\hbar)-4]\}$
		$b=\frac{1}{4}[e^2+\alpha(l+1)(l+1-\hbar)]$
M	$-\alpha(a+b)e^{-x}-\frac{1}{2}(a+b)$	$a=-\frac{1}{4\hbar\alpha}\{B^2+\alpha[B(2A+\hbar)-2\hbar^2]-2\hbar\Delta\}$
	$+\frac{2\hbar b}{\alpha(a+b)}$	$b=\frac{B}{4\hbar\alpha}[B+\alpha(2A+\hbar)]$
		$\Delta=\sqrt{4B^2+\hbar^2\alpha^2}$
E	$-(a+b)\coth x+\frac{2\hbar b}{a+b}$	$a=\frac{1}{2\hbar}[-B+2\hbar A-\frac{\alpha}{2}A(A-\hbar)]$
	$-\frac{\alpha}{2}(a+b)$	$b=\frac{1}{2\hbar}[B+\frac{\alpha}{2}A(A-\hbar)]$
RM	$-(a+b)\cot x+\frac{2\hbar b}{a+b}$	$a=\frac{1}{2\hbar}[B+2\hbar A-\frac{\alpha}{2}A(A-\hbar)]$
	$-\frac{\alpha}{2}(a+b)$	$b=\frac{1}{2\hbar}[-B+\frac{\alpha}{2}A(A-\hbar)]$

Table 2. *Cont.*

Name	W	Parameters
SHO	$(a+b)(\alpha x+\beta)$ $-\frac{2b}{\hbar^2\alpha(a+b)}$	$a=\frac{\hbar}{2}\left(-\hbar\frac{\beta}{4\alpha}\omega^2+1+\frac{\Delta}{\hbar\alpha}-\frac{d}{2}\hbar\omega\right)$ $b=\frac{1}{2}\hbar^2\omega\left(\frac{\beta}{4\alpha}\omega+\frac{d}{2}\right)$ $\Delta=\sqrt{\omega^2+\hbar^2\alpha^2}$
DRHO	$a\left(-\frac{1}{x}f-\frac{\lambda x}{f}\right)$ $+b\left(-\frac{1}{x}f+\frac{\lambda x}{f}\right)$	$a=\frac{1}{2}\left(l+1-\frac{\omega}{2\lambda}\right)$ $b=\frac{1}{2}\left(l+1+\frac{\omega}{2\lambda}\right)$
DC	$-\frac{a}{x}f+\frac{e^2}{2a}$	$a=l+1$

Table 3. Functions $g(a)$ and bound-state energies.

Name	g	$E_n^{(-)}$
PT	$(1+\alpha)a^2$	$\hbar^2(1+\alpha)n(n+1)+\hbar\Delta n$
RHO	$4\alpha a^2$	$4\hbar\alpha n\left[\left(n+\frac{1}{2}\right)\hbar+l+1\right]+2\hbar n\Delta$
S	$(1-\alpha^2)a^2$ $+2(1+\alpha^2)ab$	$\hbar^2(1-\alpha^2)n(n+1)+\hbar\left[(1+\alpha)\Delta_+ +(1-\alpha)\Delta_-\right]n$
C	$-\frac{4b^2}{(a+b)^2}$ $-\frac{\alpha^2}{4}a(a+2b)$	$\hbar n(\hbar n+2l+2)[e^2-\hbar\alpha(l+1)(n+1)]$ $\times[e^2+\alpha(l+1)(\hbar n+2l+2-\hbar)]$ $\times[4(l+1)^2(l+1+n\hbar)^2]^{-1}$
M	$-\frac{4\hbar^2b^2}{\alpha^2(a+b)^2}$ $-\frac{1}{4}a(a+2b)$	$\hbar n[\Delta+\hbar\alpha(n+1)][2B(2A+\hbar)-\hbar(\Delta+\alpha\hbar)(n+1)]$ $\times[4B^2+2\alpha B(2A+\hbar)+\hbar\alpha(\Delta+\hbar\alpha)(n+1)]$ $\times(\Delta+\hbar\alpha)^{-2}[\Delta+(2n+1)\hbar\alpha]^{-2}$
E	$-\frac{4\hbar^2b^2}{(a+b)^2}$ $-\frac{1}{4}(\alpha+2)^2a(a+2b)$	$\hbar n(2A+n\hbar)[4A^2(A+n\hbar)^2]^{-1}$ $\times\{2B+2A(A+n\hbar)+\alpha A[2A+(n-1)\hbar]\}$ $\times[2B-2A(A+n\hbar)-\hbar\alpha A(n+1)]$
RM	$-\frac{4\hbar^2b^2}{(a+b)^2}$ $-\frac{1}{4}(\alpha^2-4\beta-4)a(a+2b)$	$\hbar n(2A+n\hbar)[4A^2(A+n\hbar)^2]^{-1}$ $\times\{4A^2(A+n\hbar)^2+4B^2-4\alpha BA(A-\hbar)$ $-\hbar\alpha^2A^2[2A-\hbar+n(2A+n\hbar)]-4\beta A^2(A+n\hbar)^2\}$
SHO	$-\frac{4b^2}{\hbar^4\alpha^2(a+b)^2}$ $+(\alpha-\beta^2)a(a+2b)$	$4n[\Delta+(n+1)\hbar\alpha]$ $\times\{(\Delta+\hbar\alpha)[\Delta+\hbar\alpha(2n+1)]\}^{-2}$ $\times\{\hbar^3\alpha(\alpha-\beta^2)(n+1)[2\hbar^2\alpha^2(n+1)+\omega^2(n+2)]$ $+d\omega^2(\beta+\hbar\alpha d)+\frac{1}{4}\hbar\omega^4$ $+\hbar^2\Delta(\alpha-\beta^2)(n+1)[2\hbar^2\alpha^2(n+1)+\omega^2]\}$
DRHO	$-4\lambda a^2$	$2n\hbar\omega-4\hbar\lambda n(l+1+\hbar n)$
DC	$-\lambda a^2-\frac{e^4}{4a^2}$	$n\hbar(2l+2+n\hbar)$ $\times\left(-\lambda+\frac{e^4}{4(l+1)^2(l+1+n\hbar)^2}\right)$

4. Deformed Shape Invariance for Superpotentials with an Explicit Dependence on $\hbar$

Let us next consider the case where the superpotential contains an explicit dependence on $\hbar$. It may then be expanded in powers of $\hbar$ as

$$W(x,a,\hbar) = \sum_{n=0}^{\infty} \hbar^n W_n(x,a). \tag{32}$$

On inserting this expression in the DSI condition (15) and proceeding as in conventional SUSYQM [43,44], we arrive at the set of relations

$$\begin{aligned}&\sum_{k=0}^{n} W_k W_{n-k} + f\frac{\partial W_{n-1}}{\partial x} - \sum_{s=0}^{n}\sum_{k=0}^{s} \frac{1}{(n-s)!}\frac{\partial^{n-s}}{\partial a^{n-s}} W_k W_{s-k} \\ &+ f\sum_{k=1}^{n} \frac{1}{(k-1)!}\frac{\partial^k}{\partial a^{k-1}\partial x} W_{n-k} - \frac{1}{n!}\frac{d^n g}{da^n} = 0, \qquad n = 1,2,\ldots.\end{aligned} \tag{33}$$

The latter can be rewritten as

$$2f\frac{\partial W_0}{\partial x} - \frac{\partial}{\partial a}(W_0^2 + g) = 0, \tag{34}$$

$$f\frac{\partial W_1}{\partial x} - \frac{\partial}{\partial a}(W_0 W_1) = 0, \tag{35}$$

$$\begin{aligned}&2f\frac{\partial W_{n-1}}{\partial x} - \sum_{s=1}^{n-1}\sum_{k=0}^{s} \frac{1}{(n-s)!}\frac{\partial^{n-s}}{\partial a^{n-s}} W_k W_{s-k} + \frac{n-2}{n!} f \frac{\partial^n W_0}{\partial a^{n-1}\partial x} \\ &+ f\sum_{k=2}^{n-1} \frac{1}{(k-1)!}\frac{\partial^k}{\partial a^{k-1}\partial x} W_{n-k} = 0, \qquad n = 3,4,\ldots.\end{aligned} \tag{36}$$

In [41], two sets of rational extensions of the DRHO potential with $\lambda < 0$ considered in Section 3, referred to as type I and type II extensions, were constructed in terms of some Jacobi polynomials of degree m. The potentials belonging to these two sets were shown to be derived from superpotentials satisfying the DSI condition. The simplest potentials, corresponding to $m = 1$, turn out to be identical and given by (after changing the parameters and the variable, as explained in Appendix A)

$$\begin{aligned}V = {}& \frac{\omega(\omega - 2\hbar|\lambda|)x^2}{4(1-|\lambda|x^2)} + \frac{(l+1)(l+1-\hbar)}{x^2} + 4\hbar^2\left(\frac{\omega + 2|\lambda|(l+1-\hbar)}{[\omega - 2|\lambda|(l+1)]x^2 + 2l + 2 - \hbar}\right. \\ &\left. - \frac{2(2l+2-\hbar)(\omega - \hbar|\lambda|)}{\{[\omega - 2|\lambda|(l+1)]x^2 + 2l + 2 - \hbar\}^2}\right),\end{aligned} \tag{37}$$

with a corresponding superpotential

$$\begin{aligned}W = {}& \frac{\omega x}{2\sqrt{1-|\lambda|x^2}} - \frac{l+1}{x}\sqrt{1-|\lambda|x^2} + 2\hbar[\omega - 2(l+1)|\lambda|]x\sqrt{1-|\lambda|x^2} \\ &\times\left(\frac{1}{[\omega - 2(l+1)|\lambda|]x^2 + 2l + 2 - \hbar} - \frac{1}{[\omega - 2(l+1)|\lambda|]x^2 + 2l + 2 + \hbar}\right).\end{aligned} \tag{38}$$

Let us show that such a superpotential can be derived by the present method.

For such a purpose, we plan to prove that for

$$f = \sqrt{1-|\lambda|x^2}, \qquad a = \frac{1}{2}\left(l + 1 + \frac{\omega}{2|\lambda|}\right), \qquad b = \frac{1}{2}\left(l + 1 - \frac{\omega}{2|\lambda|}\right), \tag{39}$$

the functions

$$W_0(x,a) = -\frac{a+b}{x}f + \frac{(a-b)|\lambda|x}{f}, \tag{40}$$

$$W_{2\nu+1}(x,a) = 0, \qquad \nu = 0,1,2,\ldots, \tag{41}$$

$$W_{2\nu}(x,a) = -f\frac{16b|\lambda|x}{(4bf^2+2a-2b)^{2\nu}}, \qquad \nu = 1,2,\ldots, \tag{42}$$

provide a solution of the set of Equations (34)–(36). Note that, as in Section 3, the combinations of parameters a and b become $a+\hbar$ and b for the partner, respectively.

Let us start with Equation (34). From (40), we get

$$\frac{\partial W_0}{\partial x} = \frac{a+b}{x^2}f + \frac{(a+b)|\lambda|}{f} + \frac{(a-b)|\lambda|}{f^3}, \tag{43}$$

$$\frac{\partial W_0}{\partial a} = -\frac{1}{x}f + \frac{|\lambda|x}{f}, \tag{44}$$

from which we obtain $\frac{dg}{da} = 8a|\lambda|$ and $g = 4a^2|\lambda|$ up to some additive constant. Hence, from Equation (16),

$$E_n^{(-)} = 4\hbar|\lambda|n(n\hbar+2a) = 4\hbar|\lambda|n\left(n\hbar+l+1+\frac{\omega}{2|\lambda|}\right), \tag{45}$$

in agreement with the result obtained in [41].

Equation (35) is automatically satisfied since $W_1 = 0$.

Considering next Equation (36), we note that

$$\frac{\partial^n W_0}{\partial a^{n-1}\partial x} = 0, \qquad n = 3,4,\ldots, \tag{46}$$

and that

$$\sum_{k=0}^{s} W_k W_{s-k} = 0 \qquad \text{for odd } s. \tag{47}$$

For even s, on the other hand, we easily get

$$\sum_{k=0}^{s} W_k W_{s-k} = F_s, \tag{48}$$

with F_s defined by

$$F_s = \frac{32b|\lambda|}{(4bf^2+2a-2b)^s}\{-4b(s-2)f^4 + [2a+4b(s-2)]f^2 - a + b\}. \tag{49}$$

From this result, it is straightforward to prove that

$$\frac{\partial^{n-s}}{\partial a^{n-s}}\sum_{k=0}^{s} W_k W_{s-k} = (-2)^{n-s}\frac{(n-2)!}{(s-2)!}F_n \qquad \text{for even } s. \tag{50}$$

Equations (47) and (50) then lead to

$$\sum_{s=1}^{n-1}\sum_{k=0}^{s}\frac{1}{(n-s)!}\frac{\partial^{n-s}}{\partial a^{n-s}}W_k W_{s-k} = F_n \times \begin{cases} \frac{1}{2}(3^{n-2}-1) & \text{for even } n, \\ -\frac{1}{2}(3^{n-2}+1) & \text{for odd } n. \end{cases} \tag{51}$$

Furthermore, we obtain

$$2f\frac{\partial W_{n-1}}{\partial x} = \begin{cases} 0 & \text{for even } n, \\ -2F_n & \text{for odd } n, \end{cases}. \tag{52}$$

as well as

$$f\sum_{k=2}^{n-1} \frac{1}{(k-1)!}\frac{\partial^k}{\partial a^{k-1}\partial x} W_{n-k} = F_n \times \begin{cases} \frac{1}{2}(3^{n-2}-1) & \text{for even } n, \\ -\frac{1}{2}(3^{n-2}-3) & \text{for odd } n. \end{cases}. \tag{53}$$

On inserting Equations (46), (51), (52), and (53) in Equation (36), it is clear that the latter is satisfied, which completes the proof that Equations (40)–(42) provide a solution of Equations (34)–(36).

It now only remains to use Equations (40) and (42) in

$$W(x,a,\hbar) = \sum_{\nu=0}^{\infty} \hbar^{2\nu} W_{2\nu}(x,a) \tag{54}$$

to obtain

$$\begin{aligned} W(x,a,\hbar) = {} & \frac{(a-b)|\lambda|x}{f} - \frac{a+b}{x}f - 8\hbar b|\lambda|xf\left(\frac{1}{4bf^2+2a-2b-\hbar}\right. \\ & \left. - \frac{1}{4bf^2+2a-2b+\hbar}\right), \end{aligned} \tag{55}$$

which, after introducing the definitions of f, a, and b, given in (39), reduces to the expression (38)—i.e., the extended superpotential found in [41].

5. Conclusions

In this paper, we have shown that the approach of Gangopadhyaya, Mallow, and their coworkers of SI potentials in conventional SUSYQM can be extended to DSI ones in DSUSYQM and we have illustrated our results by considering several examples taken from [39–41]. These include both conventional potentials, for which the corresponding superpotential has no explicit dependence on $\hbar$, and a rationally-extended one, for which there is such a dependence. In all cases, it turns out that the parameter a, which is changed into $a+\hbar$ when going to the partner potential, is a combination of the potential and deforming function parameters.

An interesting open question for future investigation would be the possibility of generalizing the method to rationally-extended potentials exhibiting an "enlarged" shape invariance, for which the partner is obtained by translating some potential parameter as well as the degree m of the polynomial arising in the denominator. Such potentials are indeed known both in conventional SUSYQM [48–51] and in DSUSYQM [41].

Acknowledgments: This work was supported by the Fonds de la Recherche Scientifique—FNRS under Grant Number 4.45.10.08.

Appendix A. Going from Previously Used Conventions to the Present Ones

In this appendix, we summarize the changes that have to be carried out to go from the conventions used in [39–41] to those of the present paper. The quantities employed in the former papers are distinguished by a bar from those used here. It is worth noting too that in [39–41], the potentials used have a nonvanishing ground-state energy and must therefore be compared with $V = V_- + E_0$, where E_0 is the shift to adjust the ground-state energy of H_- to a zero value. As a consequence, $E_n^{(-)} = E_n - E_0$ corresponds to $\bar{E}_n - \bar{E}_0$.

Pöschl-Teller potential

$$\begin{aligned}
&\bar{V} = \bar{A}(\bar{A}-1)\sec^2 x, \quad \bar{f}(\bar{x}) = 1 + \bar{\alpha}\sin^2\bar{x},\\
&\bar{A} = \frac{A}{\hbar}, \quad \bar{\alpha} = \alpha, \quad \bar{x} = x,\\
&V = A(A-\hbar)\sec^2 x = \hbar^2\bar{V}, \quad f(x) = 1 + \alpha\sin^2 x = \bar{f}(\bar{x}),\\
&W = \hbar\bar{W}, \quad E_n = \hbar^2\bar{E}_n.
\end{aligned} \tag{A1}$$

Radial Harmonic Oscillator potential

$$\begin{aligned}
&\bar{V} = \frac{1}{4}\bar{\omega}^2\bar{x}^2 + \frac{\bar{l}(\bar{l}+1)}{\bar{x}^2}, \quad \bar{f}(\bar{x}) = 1 + \bar{\alpha}\bar{x}^2,\\
&\bar{\omega} = \hbar\omega, \quad \bar{l} = \frac{l+1}{\hbar} - 1, \quad \bar{\alpha} = \hbar^2\alpha, \quad \bar{x} = \frac{x}{\hbar},\\
&V = \frac{1}{4}\omega^2 x^2 + \frac{(l+1)(l+1-\hbar)}{x^2} = \bar{V}, \quad f(x) = 1 + \alpha x^2 = \bar{f}(\bar{x}),\\
&W = \bar{W}, \quad E_n = \bar{E}_n.
\end{aligned} \tag{A2}$$

Scarf I potential

$$\begin{aligned}
&\bar{V} = [\bar{A}(\bar{A}-1)] + \bar{B}^2]\sec^2\bar{x} - \bar{B}(2\bar{A}-1)\sec\bar{x}\tan\bar{x}, \quad \bar{f}(\bar{x}) = 1 + \bar{\alpha}\sin\bar{x},\\
&\bar{A} = \frac{A}{\hbar}, \quad \bar{B} = \frac{B}{\hbar}, \quad \bar{\alpha} = \alpha, \quad \bar{x} = x,\\
&V = [A(A-\hbar) + B^2]\sec^2 x - B(2A-\hbar)\sec x\tan x = \hbar^2\bar{V},\\
&f(x) = 1 + \alpha\sin x = \bar{f}(\bar{x}), \quad W = \hbar\bar{W}, \quad E_n = \hbar^2\bar{E}_n.
\end{aligned} \tag{A3}$$

Coulomb potential

$$\begin{aligned}
&\bar{V} = -\frac{2\bar{Z}}{\bar{x}} + \frac{\bar{l}(\bar{l}+1)}{\bar{x}^2}, \quad \bar{f}(\bar{x}) = 1 + \bar{\alpha}\bar{x},\\
&\bar{Z} = \frac{e^2}{2\hbar}, \quad \bar{l} = \frac{l+1}{\hbar} - 1, \quad \bar{\alpha} = \hbar\alpha, \quad \bar{x} = \frac{x}{\hbar},\\
&V = -\frac{e^2}{x} + \frac{(l+1)(l+1-\hbar)}{x^2} = \bar{V}, \quad f(x) = 1 + \alpha x = \bar{f}(\bar{x}),\\
&W = \bar{W}, \quad E_n = \bar{E}_n.
\end{aligned} \tag{A4}$$

Morse potential

$$\begin{aligned}
&\bar{V} = \bar{B}^2 e^{-2\bar{x}} - \bar{B}(2\bar{A}+1)e^{-\bar{x}}, \quad \bar{f}(\bar{x}) = 1 + \bar{\alpha}e^{-\bar{x}},\\
&\bar{A} = \frac{A}{\hbar}, \quad \bar{B} = \frac{B}{\hbar}, \quad \bar{\alpha} = \alpha, \quad \bar{x} = x,\\
&V = B^2 e^{-2x} - B(2A+\hbar)e^{-x} = \hbar^2\bar{V}, \quad f(x) = 1 + \alpha e^{-x} = \bar{f}(\bar{x}),\\
&W = \hbar\bar{W}, \quad E_n = \hbar^2\bar{E}_n.
\end{aligned} \tag{A5}$$

Eckart potential

$$\begin{aligned}
&\bar{V} = \bar{A}(\bar{A}-1)\,\mathrm{csch}^2\bar{x} - 2\bar{B}\coth\bar{x}, \quad \bar{f}(\bar{x}) = 1 + \bar{\alpha}e^{-\bar{x}}\sinh\bar{x},\\
&\bar{A} = \frac{A}{\hbar}, \quad \bar{B} = \frac{B}{\hbar^2}, \quad \bar{\alpha} = \alpha, \quad \bar{x} = x,\\
&V = A(A-\hbar)\,\mathrm{csch}^2 x - 2B\coth x = \hbar^2\bar{V}, \quad f(x) = 1 + \alpha e^{-x}\sinh x = \bar{f}(\bar{x}),\\
&W = \hbar\bar{W}, \quad E_n = \hbar^2\bar{E}_n.
\end{aligned} \tag{A6}$$

Rosen-Morse I potential

$$\begin{aligned}
&\bar{V} = \bar{A}(\bar{A}-1)\csc^2\bar{x} + 2\bar{B}\cot\bar{x}, \quad \bar{f}(\bar{x}) = 1 + \sin\bar{x}(\bar{\alpha}\cos\bar{x} + \bar{\beta}\sin\bar{x}),\\
&\bar{A} = \frac{A}{\hbar}, \quad \bar{B} = \frac{B}{\hbar^2}, \quad \bar{\alpha} = \alpha, \quad \bar{\beta} = \beta, \quad \bar{x} = x,\\
&V = A(A-\hbar)\csc^2 x + 2B\cot x = \hbar^2\bar{V}, \quad f(x) = 1 + \sin x(\alpha\cos x + \beta\sin x) = \bar{f}(\bar{x}),\\
&W = \hbar\bar{W}, \quad E_n = \hbar^2\bar{E}_n.
\end{aligned} \tag{A7}$$

Shifted Harmonic Oscillator potential

$$\begin{aligned}
&\bar{V} = \frac{1}{4}\bar{\omega}^2\left(\bar{x} - \frac{2\bar{d}}{\bar{\omega}}\right)^2, \quad \bar{f}(\bar{x}) = 1 + \bar{\alpha}\bar{x}^2 + 2\bar{\beta}\bar{x},\\
&\bar{\omega} = \hbar\omega, \quad \bar{d} = d, \quad \bar{\alpha} = \hbar^2\alpha, \quad \bar{\beta} = \hbar\beta, \quad \bar{x} = \frac{x}{\hbar},\\
&V = \frac{1}{4}\omega^2\left(x - \frac{2d}{\omega}\right)^2 = \bar{V}, \quad f(x) = 1 + \alpha x^2 + 2\beta x = \bar{f}(\bar{x}),\\
&W = \bar{W}, \quad E_n = \bar{E}_n.
\end{aligned} \tag{A8}$$

Deformed Radial Harmonic Oscillator potential

$$\begin{aligned}
&\bar{V} = \frac{\bar{\omega}(\bar{\omega} + 2\bar{\lambda})\bar{x}^2}{4(1+\bar{\lambda}\bar{x}^2)} + \frac{\bar{l}(\bar{l}+1)}{\bar{x}^2}, \quad \bar{f}(\bar{x}) = \sqrt{1+\bar{\lambda}\bar{x}^2},\\
&\bar{\omega} = \hbar\omega, \quad \bar{l} = \frac{l+1}{\hbar} - 1, \quad \bar{\lambda} = \hbar^2\lambda, \quad \bar{x} = \frac{x}{\hbar},\\
&V = \frac{\omega(\omega + 2\hbar\lambda)x^2}{4(1+\lambda x^2)} + \frac{(l+1)(l+1-\hbar)}{x^2} = \bar{V}, \quad f(x) = \sqrt{1+\lambda x^2} = \bar{f}(\bar{x}),\\
&W = \bar{W}, \quad E_n = \bar{E}_n.
\end{aligned} \tag{A9}$$

Deformed Coulomb potential

$$\begin{aligned}
&\bar{V} = -\frac{\bar{Q}}{\bar{x}}\sqrt{1+\bar{\lambda}\bar{x}^2} + \frac{\bar{l}(\bar{l}+1)}{\bar{x}^2}, \quad \bar{f}(\bar{x}) = \sqrt{1+\bar{\lambda}\bar{x}^2},\\
&\bar{Q} = \frac{e^2}{\hbar}, \quad \bar{l} = \frac{l+1}{\hbar} - 1, \quad \bar{\lambda} = \hbar^2\lambda, \quad \bar{x} = \frac{x}{\hbar},\\
&V = -\frac{e^2}{x}\sqrt{1+\lambda x^2} + \frac{(l+1)(l+1-\hbar)}{x^2} = \bar{V}, \quad f(x) = \sqrt{1+\lambda x^2} = \bar{f}(\bar{x}),\\
&W = \bar{W}, \quad E_n = \bar{E}_n.
\end{aligned} \tag{A10}$$

References

1. Cooper, F.; Khare, A.; Sukhatme, U. Supersymmetry and quantum mechanics. *Phys. Rep.* **1995**, *251*, 267–385. [CrossRef]
2. Junker, G. *Supersymmetric Methods in Quantum and Statistical Physics*; Springer: Berlin, Germany, 1996.
3. Bagchi, B. *Supersymmetry in Quantum and Classical Physics*; Chapman and Hall/CRC: Boca Raton, FL, USA, 2000.
4. Gangopadhyaya, A.; Mallow, J.; Rasinariu, C. *Supersymmetric Quantum Mechanics: An Introduction*; World Scientific: Singapore, 2010.
5. Gendenshtein, L. Derivation of exact spectra of the Schrödinger equation by means of supersymmetry. *JETP Lett.* **1983**, *38*, 356–359.
6. Darboux, G. *Leçons sur la Théorie Générale des Surfaces*, 2nd ed.; Gauthier-Villars: Paris, France, 1912.

7. Schrödinger, E. A method of determining quantum-mechanical eigenvalues and eigenfunctions. *Proc. R. Ir. Acad.* **1940**, *A46*, 9–16.
8. Schrödinger, E. Further studies on solving eigenvalue problems by factorization. *Proc. R. Ir. Acad.* **1941**, *A46*, 183–206.
9. Schrödinger, E. The factorization of the hypergeometric equation. *Proc. R. Ir. Acad.* **1941**, *A47*, 53–54.
10. Infeld, L.; Hull, T.E. The factorization method. *Rev. Mod. Phys.* **1951**, *23*, 21–68. [CrossRef]
11. Gómez-Ullate, D.; Grandati, Y.; Milson, R. Extended Krein-Adler theorem for the translationally shape invariant potentials. *J. Math. Phys.* **2014**, *55*, 043510. [CrossRef]
12. Quesne, C. Exceptional orthogonal polynomials, exactly solvable potentials and supersymmetry. *J. Phys. A* **2008**, *41*, 392001. [CrossRef]
13. Bagchi, B.; Quesne, C.; Roychoudhury, R. Isospectrality of conventional and new extended potentials, second-order supersymmetry and role of $\mathcal{PT}$ symmetry. *Pramana J. Phys.* **2009**, *73*, 337–347. [CrossRef]
14. Quesne, C. Solvable rational potentials and exceptional orthogonal polynomials in supersymmetric quantum mechanics. *SIGMA* **2009**, *5*, 084. [CrossRef]
15. Odake, S.; Sasaki, R. Infinitely many shape invariant potentials and new orthogonal polynomials. *Phys. Lett. B* **2009**, *679*, 414–417. [CrossRef]
16. Gómez-Ullate, D.; Kamran, R.; Milson, R. An extended class of orthogonal polynomials defined by a Sturm-Liouville problem. *J. Math. Anal. Appl.* **2009**, *359*, 352–367. [CrossRef]
17. Kempf, A. Uncertainty relation in quantum mechanics with quantum group symmetry. *J. Math. Phys.* **1994**, *35*, 4483–4496. [CrossRef]
18. Hinrichsen, H.; Kempf, A. Maximal localization in the presence of minimal uncertainties in positions and in momenta. *J. Math. Phys.* **1996**, *37*, 2121–2137. [CrossRef]
19. Kempf, A. Non-pointlike particles in harmonic oscillators. *J. Phys. A* **1997**, *30*, 2093–2102. [CrossRef]
20. Witten, E. Reflections on the fate of spacetime. *Phys. Today* **1996**, *49*, 24–30. [CrossRef]
21. Bastard, G. *Wave Mechanics Applied to Semiconductor Heterostructures*; Editions de Physique: Les Ulis, France, 1988.
22. Weisbuch, C; Vinter, B. *Quantum Semiconductor Heterostructures*; Academic: New York, NY, USA, 1997.
23. Serra, L.; Lipparini, E. Spin response of unpolarized quantum dots. *Europhys. Lett.* **1997**, *40*, 667–672. [CrossRef]
24. Harrison, P.; Valavanis, A. *Quantum Wells, Wires and Dots: Theoretical and Computational Physics of Semiconductor Nanostructures*; Wiley: Chichester, UK, 2016.
25. Barranco, M.; Pi, M.; Gatica, S.M.; Hernández, E.S.; Navarro, J. Structure and energetics of mixed ^{4}He-^{3}He drops. *Phys. Rev. B* **1997**, *56*, 8997–9003. [CrossRef]
26. Geller, M.R.; Kohn, W. Quantum mechanics of electrons in crystals with graded composition. *Phys. Rev. Lett.* **1993**, *70*, 3103–3106. [CrossRef]
27. Arias de Saavedra, F.; Boronat, J.; Polls, A.; Fabrocini, A. Effective mass of one ^{4}He atom in liquid ^{3}He. *Phys. Rev. B* **1994**, *50*, 4248(R)–4251(R). [CrossRef]
28. Puente, A.; Serra, L.; Casas, M. Dipole excitation of Na clusters with a non-local energy density functional. *Z. Phys. D* **1994**, *31*, 283–286. [CrossRef]
29. Ring, P.; Schuck, P. *The Nuclear Many Body Problem*; Springer: New York, NY, USA, 1980.
30. Bonatsos, D.; Georgoudis, P.E.; Lenis, D.; Minkov, N.; Quesne, C. Bohr Hamiltonian with a deformation-dependent mass term for the Davidson potential. *Phys. Rev. C* **2011**, *83*, 044321. [CrossRef]
31. Willatzen, M.; Lassen B. The Ben Daniel-Duke model in general nanowire structures. *J. Phys. Condens. Matter* **2007**, *19*, 136217. [CrossRef]
32. Chamel, N. Effective mass of free neutrons in neutron star crust. *Nucl. Phys. A* **2006**, *773*, 263–278. [CrossRef]
33. Infeld, L. On a new treatment of some eigenvalue problems. *Phys. Rev.* **1941**, *59*, 737–747. [CrossRef]
34. Stevenson, A.F. Note on the "Kepler problem" in a spherical space, and the factorization method of solving eigenvalue problems. *Phys. Rev.* **1941**, *59*, 842. [CrossRef]
35. Infeld, L.; Schild, A. A note on the Kepler problem in a space of constant negative curvature. *Phys. Rev.* **1945**, *67*, 121. [CrossRef]
36. Kalnins, E.G.; Miller, W., Jr.; Pogosyan, G.S. Superintegrability and associated polynomial solutions: Euclidean space and the sphere in two dimensions. *J. Math. Phys.* **1996**, *37*, 6439–6467. [CrossRef]
37. Kalnins, E.G.; Miller, W., Jr.; Pogosyan, G.S. Superintegrability on the two-dimensional hyperboloid. *J. Math. Phys.* **1997**, *38*, 5416–5433. [CrossRef]

38. Quesne, C.; Tkachuk, V.M. Deformed algebras, position-dependent effective masses and curved spaces: An exactly solvable Coulomb problem. *J. Phys. A* **2004**, *37*, 4267–4281. [CrossRef]
39. Bagchi, B.; Banerjee, A.; Quesne, C.; Tkachuk, V.M. Deformed shape invariance and exactly solvable Hamiltonians with position-dependent effective mass. *J. Phys. A* **2005**, *38*, 2929–2945. [CrossRef]
40. Quesne, C. Point canonical transformations versus deformed shape invariance for position-dependent mass Schödinger equations. *SIGMA* **2009**, *5*, 046.
41. Quesne, C. Quantum oscillator and Kepler-Coulomb problems in curved spaces: Deformed shape invariance, point canonical transformations, and rational extensions. *J. Math. Phys.* **2016**, *57*, 102101. [CrossRef]
42. Gangopadhyaya, A.; Mallow, J.V. Generating shape invariant potentials. *Int. J. Mod. Phys. A* **2008**, *23*, 4949–4978. [CrossRef]
43. Bougie, J.; Gangopadhyaya, A.; Mallow, J.V. Generation of a complete set of additive shape-invariant potentials from an Euler equation. *Phys. Rev. Lett.* **2010**, *105*, 210402. [CrossRef]
44. Bougie, J.; Gangopadhyaya, A.; Mallow, J.; Rasinariu, C. Supersymmetric quantum mechanics and solvable models. *Symmetry* **2012**, *4*, 452–473. [CrossRef]
45. Bougie, J.; Gangopadhyaya, A.; Mallow, J.V.; Rasinariu, C. Generation of a novel exactly solvable potential. *Phys. Lett. A* **2015**, *379*, 2180–2183. [CrossRef]
46. Mallow, J.V.; Gangopadhyaya, A.; Bougie, J.; Rasinariu, C. Inter-relations between additive shape invariant superpotentials. *Phys. Lett. A* **2020**, *384*, 126129. [CrossRef]
47. Mustafa, O.; Mazharimousavi, S.H. Ordering ambiguity revisited via position dependent mass pseudo-momentum operators. *Int. J. Theor. Phys.* **2007**, *46*, 1786–1796. [CrossRef]
48. Quesne, C. Revisiting (quasi-)exactly solvable rational extensions of the Morse potential. *Int. J. Mod. Phys. A* **2012**, *27*, 1250073. [CrossRef]
49. Quesne, C. Novel enlarged shape invariance property and exactly solvable rational extensions of the Rosen-Morse II and Eckart potentials. *SIGMA* **2012**, *8*, 080. [CrossRef]
50. Grandati, Y. New rational extensions of solvable potentials with finite bound state spectrum. *Phys. Lett. A* **2012**, *376*, 2866–2872. [CrossRef]
51. Grandati, Y.; Quesne, C. Confluent chains of DBT: Enlarged shape invariance and new orthogonal polynomials. *SIGMA* **2015**, *11*, 061. [CrossRef]

7

Symmetry, Transactions and the Mechanism of Wave Function Collapse

John Gleason Cramer [1,*] and Carver Andress Mead [2]

1 Department of Physics, University of Washington, Seattle, WA 98195, USA
2 California Institute of Technology, Pasadena, CA 91125, USA; carver@caltech.edu
* Correspondence: jcramer@uw.edu

Abstract: The Transactional Interpretation of quantum mechanics exploits the intrinsic time-symmetry of wave mechanics to interpret the ψ and ψ^* wave functions present in all wave mechanics calculations as representing retarded and advanced waves moving in opposite time directions that form a quantum "handshake" or transaction. This handshake is a 4D standing-wave that builds up across space-time to transfer the conserved quantities of energy, momentum, and angular momentum in an interaction. Here, we derive a two-atom quantum formalism describing a transaction. We show that the bi-directional electromagnetic coupling between atoms can be factored into a matched pair of vector potential Green's functions: one retarded and one advanced, and that this combination uniquely enforces the conservation of energy in a transaction. Thus factored, the single-electron wave functions of electromagnetically-coupled atoms can be analyzed using Schrödinger's original wave mechanics. The technique generalizes to any number of electromagnetically coupled single-electron states—no higher-dimensional space is needed. Using this technique, we show a worked example of the transfer of energy from a hydrogen atom in an excited state to a nearby hydrogen atom in its ground state. It is seen that the initial exchange creates a dynamically unstable situation that avalanches to the completed transaction, demonstrating that wave function collapse, considered mysterious in the literature, can be implemented with solutions of Schrödinger's original wave mechanics, coupled by this unique combination of retarded/advanced vector potentials, without the introduction of any additional mechanism or formalism. We also analyze a simplified version of the photon-splitting and Freedman–Clauser three-electron experiments and show that their results can be predicted by this formalism.

Keywords: quantum mechanics; transaction; Wheeler–Feynman; transactional interpretation; handshake; advanced; retarded; wave function collapse; collapse mechanism; EPR; HBT; Jaynes; NCT; split photon; Freedman–Clauser; nonlocality; entanglement

---

1. Introduction

Quantum mechanics (QM) was never properly finished. Instead, it was left in an exceedingly unsatisfactory state by its founders. Many attempts by highly qualified individuals to improve the situation have failed to produce any consensus about either (a) the precise nature of the problem, or (b) what a better form of QM might look like.

At the most basic level, a simple observation illustrates the central conceptual problem:

An excited atom somewhere in the universe transfers *all* of its excitation energy to another single atom, independent of the presence of the vast number of alternative atoms that could have received all or part of the energy. The obvious "photon-as-particle" interpretation of this situation has a one-way symmetry: The excited source atom is depicted as emitting a particle, a *photon* of electromagnetic energy that is somehow oscillating with angular frequency ω while moving in a particular direction.

The photon is depicted as carrying a quantum of energy $\hbar\omega$, a momentum $\hbar\omega/c$, and an angular momentum $\hbar$ through space, until it is later absorbed by some unexcited atom. The emission and absorption are treated as independent isolated events without internal structure. It is insisted that the only real and meaningful quantities describing this process are *probabilities*, since these are measurable. The necessarily abrupt change in the quantum wave function of the system when the photon arrives (and an observer potentially gains information) is called "wave function collapse" and is considered to be a mysterious process that the founders of QM found it necessary to "put in by hand" without providing any mechanism. [The missing mechanism behind wave function collapse is sometimes called "the measurement problem", particularly by acolytes of Heisenberg's knowledge interpretation. In our view, measurement requires wave function collapse *but does not cause it.*] [Side comments will be put in square brackets]

Referring to statistical quantum theory, which is reputed to apply only to *ensembles* of similar systems, Albert Einstein [1] had this to say:

> *"I do not believe that this fundamental concept will provide a useful basis for the whole of physics."*
>
> *"I am, in fact, firmly convinced that the essentially statistical character of contemporary quantum theory is solely to be ascribed to the fact that this [theory] operates with an incomplete description of physical systems."*
>
> *"One arrives at very implausible theoretical conceptions, if one attempts to maintain the thesis that the statistical quantum theory is in principle capable of producing a complete description of an individual physical system ..."*
>
> *"Roughly stated, the conclusion is this: Within the framework of statistical quantum theory, there is no such thing as a complete description of the individual system. More cautiously, it might be put as follows: The attempt to conceive the quantum-theoretical description as the complete description of the individual systems leads to unnatural theoretical interpretations, which become immediately unnecessary if one accepts the interpretation that the description refers to ensembles of systems and not to individual systems. In that case, the whole 'egg-walking' performed in order to avoid the 'physically real' becomes superfluous. There exists, however, a simple psychological reason for the fact that this most nearly obvious interpretation is being shunned—for, if the statistical quantum theory does not pretend to describe the individual system (and its development in time) completely, it appears unavoidable to look elsewhere for a complete description of the individual system. In doing so, it would be clear from the very beginning that the elements of such a description are not contained within the conceptual scheme of the statistical quantum theory. With this. one would admit that, in principle, this scheme could not serve as the basis of theoretical physics. Assuming the success of efforts to accomplish a complete physical description, the statistical quantum theory would, within the framework of future physics, take an approximately analogous position to the statistical mechanics within the framework of classical mechanics. I am rather firmly convinced that the development of theoretical physics will be of this type, but the path will be lengthy and difficult."*
>
> *"If it should be possible to move forward to a complete description, it is likely that the laws would represent relations among all the conceptual elements of this description which, per se, have nothing to do with statistics."*

In what follows we put forth a simple approach to *describing the individual system (and its development in time)*, which Einstein believed was missing from statistical quantum theory and which must be present before any theory of physics could be considered to be complete.

The way forward was suggested by the phenomenon of *entanglement*. Over the past few decades, many increasingly exquisite Einstein–Podolsky–Rosen [2] (EPR) experiments [3–11] have demonstrated that multi-body quantum systems with separated components that are subject to conservation laws exhibit a property called "quantum entanglement" [12]: Their component wave functions are inextricably locked together, and they display a nonlocal correlated behavior enforced over an arbitrary interval of space-time without any hint of an underlying mechanism or any show of respect for our cherished classical "arrow of time." Entanglement is the most mysterious of the many so-called "quantum mysteries."

It has thus become clear that the quantum transfer of energy must have quite a different symmetry from that implied by this simple "photon-as-particle" interpretation. Within the framework of statistical QM, the intrinsic symmetry of the energy transfer and the mechanisms behind wave function collapse and entanglement have been greatly clarified by the **Transactional Interpretation of quantum mechanics** (TI), developed over several decades by one of us and recently described in some detail in the book ***The Quantum Handshake*** [12]. [We note that Ruth Kastner has extended her "probabilist" variant of the TI, which embraces the Heisenberg/probability view and characterizes transactions as events in many-dimensional Hilbert space, into the quantum-relativistic domain [13,14] and has used it to extend and enhance the "decoherence" approach to quantum interpretation [15]].

This paper begins with a tutorial review of the TI approach to a credible photon mechanism developed in the book ***Collective Electrodynamics*** [16], followed by a deeper dive into the electrodynamics of the quantum handshake, and finally includes descriptions of several historic experiments that have excluded entire classes of theories. We conclude that the approach described here has *not* been excluded by any experiment to date.

1.1. Wheeler–Feynman Electrodynamics

The Transactional Interpretation was inspired by classical time-symmetric Wheeler–Feynman electrodynamics [17,18] (WFE), sometimes called "absorber theory." Basically, WFE assumes that electrodynamics must be time-symmetric, with equally valid retarded waves (that arrive *after* they are emitted) and advanced waves (that arrive *before* they are emitted). WFE describes a "handshake" process accounting for emission recoil in which the emission of a retarded wave stimulates a future absorber to produce an advanced wave that arrives back at the emitter at the instant of emission. WFE is based on electrodynamic time symmetry and has been shown to be completely interchangeable with conventional classical electrodynamics in its predictions.

WFE asserts that the breaking of the intrinsic time-symmetry to produce the electromagnetic arrow of time, i.e., the observed dominance of retarded radiation and absence of advanced radiation in the universe, arises from the presence of more absorption in the future than in the past. In an expanding universe, that assertion is questionable. One of us has suggested an alternative cosmological explanation [19], which employs advanced-wave termination and reflection from the singularity of the Big Bang.

1.2. The Transactional Interpretation of Quantum Mechanics

The Transactional Interpretation of quantum mechanics [12] takes the concept of a WFE handshake from the classical regime into the quantum realm of photons and massive particles. The retarded and advanced waves of WFE become the quantum wave functions ψ and ψ^*. Note that the complex conjugation of ψ^* is in effect the application of the Wigner time-reversal operator, thus representing an advanced wave function that carries negative energy from the present to the past.

Let us here clarify what an *interpretation* of quantum mechanics actually is. An interpretation serves the function of explaining and clarifying the formalism and procedures of its theory. In our view, the mathematics is (and should be) exclusively contained in the formalism itself. The interpretation should not introduce additional variant formalism. [We note, however, that this principle is violated by the Bohm–de Broglie "interpretation" with its "quantum potentials" and uncertainty-principle-violating trajectories, by the Ghirardi–Rimini–Weber "interpretation" with its nonlinear stochastic term, and by many other so-called interpretations that take the questionable liberty of modifying the standard QM formalism. In that sense, these are alternative variant quantum theories, *not* interpretations at all.]

The present work is a calculation describing the formation of a transaction that was inspired by the Transactional Interpretation but has not previously been a part of it. In Section 12 below, we discuss how the TI is impacted by this work. We use Schrödinger's original wave mechanics formalism *with the inclusion of retarded and advanced electromagnetic four-potentials* to describe and illuminate the processes of transaction formation and the collapse of the wave function. We show that this approach can provide a detailed mathematical description of a "quantum-jump" in which what seems to be a photon is emitted by one hydrogen atom in an excited state and excites another hydrogen atom initially in its ground state. Thus, the mysterious process of wave function collapse becomes just a phenomenon involving an exchange of advanced/retarded electromagnetic waves between atomic wave functions described by the Schrödinger formalism.

As illustrated schematically in Figure 1, the process described involves the initial existence in each atom of a very small admixture of the wave function for the opposite state, thereby forming two-component states in both atoms. This causes them to become weak dipole radiators oscillating at the same difference-frequency ω_0. The interaction that follows, characterized by a retarded-advanced exchange of 4-vector potentials, leads to an exponential build-up of a transaction, resulting in the complete transfer of one photon worth of energy $\hbar\omega_0$ from one atom to the other. This process is described in more detail below.

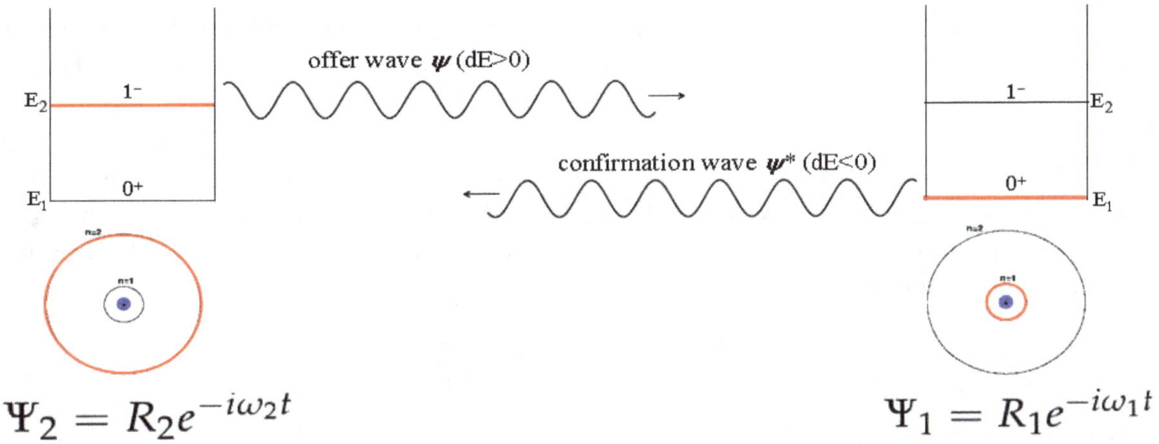

Figure 1. Model of transaction formation: An emitter atom 2 in a space-antisymmetric excited state of energy E_2 and an absorber atom 1 in a space-symmetric ground state of energy E_1 both have slight admixtures of the other state, giving both atoms dipole moments that oscillate with the same difference-frequency $\omega_0 = \omega_2 - \omega_1$. If the relative phase of the initial small offer wave ψ and confirmation wave $\psi*$ is optimal, this condition initiates energy transfer, which avalanches to complete transfer of one photon-worth of energy $\hbar\omega_0$.

2. Physical Mechanism of the Transfer

The standard formalism of QM consists of a set of arbitrary rules, conventionally viewed as dealing only with probabilities. When illuminated by the TI, that formalism hints at an underlying physical mechanism that might be understood, in the usual sense of the concept *understood*. The first glimpse of such an understanding, and of the physical nature of the transactional symmetry, was suggested by Gilbert N. Lewis in 1926 [20,21], the same year he gave electromagnetic energy transfer the name "photon":

> *"It is generally assumed that a radiating body emits light in every direction, quite regardless of whether there are near or distant objects which may ultimately absorb that light; in other words that it radiates 'into space'..."*
>
> *"I am going to make the contrary assumption that an atom never emits light except to another atom..."*

"I propose to eliminate the idea of mere emission of light and substitute the idea of transmission, or a process of exchange of energy between two definite atoms... Both atoms must play coordinate and symmetrical parts in the process of exchange..."

"In a pure geometry it would surprise us to find that a true theorem becomes false when the page upon which the figure is drawn is turned upside down. A dissymmetry alien to the pure geometry of relativity has been introduced by our notion of causality."

In what follows, we demonstrate that the pair of coupled Schrödinger equations describing the two atoms, as coupled by a relativistically correct description of the electromagnetic field, exhibit a unique solution. This solution has exactly the symmetry of the TI and thus provides a *physically understandable* mechanism for the experimentally observed behavior: Both atoms, in the words of Lewis, *"play coordinate and symmetrical parts in the process of exchange"*.

The solution gives a smooth transition in each of the atomic wave functions, brought to abrupt closure by the highly nonlinear increase in coupling as the transition proceeds. The origin of statistical behavior and "quantum randomness" can be understood in terms of the random distribution of wave-function amplitudes and phases provided by the perturbations of the many other potential recipient atoms; no "hidden variables" are required. Although much remains to be done, these findings might be viewed as a next step towards a physical understanding of the process of quantum energy transfer.

We will close by indicating the deep, fundamental questions that we have not addressed, and that must be understood before anything like a complete physical understanding of QM is in hand.

3. Quantum States

In 1926, Schrödinger, seeking a wave-equation description of a quantum system with mass, adopted Planck's notion that energy was somehow proportional to frequency together with deBroglie's idea that momentum was the propagation vector of a wave and crafted his wave equation for the time evolution of the wave function Ψ [22]:

$$-\frac{\hbar}{2m\,i}\nabla^2\Psi + \frac{q\,V}{i\,\hbar}\Psi = \frac{\partial\Psi}{\partial t}. \tag{1}$$

Here, V is the electrical potential, m is the electron mass, and q is the (negative) charge on the electron. Thus, what is the meaning of the wave function Ψ that is being characterized? In modern treatments, Ψ is called a "probability amplitude", which has only a probabilistic interpretation. In what follows, however, we return to Schrödinger's original vision, which provides a detailed physical picture of the wave function and how it interacts with other charges:

"The hypothesis that we have to admit is very simple, namely that the square of the absolute value of Ψ is proportional to an electric density, which causes emission of light according to the laws of ordinary electrodynamics."

That vision has inspired generations of talented conceptual thinkers to invent solutions to technical problems using Schrödinger's approach. Foremost among these was Ed Jaynes who, with a number of students and collaborators, attacked a host of quantum problems in this manner [23–30]. A great deal of physical understanding was obtained, in particular concerning lasers and the coherent optics made possible by them. The theory evolved rapidly and had an enabling role in the explosive progress of that field. Indeed, the continued rapid technical progress into the present is due, in no small part, to the understanding gained through application of the Jaynes way of thinking. A detailed review of the progress up to 1972 was reported in a conference that year [30]. By then this class of theory was called **neoclassical** (NCT) because of its use of Maxwell's equations. While there was no question about the utility of NCT in the conceptualization and technical realization of amazing quantum-optics devices and their application, there was a deep concern about whether it could possibly be correct *at* *the fundamental level*—maybe it was just a clever bunch of hacks. The tension over this concern was a major focus of the 1972 *Third Rochester Conference on Coherence and Quantum Optics* [31], and several experiments testing NCT predictions were discussed there by Jaynes [30].

He ended his presentation this way:

> *"We have not commented on the beautiful experiment reported here by Clauser [32], which opens up an entirely new area of fundamental importance to the issues facing us..."*
>
> *"What it seems to boil down to is this: a perfectly harmless looking experimental fact (nonoccurence of coincidences at 90°), which amounts to determining a single experimental point—and with a statistical measurement of unimpressive statistical accuracy—can, at a single stroke, throw out a whole infinite class of alternative theories of electrodynamics, namely all local causal theories."*
>
> *"...if the experimental result is confirmed by others, then this will surely go down as one of the most incredible intellectual achievements in the history of science, and my own work will lie in ruins."*

The experiment he was alluding to was that of Freedman and Clauser [6], and in particular to their observaton of an essentially zero coincidence rate with crossed polarizers. The Freedman–Clauser experiment (see Section 13.3 below), with its use of entangled photon pairs, was the vanguard of an entire new direction in quantum physics that now goes under the rubric of **Tests of Bell's Inequality** [3,4] and/or **EPR experiments** [6–11]. Both the historic EPR experiment and its analysis have been repeated many times with ever-increasing precision, and always with the same outcome: a difinitive violaton of Bell's inequalities. Local causal theories were dead! [Much of the literature on violations of Bell's inequalities in EPR experiments has unfortunately emphasized the refutation of *local hidden-variable theories*. In our view, this is a regrettable historical accident attributable to Bell. *Nonlocal* hidden-variable theories have been shown to be compatible with EPR results. It is *locality* that has been refuted. Entangled systems exhibit correlations that can only be accomodated by quantum nonlocality. The TI supplies the mechanism for that nonlocality.]

In fact, it was the manifest quantum nonlocality evident in the early EPR experiments of the 1970s that led to the synthesis of the transactional interpretation in the 1980s [19,33,34], designed to compactly explain entanglement and nonlocality. This in turn led to the search for an underlying transaction mechanism, as reported in 2000 in ***Collective Electrodynamics*** [16]. As we detail below, the quantum handshake, as mediated by advanced/retarded electromagnetic four-potentials, provides the effective non-locality so evident in modern versions of these EPR experiments. In Section 13, we analyze the Freedman–Clauser experiment in detail and show that their result is a natural outcome of our approach. Jaynes' work does not lie in ruins—all that it needed for survival was the non-local quantum handshake! What follows is an extension and modification of NCT using a different non-Maxwellian form of E&M [16] and including our non-local Transactional approach. We illustrate the approach with the simplest possible physical arrangements, described with the major goal of conceptual understanding rather than exhaustion. Obviously, much more work needs to be done, which we point out where appropriate.

Atoms

We will begin by visualizing the electron as Schrödinger and Jaynes did: as having a smooth charge distribution in three-dimensional space, whose density is given by $\Psi^*\Psi$. There is no need for statistics and probabilities at any point in these calculations, and none of the quantities have statistical meaning. The probabilistic outcome of quantum experiments has the same origin as it does in all other experiments—random perturbations beyond the control of the experimenter. We return to the topic of probability after we have established the nature of the transaction.

For a local region of positive potential V, for example near a positive proton, the negative electron's wave function has a local potential energy (qV) minimum in which the electron's wave function can form local **bound states**. The spatial shape of the wave function amplitude is a trade-off between getting close to the proton, which lowers its potential energy, and bunching together too much, which increases its ∇^2 "kinetic energy." Equation (1) is simply a mathematical expression of this trade-off, a statement of the physical relation between mass, energy, and momentum in the form of a wave equation.

A discrete set of states called **eigenstates** are standing-wave solutions of Equation (1) and have the form $\Psi = Re^{-i\omega t}$, where R and V are functions of only the spatial coordinates, and the angular frequency ω is itself independent of time. For the hydrogen atom, the potential $V = \epsilon_0 q_p / r$, where q_p is the positive charge on the nucleus, equal in magnitude to the electron charge q. Two of the lowest-energy solutions to Equation (1) with this potential are:

$$\Psi_{100} = \frac{e^{-r}}{\sqrt{\pi}} e^{-i\omega_1 t} \qquad \Psi_{210} = \frac{r\, e^{-r/2} \cos(\theta)}{4\sqrt{6\pi}} e^{-i\omega_2 t}, \tag{2}$$

where the dimensionless radial coordinate r is the radial distance divided by the **Bohr radius** a_0:

$$a_0 \equiv \frac{4\pi\epsilon_0 \hbar^2}{mq^2} = 0.0529\ \text{nm}, \tag{3}$$

and θ is the azimuthal angle from the North Pole ($+z$ axis) of the spherical coordinate system.

The spatial "shape" of the two lowest energy eigenstates for the hydrogen atom is shown in Figure 2. Here, we focus on the excited-state wave function Ψ_{210} that has no angular momentum projection on the z-axis. For the moment, we set aside the wave functions $\Psi_{21\pm1}$, which have $+1$ and -1 angular momentum z-axis projections. Because, for any single eigenstate, the electron density is $\Psi^*\Psi = Re^{i\omega t} Re^{-i\omega t} = R^2$, the charge density is not a function of time, so none of the other properties of the wave function change with time. The individual eigenstates are thus **stationary states**. The lowest energy bound eigenstate for a given form of potential minimum is called its **ground state**, shown on the left in Figure 3. The corresponding charge densities are shown in Figure 4.

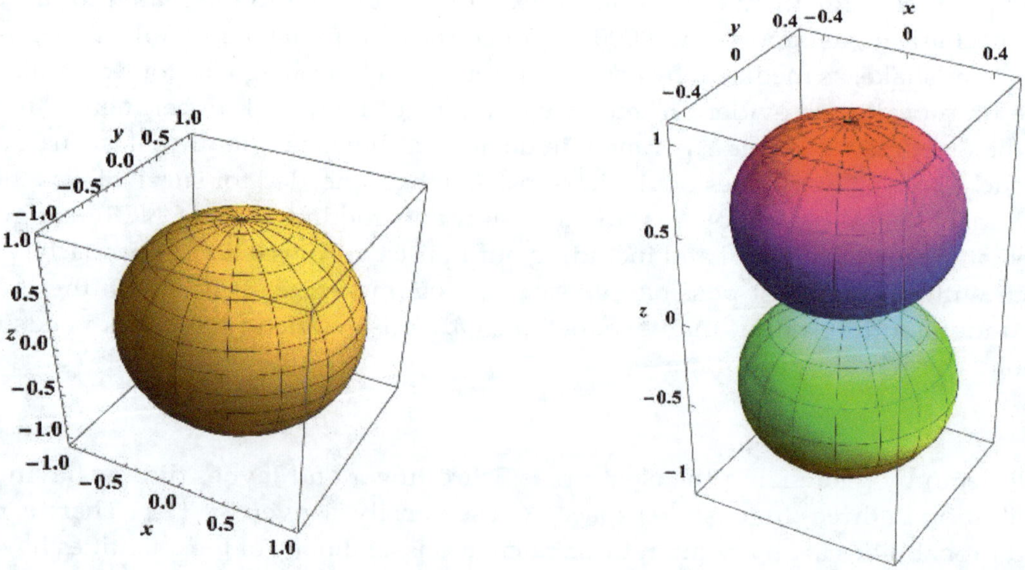

Figure 2. Angular dependence of the spatial wave function amplitudes for the lowest (100, **left**) and next higher (210, **right**) states of the hydrogen atom, plotted as unit radius in spherical coordinates from Equation (2). The 100 wave function has spherical symmetry: positive in all directions. The 210 wave function is antisymmetric along the z-axis, as indicated by the color change. In practice, the direction of the z-axis will be established by an external electromagnetic field, as we shall analyze shortly.

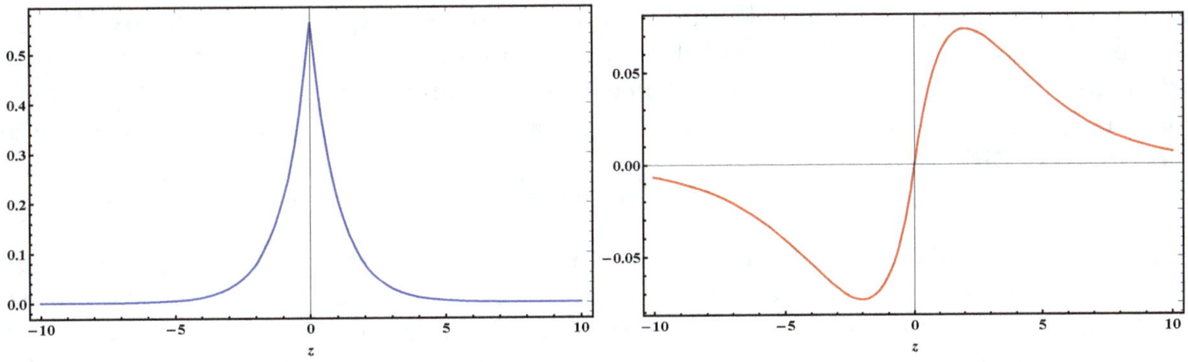

Figure 3. Wave function amplitudes Ψ for the 100 and 210 states, along the z-axis of the hydrogen atom. The horizontal axis in all plots is the position along the z-axis in units of the Bohr radius.

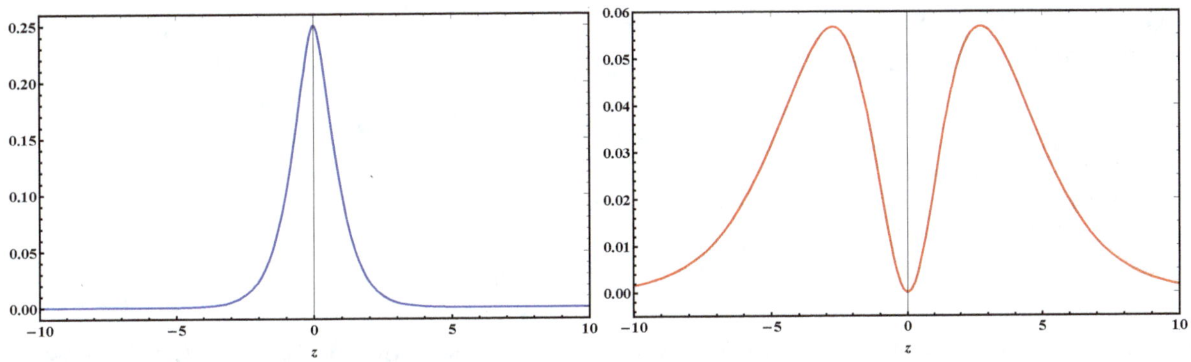

Figure 4. Contribution of $x - y$ "slices" at position z of wave function density $\Psi^*\Psi$ to the total charge or mass of the 100 and 210 states of the hydrogen atom. Both curves integrate to 1.

In 1926, Schrödinger had already derived the energies and wave functions for the stationary solutions of his equation for the hydrogen atom. His physical insight that the absolute square $\Psi^*\Psi$ of the wave function was the *electron density* had enabled him to work out the energy shifts of these levels caused by external applied electric and magnetic fields, the expected strengths of the transitions between pairs of energy levels, and the polarization of light from certain transitions.

These predictions could be compared directly with experimental data, which had been previously observed but not understood. He reported that these calculations were:

> *"...not at all difficult, but very tedious. In spite of their tediousness, it is rather fascinating to see all the well-known but not understood "rules" come out one after the other as the result of familiar elementary and absolutely cogent analysis, like e.g., the fact that* $\int_0^{2\pi} \cos m\phi \, \cos n\phi \, d\phi$ *vanishes unless* $n = m$*. Once the hypothesis about* $\Psi^*\Psi$ *has been made, no accessory hypothesis is needed or is possible; none could help us if the "rules" did not come out correctly. However, fortunately they do [22,35]."*

The Schrödinger/Jaynes approach enables us to describe continuous quantum transitions in an intuitively appealing way: We extend the electromagnetic coupling described in Collective Electrodynamics [16] to the wave function of a single electron, and require only the most rudimentary techniques of Schrödinger's original quantum theory.

4. The Two-State System

The first two eigenstates of the Hydrogen atom, from Equation (2), form an ideal two-state system. We refer to the 100 ground state as State 1, with wave function Ψ_1 and energy E_1, and to the 210 excited state as State 2, with wave function Ψ_2 and energy $E_2 > E_1$:

$$\Psi_1 = R_1 e^{-i\omega_1 t} \qquad\qquad \Psi_2 = R_2 e^{-i\omega_2 t}, \tag{4}$$

where $\omega_1 = E_1/\hbar$, $\omega_2 = E_2/\hbar$, and R_1 and R_2 are real, time-independent functions of the space coordinates. The wave functions represent totally continuous matter waves, and all of the usual operations involving the wave function are methods for computing properties of this continuous distribution. The only particularly quantal assumption involved is that the wave function obeys a **normalization condition**:

$$\int \Psi^* \Psi \, d\text{vol} = 1, \tag{5}$$

where the integral is taken over the entire three-dimensional volume where Ψ is non-vanishing. [Envelope functions like R_1 and R_2 generally die out exponentially with distance sufficiently far from the "region of interest", such as an atom. Integrals like this one and those that follow in principle extend to infinity but in practice are taken out far enough that the part being neglected is within the tolerance of the calculation.].

Equation (5) ensures that the total charge will be a single electronic charge, and the total mass will be a single electronic mass.

By construction, the eigenstates of the Schrödinger equation are real and orthogonal:

$$\int R_1 R_2 \, d\text{vol} = 0. \tag{6}$$

The first moment $\langle z \rangle$ of the electron distribution along the atom's z-axis is:

$$\langle z \rangle \equiv \int \Psi^* z \, \Psi \, d\text{vol}, \tag{7}$$

In statistical treatments, $\langle z \rangle$ would be called the "expectation value of z", whereas for our continuous distribution it is called the "average value of z" or the "first moment of z." The electron wave function is a *wave packet* and is subject to all the Fourier properties of one, as treated at some length in Ref. [12]. Statistical QM insisted that electrons were "point particles", so one was no longer able to visualize how they could exhibit interference or other wave properties, so a set of rules was constructed to make the outcomes of statistical experiments come out right. Among these was the **uncertainty principle**, which simply restated the Fourier properties of an object described by waves in a statistical context. No statistical attributes are attached to any properties of the wave function in this treatment.

Equation (7) gives the position of the center of negative charge of the electron wave function relative to the positive charge on the nucleus. When multiplied by the electronic charge q, it is called *the electric dipole moment* $q\langle z \rangle$ of the charge distribution of the atom:

$$q \langle z \rangle = q \int \Psi^* z \, \Psi \, d\text{vol}. \tag{8}$$

From Equations (7) and (4), the dipole moment for the ith eigenstate is:

$$q \langle z_i \rangle = q \int \Psi_i^* z \, \Psi_i \, d\text{vol} = q \int R_i^* z \, R_i \, d\text{vol} = q \int R_i^2 \, z \, d\text{vol}. \tag{9}$$

Pure eigenstates of the system will have a definite parity, i.e., they will have wave functions with either even symmetry $[\Psi(z) = \Psi(-z)]$, or odd symmetry $[\Psi(z) = -\Psi(-z)]$. For either symmetry, the integral over $R^2 z$ vanishes, and the dipole moment is zero. We note that, even if the wave function did not have even or odd symmetry, the dipole moment, and all higher moments as well, would be independent of time. By their very nature, eigenstates are stationary states and can be visualized as standing-waves—none of their physical spatial attributes can be functions of time. In order to radiate electromagnetic energy, the charge distribution *must change with time*.

The notion of *stationarity* is the quantum answer to the original question about atoms depicted as electrons orbiting a central nucleus like a tiny Solar System:

Why doesn't the electron orbiting the nucleus radiate its energy away?
In his 1917 book, ***The Electron***, R.A. Millikan [36] anticipates the solution in his comment about the

> *"...apparent contradiction involved in the non-radiating electronic orbit—a contradiction which would disappear, however, if the negative electron itself, when inside the atom, were a ring of some sort, capable of expanding to various radii, and capable, only when freed from the atom, of assuming essentially the properties of a point charge."*

Millikan was the first researcher to directly observe and measure the quantized charge on the electron with his famous oil-drop experiment, for which he later received the Nobel prize. Ten years before the statistical quantum theory was put in place, he had clearly seen that a continuous, symmetric electronic charge distribution would not radiate, and that the real problem was the assumption of a point charge.

5. Transitions

The eigenstates of the system form a complete basis set, so any behavior of the system can be expressed by forming a linear combination (superposition) of its eigenstates.

The general form of such a superposition of our two chosen eigenstates is:

$$\Psi = ae^{i\phi_a} R_1 e^{-i\omega_1 t} + be^{i\phi_b} R_2 e^{-i\omega_2 t}, \tag{10}$$

where a and b are real amplitudes, and ϕ_a and ϕ_b are real constants that determine the phases of the oscillations ω_1 and ω_2.

Using $\omega_0 = \omega_2 - \omega_1$ and $\phi = \phi_a - \phi_b$, the charge density ρ of the two-component-state wave function is:

$$\begin{aligned}
\rho &= q\Psi^* \Psi \\
\frac{\rho}{q} &= \left(ae^{-i\phi_a} R_1 e^{i\omega_1 t} + be^{-i\phi_b} R_2 e^{i\omega_2 t}\right)\left(ae^{i\phi_a} R_1 e^{-i\omega_1 t} + be^{i\phi_b} R_2 e^{-i\omega_2 t}\right) \\
&= a^2 R_1^2 + b^2 R_2^2 + \left(ae^{-i\phi_a} be^{i\phi_b}\, e^{-i\omega_0 t} + be^{-i\phi_b} ae^{i\phi_a}\, e^{i\omega_0 t}\right) R_1 R_2 \\
&= a^2 R_1^2 + b^2 R_2^2 + ab\left(e^{i(\phi_b - \phi_a)}\, e^{-i\omega_0 t} + e^{i(\phi_a - \phi_b)}\, e^{i\omega_0 t}\right) R_1 R_2 \\
&= a^2 R_1^2 + b^2 R_2^2 + ab\left(e^{-i(\omega_0 t - \phi)} + e^{i(\omega_0 t + \phi)}\right) R_1 R_2 \\
&= a^2 R_1^2 + b^2 R_2^2 + 2abR_1 R_2 \cos(\omega_0 t + \phi).
\end{aligned} \tag{11}$$

Thus, the charge density of the two-component wave function is made up of the charge densities of the two separate wave functions, shown in Figure 4, plus a term proportional to the product of the two wave function amplitudes. It reduces to the individual charge density of the ground state when $a = 1, b = 0$ and to that of the excited state when $a = 0, b = 1$. The product term, shown in green in Figure 5, is the only non-stationary term; it oscillates at the transition frequency ω_0. The integral of the total density shown in the right-hand plot is equal to 1 for any phase of the cosine term, since there is only one electron in this two-component state.

All the $\Psi^*\Psi$ plots represent the density of negative charge of the electron. The atom as a whole is neutral because of the equal positive charge on the nucleus. The dipole is formed when the center of charge of the electron wave function is displaced from the central positive charge of the nucleus.

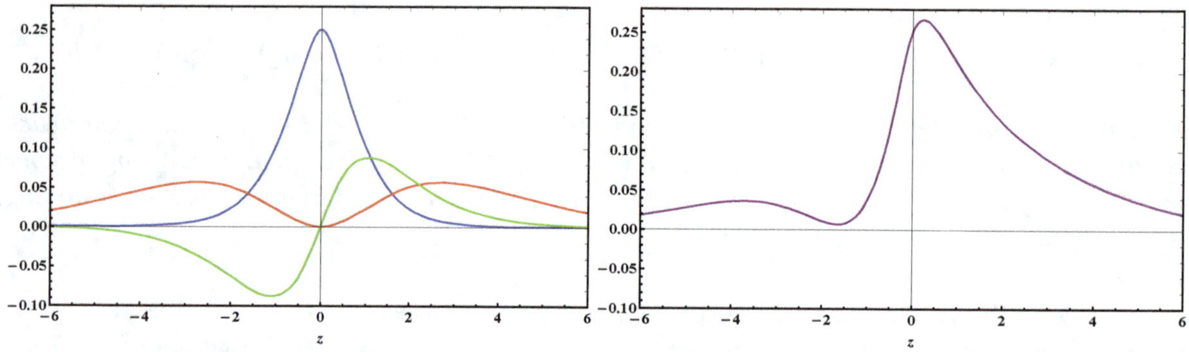

Figure 5. Left: Plot of the three terms in the wave-function density in Equation (11) for an equal $\left(a = b = 1/\sqrt{2}\right)$ superposition of the ground state (R_1^2, blue) and first excited state (R_2^2, red) of the hydrogen atom. The green curve is a snapshot of the time-dependent $R_1 R_2$ product term, which oscillates at difference frequency ω_0. **Right**: Snapshot of the total charge density, which is the sum of the three curves in the left plot. The magnitudes plotted are the contribution to the total charge in an $x - y$ "slice" of $\Psi^*\Psi$ at the indicated z coordinate. All plots are shown for the time such that $\cos(\omega_0 t + \phi) = 1$. The horizontal axis in each plot is the spatial coordinate along the z-axis of the atom, given in units of the Bohr radius a_0. Animation here [37] (see Supplementary Materials).

The two-component wave function must be normalized, since it is the state of one electron:

$$\begin{aligned}
&\int \Psi^* \, \Psi \; d\text{vol} = 1 \\
&= \int \left(ae^{-i\phi_a} R_1 e^{i\omega_1 t} + be^{-i\phi_b} R_2 e^{i\omega_2 t}\right) \\
&\quad \left(ae^{i\phi_a} R_1 e^{-i\omega_1 t} + be^{i\phi_b} R_2 e^{-i\omega_2 t}\right) \; d\text{vol} \\
&= a^2 \int R_1^2 \, d\text{vol} + b^2 \int R_2^2 d\text{vol} \\
&+ \left(ae^{-i\phi_a} be^{i\phi_b} \, e^{-i\omega_0 t} + be^{-i\phi_b} ae^{i\phi_a} \, e^{i\omega_0 t}\right) \int R_1 R_2 d\text{vol}.
\end{aligned} \tag{12}$$

Recognizing from Equations (5) and (6) that the individual eigenfunctions are normalized and orthogonal:

$$\int R_1^2 \, d\text{vol} = 1 \qquad \int R_2^2 \, d\text{vol} = 1 \qquad \int R_1 R_2 \; d\text{vol} = 0. \tag{13}$$

Equation (12) becomes

$$\int \Psi^* \, \Psi \; d\text{vol} = 1 = a^2 + b^2. \tag{14}$$

Thus, a^2 represents the fraction of the two-component wave function made up of the lower state Ψ_1, and b^2 represents the fraction made up of the upper state Ψ_2. The total energy E of a system whose wave function is a superposition of two eigenstates is:

$$E = a^2 E_1 + b^2 E_2. \tag{15}$$

Using the normalization condition $a^2 + b^2 = 1$ and solving Equation (15) for b^2, we obtain:

$$b^2 = \frac{E - E_1}{E_2 - E_1}. \tag{16}$$

In other words, b^2 is just the energy of the wave function, normalized to the transition energy, and using E_1 as its reference energy. Taking E_1 as our zero of energy and $E_0 = E_2 - E_1$, Equation (16) becomes:

$$E = E_0 b^2 \quad \Rightarrow \quad \frac{\partial E}{\partial t} = E_0 \frac{\partial \left(b^2\right)}{\partial t}. \tag{17}$$

Defining:

$$d_{12} \equiv 2q \int R_1 R_2 \, z \, dvol = 2q \langle z \rangle_{max}, \tag{18}$$

the dipole moment $q \langle z \rangle$ of such a superposition can, from Equation (11), be written:

$$q \langle z \rangle = d_{12} ab \cos(\omega_0 t + \phi). \tag{19}$$

The factor d_{12} we call the **dipole strength** for the transition. When one R is an even function of z and the other is an odd function of z, as in the case of the 100 and 210 states of the hydrogen atom, then d_{12} is nonzero, and the transition is said to be **electric dipole allowed**. When both R_1 and R_2 are either even or odd functions of z, $d_{12} = 0$, and the transition is said to be **electric dipole forbidden**.

Even in this case, some other moment of the distribution generally will be nonzero, and the transition can proceed by magnetic dipole, magnetic quadrupole, or other higher-order moments.

For now, we will concentrate on transitions that are electric dipole allowed.

We have the time dependence of the electron dipole moment $q \langle z \rangle$ from Equation (19), from which we can derive the velocity and acceleration of the charge:

$$\begin{aligned} q \langle z \rangle &= d_{12} ab \cos(\omega_0 t + \phi) \\ q \frac{\partial \langle z \rangle}{\partial t} &= -\omega_0 d_{12} ab \sin(\omega_0 t + \phi) + d_{12} \cos(\omega_0 t + \phi) \frac{\partial (ab)}{\partial t} \\ &\approx -\omega_0 d_{12} ab \sin(\omega_0 t + \phi) \\ q \frac{\partial^2 \langle z \rangle}{\partial t^2} &\approx -\omega_0^2 d_{12} ab \cos(\omega_0 t + \phi), \end{aligned} \tag{20}$$

where the approximation arises because we will only consider situations where the coefficients a and b change slowly with time over a large number of cycles of the transition frequency: $\left(\frac{\partial (ab)}{\partial t} \ll ab \, \omega_0 \right)$.

The motion of the electron mass density endows the electron with a momentum $\vec{p}$:

$$\vec{p} = m\vec{v} \quad \Rightarrow \quad p_z = m \frac{\partial \langle z \rangle}{\partial t} \approx -\frac{m}{q} \omega_0 d_{12} ab \sin(\omega_0 t + \phi). \tag{21}$$

6. Atom in an Applied Field

Schrödinger had a detailed physical picture of the wave function, and he gave an elegant derivation of the process underlying the change of atomic state mediated by electromagnetic coupling. [The original derivation in Ref. [38], p. 137, is not nearly as readable as that in Schrödinger's second and third 1928 lectures [39], where the state transition is described in Section 9 starting at the bottom of page 31, for which the second lecture is preparatory. There he solved the problem more generally, including the effect of a slight detuning of the field frequency from the atom's transition frequency.] Instead of directly tackling the transfer of energy between two atoms, he considered the response of a single atom to a small externally applied vector potential field $\vec{A}$. He found that the immediate effect of an applied vector potential is to change the momentum p of the electron wave function:

$$\begin{aligned} p_z &= m \frac{\partial \langle z \rangle}{\partial t} - q A_z \\ \frac{\partial p_z}{\partial t} &= m \frac{\partial^2 \langle z \rangle}{\partial t^2} - q \frac{\partial A_z}{\partial t}. \end{aligned} \tag{22}$$

Thus, the quantity $-q \frac{\partial A_z}{\partial t}$ acts as a *force*, causing an *acceleration* of the electron wave function.

This is the physical reason that $-\frac{\partial A_z}{\partial t}$ can be treated as an *electric field* $\mathcal{E}_z$. [At a large distance from an overall charge-neutral charge distribution like an atom, the longitudinal gradient of the scalar potential just cancels the longitudinal component of $\partial \vec{A}/\partial t$, so what is left is $\vec{\mathcal{E}} = -\partial A_{\perp}/\partial t$, which is purely transverse.].

$$\mathcal{E}_z = -\frac{\partial A_z}{\partial t}. \tag{23}$$

In the simplest case, the qA_z term makes only a tiny perturbation to the momentum over a single cycle of the ω_0 oscillation, so its effects will only be appreciable over many cycles.

We consider an additional simplification, where the frequency of the applied field is exactly equal to the transition frequency ω_0 of the atom:

$$A_z = A\cos(\omega_0 t) \quad \Rightarrow \quad -\frac{\partial A_z}{\partial t} = \mathcal{E}_z = \omega_0 A \sin(\omega_0 t). \tag{24}$$

In such evaluations, we need to be very careful to identify *exactly which energy* we are calculating: The electric field is merely a bookkeeping device to keep track of the energy that an electron in one atom exchanges with another electron in another atom, in such a way that the total energy is conserved. We will evaluate how much energy a given electron gains from or loses to the field, recognizing that the negative of that energy represents work done by the electron on the source electron responsible for the field. The force on the electron is just $q\mathcal{E}_z$. Because $\mathcal{E}_z = \omega_0 A \sin(\omega_0 t)$, for a stationary charge, the force is in the $+z$ direction as much as it is in the $-z$ direction, and, averaged over one cycle of the electric field, the work averages to zero. However, if the charge itself oscillates with the electric field, it will gain energy ΔE from the work done by the field on the electron over one cycle:

$$\frac{\Delta E}{\text{cycle}} = \int q\mathcal{E}_z \, dz = \int_0^{2\pi/\omega_0} q\mathcal{E}_z \frac{\partial \langle z \rangle}{\partial t} \, dt, \tag{25}$$

where $\langle z \rangle$ is the z position of the electron center of charge from Equation (20).

When the electron is "coasting downhill" *with* the electric field, it gains energy and ΔE is positive. When the electron is moving "uphill" *against* the electric field, the electron loses energy and ΔE is negative.

As long as the energy gained or lost in each cycle is small compared with E_0, we can define a continuous **power** (rate of change of electron energy), which is understood to be an average over many cycles. The time required for one cycle is $2\pi/\omega_0$, so Equation (25) becomes:

$$\frac{\partial E}{\partial t} = \frac{\omega_0 \Delta E}{2\pi} = \frac{\omega_0}{2\pi} \int_0^{2\pi/\omega_0} q\mathcal{E}_z \frac{\partial \langle z \rangle}{\partial t} \, dt = \frac{1}{2\pi} \int_0^{2\pi} q\mathcal{E}_z \frac{\partial \langle z \rangle}{\partial t} \, d(\omega_0 t). \tag{26}$$

7. Electromagnetic Coupling

Because our use of electromagnetism is conceptually quite different from that in traditional Maxwell treatments (including Jaynes' NCT), we review here the foundations of that discipline from the present perspective. [A more detailed discussion from the present viewpoint is given in Mead, ***Collective Electrodynamics*** [16]. The standard treatment is given in Jackson, ***Classical Electrodynamics, 3rd Edition,*** Chapter 8 [40].] It is shown in Ref. [16] that electromagnetism is of totally quantum origin. We saw in Equation (22) that it is the vector potential $\vec{A}$ that appears as part of the momentum of the wave function, signifying the coupling of one wave function to one or more other wave functions. Thus, to stay in a totally quantum context, we must work with electromagnetic relations based on the vector potential and related quantities. The entire content of electromagnetism is contained in the relativistically-correct Riemann–Sommerfeld second-order differential equation:

$$\left(\nabla^2 - \frac{\partial^2}{\partial t^2}\right) \mathbf{A} = -\mu_0 \, \mathbf{J}, \tag{27}$$

where $\mathbf{A} = [\vec{A}, V/c]$ is the four-potential and $\mathbf{J} = [\vec{J}, c\rho]$ is the four-current, $\vec{A}$ is the vector potential, V is the scalar potential, $\vec{J}$ is the physical current density (no displacement current), and ρ is the physical charge density, all expressed in the same inertial frame.

Connection with the usual electric and magnetic field quantities $\vec{\mathcal{E}}$ and $\vec{B}$ is as follows:

$$\vec{\mathcal{E}} = -\nabla V - \frac{\partial \vec{A}}{\partial t} \qquad \vec{B} = \nabla \times \vec{A}. \tag{28}$$

Thus, once we have the four-potential $\mathbf{A}$, we can derive any electromagnetic relations we wish.

Equation (27) has a completely general Green's Function solution for the four-potential $\mathbf{A}(t)$ at a point in space, from four-current density $\mathbf{J}(r,t)$ in volume elements $d$vol at at distance r from that point:

$$\mathbf{A}(t) = \frac{\mu_0}{4\pi} \int \frac{\mathbf{J}(r,t')}{r} \bigg|^{t'=t\pm\frac{r}{c}} d\mathrm{vol}, \tag{29}$$

where r is the distance from element $d$vol to the point where $\mathbf{A}$ is evaluated, assumed large compared to the size of the atomic wave functions, and c is the speed of light.

Equation (29) is the first fundamental equation of electromagnetic coupling: The vector potential, which will appear as part of an electron's momentum, is simply the sum of all current elements on that electron's light cone, each weighted inversely with its distance from that electron. The second-order nature of derivatives in Equation (27) does not favor any particular sign of space or time. Thus, the four-potential from a current element on the *past* light cone of the electron $(t - r/c)$ will be "felt" by the electron at later time t, and is termed a **retarded** field. Conversely, the four-potential from a current element on the *future* light cone of the electron $(t + r/c)$ will be "felt" by the electron at earlier time t, and is termed an **advanced** field. Historically, with rare exception, advanced fields have been discarded as non-physical because evidence for them has been explained in other ways. We shall see that modern quantum experiments provide overwhelming evidence for their active role in **quantum entanglement.**

Equation (29) can be expressed in terms of more familiar E&M quantities:

$$\vec{A}(t) = \frac{\mu_0}{4\pi} \int \frac{\vec{J}(r,t')}{r} \bigg|^{t'=t\pm\frac{r}{c}} d\mathrm{vol} \qquad V(t) = \frac{\mu_0 c^2}{4\pi} \int \frac{\rho(r,t')}{r} \bigg|^{t'=t\pm\frac{r}{c}} d\mathrm{vol} \tag{30}$$

If the current density $\vec{J}$ is due to the movement of a small, unified "cloud" of charge, as is the case for the wave function of an atomic electron, and the motion of the wave function is non-relativistic, the $\vec{J}$ integral just becomes the movement of the center of charge relative to its average position at the nucleus:

$$\vec{A}(t) \approx \frac{\mu_0}{4\pi} \int \frac{\rho\vec{v}(r,t')}{r} \bigg|^{t'=t\pm\frac{r}{c}} d\mathrm{vol} \approx \frac{\mu_0}{4\pi} \frac{q\,\vec{v}(r,t')}{r} \bigg|^{t'=t\pm\frac{r}{c}} \tag{31}$$

If, as we have chosen previously, the motion is in the z direction,

$$A_z(t) \approx \frac{q\mu_0}{4\pi r} \frac{\partial \langle z(r,t') \rangle}{\partial t'} \bigg|^{t'=t\pm\frac{r}{c}} \tag{32}$$

If we use the current element as our origin of time, the signs are reversed:

$$A_z(t')\big|^{t'=t\mp\frac{r}{c}} \approx \frac{q\mu_0}{4\pi r} \frac{\partial \langle z(r,t) \rangle}{\partial t} \tag{33}$$

In this case, $t + r/c$ represents the retarded field and $t - r/c$ represents the advanced field. We shall use these two forms for the simple examples presented below.

An important difference between standard Maxwell E&M practice and our use of the four-potential to couple atomic wave functions is highlighted by Wheeler and Feynman [17]:

"There is no such concept as 'the' field, an independent entity with degrees of freedom of its own."

The field has no degrees of freedom of its own. It is simply a convenient bookkeeping device for keeping track of the total effect of an arbitrary number of charges on a particular charge distribution in some region of space. The general form of interaction energy E is given by:

$$E = \int (\vec{A} \cdot \vec{J} + \rho V)\, d\text{vol} \tag{34}$$

Equation (34) is the second fundamental equation of electromagnetic coupling. The interaction described in Section 6 is a simplified case of this relation for a single atom. The energy minimum created by the positive nucleus is the V, and the ρ is the charge distribution of the electron wave function. The $\vec{A}$ from a second atom is very small, and is assumed to not change the eigenfunctions.

From Equations (34) and (29), we see that, when applied to two atoms, what is being described is a way to factor a bi-directional connection between them so that each can be analyzed separately by Schrödinger's equation as a one-electron problem, using the vector potential from the other as part of its energy.

The fact that the four-potential field from a charge is defined everywhere on its light cone does not imply that it is "radiating into space", carrying energy with it. Energy is only transferred at the position of another charge. Since all charges are the finite charge densities of wave functions, **there are no self-energy infinities in this formulation**.

One widely-held viewpoint treats the "quantum vacuum" as being made up of an infinite number of quantum harmonic oscillators. The problem with this view is that each such oscillator would have a zero-point energy that would contribute to the energy density of space in any gravitational treatment of cosmology. Even when the energies of the oscillators are cut off at some high value, the contribution of this "dark energy" is 120 orders of magnitude larger than that needed to agree with astrophysical observations. Such a disagreement between theory and observation (called the "cosmological constant problem"), even after numerous attempts to reduce it, is *many orders of magnitude worse than any other theory-vs-observation discrepancy in the history of science*! However, somehow this viewpoint remains a central part of the standard model of particle physics and standard practice in QM.

Our approach does not suffer from this serious defect, since its vacuum has no degrees of freedom of its own. Where, then, is radiated energy going if an atom's excitation decays and does not interact locally? The obvious candidate is the enormous continuum of states of matter in the early universe, source of the **cosmic microwave background**, to which atoms here and now are coupled by the quantum handshake. For independent discussions from the two of us, see Ref. [16], p. 94 and Ref. [19].

8. Two Coupled Atoms

The central point of this paper is to understand the *photon* mechanism by which energy is transferred from a *single* excited atom (atom α) to another *single* atom (atom β) initially in its ground state.

We proceed with the simplest and most idealized case of two identical atoms, where:

(1) Excited atom α will start in a state where $b \approx 1$ and a is very small, but never zero because of its ever-present random statistical interactions with a vast number of other atoms in the universe, and
(2) Likewise, atom β will start in a state where $a \approx 1$ and b is very small, but never zero for the same reason.

Thus, each atom starts in a two-component state that has an oscillating electrical current described by Equation (20):

$$
\begin{aligned}
q\frac{\partial \langle z_\alpha \rangle}{\partial t} &\approx -\omega_0 d_{12} a_\alpha b_\alpha \sin(\omega_0 t) \\
q\frac{\partial \langle z_\beta \rangle}{\partial t} &\approx -\omega_0 d_{12} a_\beta b_\beta \sin(\omega_0 t + \phi),
\end{aligned} \tag{35}
$$

where we have taken excited atom α as our reference for the phase of the oscillations ($\phi_\alpha = 0$), and the approximation assumes that a and b are changing slowly on the scale of ω_0.

Although that random starting point will contain small excitations of a wide range of phases, we simplify the problem by assuming the following:

All of the vector potential A_β at atom α is supplied by atom β,
All of the vector potential A_α at atom β is supplied by atom α,
The dipole moments of both atoms are in the z direction,
The atoms are separated by a distance r in a direction orthogonal to z,

The vector potential at distance r from a small charge distribution oscillating in the z-direction is from Equation (32):

$$
A_z(t) = \frac{q\mu_0}{4\pi r} \left. \frac{\partial \langle z(r,t') \rangle}{\partial t'} \right|^{t'=t\pm\frac{r}{c}}. \tag{36}
$$

Since all electron motions and fields are in the z direction, we can henceforth omit the z subscript.

When the distance r is small compared with the wavelength, i.e., $r \ll 2\pi c/\omega_0$, the delay r/c can be neglected. Since atomic dimensions are of the order of 10^{-10} m and the wavelength is of the order of 10^{-7} m, this case can be accommodated. We shall find that the results we arrive at here are directly adaptable to the centrally important case in which the atoms are separated by an arbitrarily distance, which will be analyzed in Section 9. Using Equation (23) and (32), the vector potentials, and hence the electric fields, from the two atoms become:

$$
\begin{aligned}
A_\alpha &\approx \frac{q\mu_0}{4\pi r}\frac{\partial \langle z_\alpha \rangle}{\partial t} \quad \Rightarrow \quad \mathcal{E}_\alpha = -\frac{\partial A_\alpha}{\partial t} \approx -\frac{q\mu_0}{4\pi r}\frac{\partial^2 \langle z_\alpha \rangle}{\partial t^2} \\
A_\beta &\approx \frac{q\mu_0}{4\pi r}\frac{\partial \langle z_\beta \rangle}{\partial t} \quad \Rightarrow \quad \mathcal{E}_\beta = -\frac{\partial A_\beta}{\partial t} \approx -\frac{q\mu_0}{4\pi r}\frac{\partial^2 \langle z_\beta \rangle}{\partial t^2}.
\end{aligned} \tag{37}
$$

When atom α is subject to electric field $\mathcal{E}_\beta$ and atom β is subject to electric field $\mathcal{E}_\alpha$, the energy of both atoms will change with time in such a way that the total energy is conserved. Thus, the superposition amplitudes a and b of both atoms change with time, as described by Equation (17) and (26), from which:

$$
\begin{aligned}
\frac{\partial E_\alpha}{\partial t} &\approx \frac{1}{2\pi}\int_0^{2\pi} q\mathcal{E}_\beta \frac{\partial \langle z_\alpha \rangle}{\partial t}\, d(\omega_0 t) = -\frac{q^2\mu_0}{8\pi^2 r}\int_0^{2\pi} \frac{\partial^2 \langle z_\beta \rangle}{\partial t^2}\frac{\partial \langle z_\alpha \rangle}{\partial t}\, d(\omega_0 t) \\
\frac{\partial E_\beta}{\partial t} &\approx \frac{1}{2\pi}\int_0^{2\pi} q\mathcal{E}_\alpha \frac{\partial \langle z_\beta \rangle}{\partial t}\, d(\omega_0 t) = -\frac{q^2\mu_0}{8\pi^2 r}\int_0^{2\pi} \frac{\partial^2 \langle z_\alpha \rangle}{\partial t^2}\frac{\partial \langle z_\beta \rangle}{\partial t}\, d(\omega_0 t).
\end{aligned} \tag{38}
$$

From Equation (38), using the $\langle z \rangle$ derivatives from Equation (20):

$$
\begin{aligned}
\frac{\partial E_\alpha}{\partial t} &\approx -\frac{\mu_0}{8\pi^2 r}\int_0^{2\pi} \left(-\omega_0^2 d_{12} a_\beta b_\beta \cos(\omega_0 t + \phi)\right)\left(-\omega_0 d_{12} a_\alpha b_\alpha \sin(\omega_0 t)\, d(\omega_0 t)\right) \\
&\approx -\frac{\mu_0 \omega_0^3 d_{12}^2 a_\beta b_\beta a_\alpha b_\alpha}{8\pi^2 r}\int_0^{2\pi} \cos(\omega_0 t + \phi)\sin(\omega_0 t)\, d(\omega_0 t) = \frac{\mu_0 \omega_0^3 d_{12}^2 a_\beta b_\beta a_\alpha b_\alpha}{8\pi r}\sin(\phi) \\
\frac{\partial E_\beta}{\partial t} &\approx -\frac{\mu_0}{8\pi^2 r}\int_0^{2\pi} \left(-\omega_0^2 d_{12} a_\alpha b_\alpha \cos(\omega_0 t)\right)\left(-\omega_0 d_{12} a_\beta b_\beta \sin(\omega_0 t + \phi)\, d(\omega_0 t)\right) \\
&\approx -\frac{\mu_0 \omega_0^3 d_{12}^2 a_\alpha b_\alpha a_\beta b_\beta}{8\pi^2 r}\int_0^{2\pi} \cos(\omega_0 t)\sin(\omega_0 t + \phi)\, d(\omega_0 t) = -\frac{\mu_0 \omega_0^3 d_{12}^2 a_\alpha b_\alpha a_\beta b_\beta}{8\pi r}\sin(\phi).
\end{aligned} \tag{39}
$$

These equations describe energy transfer between the two atoms in either direction, depending on the sign of $\sin(\phi)$. For transfer from atom α to atom β, $\partial E_\alpha/\partial t$ is negative. Since this transaction dominated all competing potential transfers, its amplitude will be maximum, so $\sin(\phi) = -1$.

If the starting state had been atom β in the excited ($b \approx 1$) state, the $\sin(\phi) = +1$ choice would have been appropriate:

$$\text{Using}: \qquad \sin(\phi) = -1 \qquad \text{and} \qquad P_{\alpha\beta} \equiv \frac{\mu_0 \omega_0^3 d_{12}^2}{8\pi r}, \tag{40}$$

This rate of transferred energy is calculated for two isolated atoms suspended in space. We have no experimental data whatsoever for such a situation. All optical experiments are done with some **optical system** between the two atoms. Even the simplest such arrangement couples the two atoms *orders of magnitude* better than the simple $1/r$ dependence in Equation (40) would indicate. We take up the enhancement due to an intervening optical system in Section 10. Any such enhancement merely provides a constant multiplier in $P_{\alpha\beta}$. In any case, Equation (39) becomes:

$$\frac{\partial E_\alpha}{\partial t} = E_0 \frac{\partial b_\alpha^2}{\partial t} = -P_{\alpha\beta}\, a_\beta b_\beta a_\alpha b_\alpha \qquad \frac{\partial E_\beta}{\partial t} = E_0 \frac{\partial b_\beta^2}{\partial t} = P_{\alpha\beta}\, a_\alpha b_\alpha a_\beta b_\beta. \tag{41}$$

Remembering that our total starting energy was E_0 that b^2 is the energy of the electron in units of E_0 referred to the ground state, that energy is conserved by the two atoms during the transfer, and that the wave functions are normalized:

$$\begin{aligned} E_0\left(b_\alpha^2 + b_\beta^2\right) = E_0 \quad \Rightarrow \quad b_\alpha^2 + b_\beta^2 &= 1 \quad \Rightarrow \quad b_\beta = \sqrt{1 - b_\alpha^2} \\ a_\alpha^2 + b_\alpha^2 &= 1 \quad \Rightarrow \quad a_\alpha = \sqrt{1 - b_\alpha^2} = b_\beta \\ a_\beta^2 + b_\beta^2 &= 1 \quad \Rightarrow \quad a_\beta = \sqrt{1 - b_\beta^2} = b_\alpha, \end{aligned} \tag{42}$$

after which substitutions Equation (41) becomes:

$$\frac{\partial (b_\alpha^2)}{\partial t} = b_\alpha^2 (1 - b_\alpha^2)/\tau, \quad \text{where the transition time scale is } \tau \equiv \frac{E_0}{P_{\alpha\beta}}. \tag{43}$$

This has the solution plotted in Figure 6:

$$b_\alpha^2 = a_\beta^2 = \frac{1}{e^{t/\tau} + 1} \qquad a_\alpha^2 = b_\beta^2 = \frac{1}{e^{-t/\tau} + 1}. \tag{44}$$

Note that this waveform is that of an *individual interaction* and has no probabilistic meaning. It was the subject of many intense discussions about NCT in general, including a quite detailed one in [30]. We refrain from such discussion here because the dependence of the dipole moment with superposition makeup is not the only nonlinearity in the problem. The self-focusing nature of the matched advanced/retarded electromagnetic solutions, described in Section 11, may be an even larger nonlinearity in many cases. Although its time dependence is much more difficult to estimate, it will almost certainly make the individual quantum transition much more abrupt.

The direction and magnitude of the entire energy-transfer process is governed by the relative phase ϕ of the electric field and the electron motion in *both atoms*: When the electron motion of *either atom* is *in phase* with the field, the field transfers energy to the electron, and the field is said to **excite** the atom. When the the electron motion has *opposite phase* from the field, the electron transfers energy "to the field", and the process is called **stimulated emission**.

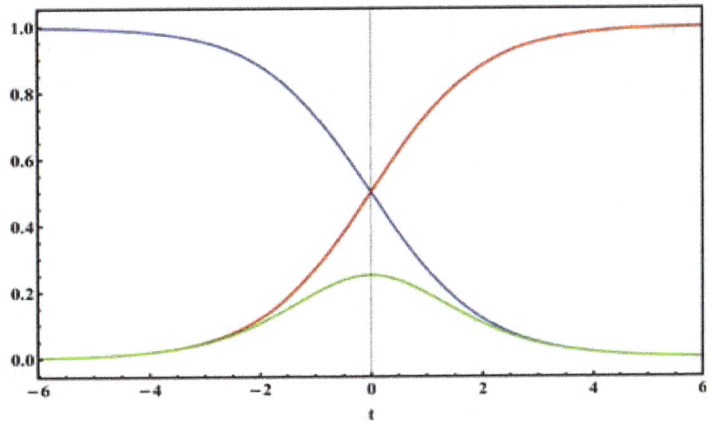

Figure 6. Squared state amplitudes for atom α: b_α^2 (blue) and $a_\alpha^2 = b_\beta^2$ (red) for the Photon transfer of energy $E_0 = \hbar\omega_0$ from atom α to atom β, from Equation (44). Using the lower state energy as the zero reference, $E_0 b^2$ is the energy of the state. The green curve shows the normalized power radiated by the atom α and absorbed by atom β, from Equation (43). The optical oscillations at ω_0 are not shown, as they are normally many orders of magnitude faster than the transition time scale τ. The time t is in units of τ. In the next section, we will find that atoms spaced by an arbitrary distance exhibit transactions of exactly the same form.

Therefore, for the photon transaction to proceed the field from atom α must have a phase such that it "excites" atom β, while the field from atom β must have a phase such that it absorbs energy and "de-excites" atom α. In the above example, that unique combination occurs when $\sin(\phi) = -1$.

This dependence on phase makes a transaction exquisitely sensitive to the frequency match between atoms α and β. The frequency $\omega_0 \approx 10^{16}/$ s, so for transitions in the nanosecond range, a mismatch of one part in 10^7 can cause a transaction to fail.

8.1. Competition between Recipient Atoms

As discussed at the end of Section 7, the bi-directional vector potential coupling has allowed us to analyze the problem of two coupled atoms, which looks like a two-electron problem, as two coupled one-electron problems, and therefore is treatable using Schrödinger's wave mechanics. The approach generalizes, so we can now examine the important three-atom case of a single excited atom α that is equally coupled to two ground-state atoms $\beta 1$ and $\beta 2$. Recipient atoms $\beta 1$ and $\beta 2$ have the same level spacing as source atom α. For atom $\beta 1$, this corresponds to transition frequency ω_0, and we assume a relative phase of $\sin(\phi) = -1$. However, atom $\beta 2$ is moving and has a slightly Doppler-shifted transition frequency $\omega_0 + \Delta\omega$. We assume that $\beta 2$ has the same structure and intial phase as $\beta 1$ at time $t = 0$. Thence, Equation (41), with the transition time scale $\tau \equiv E_0/P_{0\beta 1} = E_0/P_{0\beta 2}$, becomes:

$$\tau\frac{\partial b_{\beta 1}^2}{\partial t} = a_\alpha b_\alpha (a_{\beta 1} b_{\beta 1}) \qquad \tau\frac{\partial b_{\beta 2}^2}{\partial t} = a_\alpha b_\alpha (a_{\beta 2} b_{\beta 2}\cos(\Delta\omega t)) \qquad \tau\frac{\partial b_\alpha^2}{\partial t} = -a_\alpha b_\alpha (a_{\beta 1} b_{\beta 1} + a_{\beta 2} b_{\beta 2}\cos(\Delta\omega t)) \tag{45}$$

As with Equation (42), wave functions are normalized and energy is conserved:

$$a_\alpha^2 + b_\alpha^2 = 1 \qquad a_{\beta 1}^2 + b_{\beta 1}^2 = 1 \qquad a_{\beta 2}^2 + b_{\beta 2}^2 = 1 \qquad b_\alpha^2 + b_{\beta 1}^2 + b_{\beta 2}^2 = 1. \tag{46}$$

After these substitutions, we obtain two simultaneous differential equations in $b_\beta^2 1$ and $b_\beta^2 2$:

$$\begin{aligned}
\tau\frac{\partial b_{\beta 1}^2}{\partial t} &= \sqrt{b_{\beta 1}^2\left(1 - b_{\beta 1}^2\right)\left(1 - b_{\beta 1}^2 - b_{\beta 2}^2\right)\left(b_{\beta 1}^2 + b_{\beta 2}^2\right)} \\
\tau\frac{\partial b_{\beta 2}^2}{\partial t} &= \sqrt{b_{\beta 2}^2\left(1 - b_{\beta 2}^2\right)\left(1 - b_{\beta 1}^2 - b_{\beta 2}^2\right)\left(b_{\beta 1}^2 + b_{\beta 2}^2\right)}\cos(\Delta\omega\, t)
\end{aligned} \tag{47}$$

The solutions of Equation (47) are shown in Figure 7 for two very small values of $\Delta\omega$:

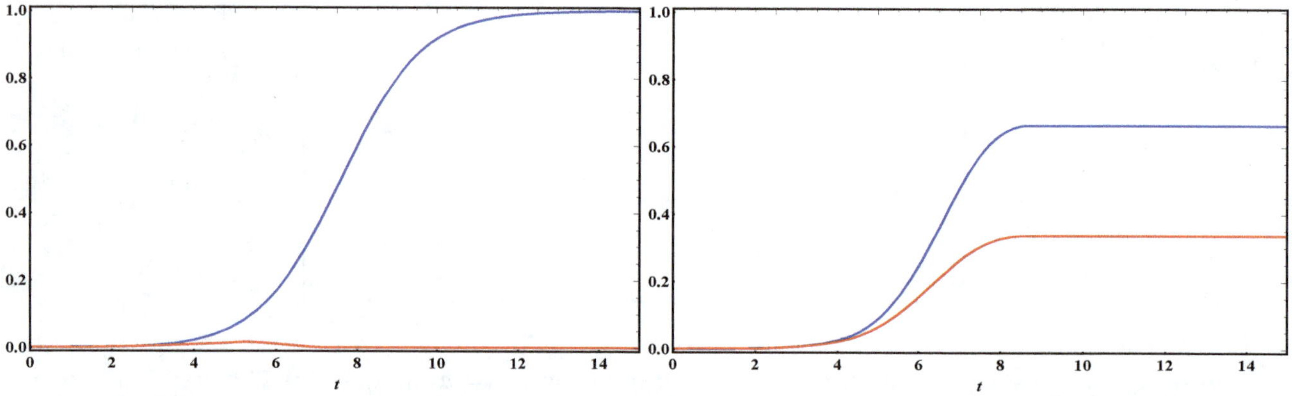

Figure 7. Squared excited state amplitudes for recipient atoms $\beta 1$ and $\beta 2$: $b_{\beta 1}^2$ (blue) and $b_{\beta 2}^2$ (red), for the photon transfer of energy $E_0 = \hbar\omega_0$ from atom α. The time t is in units of τ. The left plot, for $\Delta\omega = 0.3/\tau$, shows both recipient atoms being equally excited at the beginning, but the slip in phase of the red $\beta 2$ atom causes it to rapidly lose out, so the blue $\beta 1$ atom hogs all the energy and proceeds to become fully excited, much as if it were the "only atom in town". Its curve is nearly identical with that for the isolated recipient atom in Figure 6. The right plot, for $\Delta\omega = 0.15/\tau$, shows a totally different story. Its red $\beta 2$ atom transition frequency is just close enough to that of source atom 0 that it "hangs in there" during the transition period and ends up partially excited, leaving blue $\beta 1$ atom partially excited as well. The smaller the "slip" frequency $\Delta\omega$, the closer are the post-transition excitations of the two recipient atoms. "Split-photons" of this kind are observed in the Hanbury–Brown–Twiss Effect described in Section 13. As noted above, the frequency window for such events is *extraordinarily narrow*, typically of order $10^{-7}\omega_0$. Doppler shift of this magnitude requires velocity ≈ 30 m/s. Room temperature thermal velocities of gasses are typically tens to hundreds of times this value, which would eliminate such competition. Thus, complete transactions are the most common, with "split" transactions relatively rare and are likely to end as HBT-type four-atom events.

Here, again the transaction time scale τ is not to be confused with the time scale for initiating a transaction after atom α is excited, which is a question of probability. We make no pretense here of deriving probability results for a random population of atoms, but, from these results, we can imagine what one might look like: The random starting point for the transaction involving one excited atom will contain small excitations of a wide range of phases. Equation (43) is a highly nonlinear equation—the amplitude of each of those excitations will initially grow exponentially at a rate proportional to its own phase match. Thus, the excitation of a random recipient atom that happens to have $\sin(\phi) \approx -1$ will win in the race and become the dominant partner in the coordinated oscillation of both atoms. *Thus, we have conceptualized the source of the intrinsic randomness within quantum mechanics, an aspect of statistical QM that has been considered mysterious since its inception in the 1920s.*

Each wave function will thus evolve its motion to follow the applied field to its maximum resonant coupling and we can take $\sin(\phi) = -1$ in these expressions, which we have done in Equation (41), Figure 6 and Equation (45). [What we have not done is to derive the full second-order nature of phase locking in this arrangement. That analysis is rendered much more difficult by the potentially *huge* amplification due to the self-focusing nature of the bi-directional electromagnetic coupling described in Section 10. Thus, a full derivation remains open to future generations.]

From the TI point of view, all three atoms start in stable states, with each having extremely small admixtures of the other state, so that they have very small dipole moments oscillating with angular frequency $\omega_0 \approx (E_2 - E_1)/\hbar$. We assume that in source atom α this admixture creates an initial positive-energy offer wave that interacts with the small dipole moments of absorber atoms $\beta 1$ and $\beta 2$ to transfer positive energy, and that in atoms $\beta 1$ and $\beta 2$ this admixture creates initial negative-energy confirmation waves to the excited emitter atom α that interact with the small dipole moment of emitter atom α to transfer negative energy, as shown schematically in Figure 1. As a result of the mixed-energy superposition of states as shown in Figure 5, all three atoms oscillate with very nearly the same frequency ω_0 and act as coupled dipole resonators.

Energy transferred from source atom α to both recipient atoms $\beta 1$ and $\beta 2$ causes an increase in both minority states of the superposition, thus increasing the dipole moment of all three states, thereby increasing the coupling and, hence, the initial rate of energy transfer. This behavior is self-reinforcing for any atom that can stay in phase, giving the transition its initial exponential character. In the usual case, only one atom is sufficiently well frequency matched to stay in phase for the entire duration of the transition, the unfortunate runner-up is rudely driven out of the competition, and the winner drives the transaction to its conclusion, as shown in the left panel of Figure 7.

In the presence of near-equal competition, one competitor either loses out to a competing transaction, or, in case of a tie, results in a "split-photon", as shown in the right panel of Figure 7. That situation represents an *intermediate* state with actively oscillating dipole moments, as discussed above, in which the two confirmation waves, in the phase at the source, can precipitate the de-excitation of an additional excited atom to create a final HBT four-atom event.

The universe is full of similar atoms, all with slightly different transition frequencies due to random velocities. There are also random perturbation by waves from other systems that can randomly drive the exponential instability in either direction. This random environment is the source of the intrinsic randomness in quantum processes. Ruth Kastner [41] attributes intrinsic randomness to "spontaneous symmetry breaking", which could split a "tie" in the absence of environmental factors.

We note here that the probability of the transition must depend on two things: the strength of the electromagnetic coupling between the two states, and the degree to which the wave functions of the initial states are superposed. The magnitude of the latter must depend on the environment, in which many other atoms are "chattering" and inducing state-mixing perturbations. The more potential partner atoms there are per unit energy, the greater the probability of a perfect match. Thus, we see the emergence of **Fermi's "Golden Rule" [42]**, the assertion that a transition probability in a coupled quantum system depends on the strength of the coupling and the density of states present to which the transition can proceed. The emergence of Fermi's Golden Rule is an unexpected gift delivered to us by the logic of the present formalism.

It is certainly not obvious a priori that the Schrödinger recipe for the vector potential in the momentum (Equation (22)), together with the radiation law from a charge in motion (Equation (33)), would conspire to enable the composition of the superposed states of two electromagnetically coupled wave functions to reinforce in such a way that, from the asymmetrical starting state, the energy of one excited atom could *explosively and completely* transfer to the unexcited atom, as shown in Figure 6 and Figure 7.

If nature had worked a slightly different way, an interaction between those atoms might have resulted in a different phase, and no full transaction would have been possible. The fact that transfer of energy between two atoms has this nonlinear, self-reinforcing character makes possible arrangements like a *laser*, where many atoms in various states of excitation participate in a glorious dance, all participating at exactly the same frequency and locked in phase.

Why do the signs come out that way? No one has the slightest idea, but the behavior is so remarkable that it has been given a name: Photons are classified as *bosons*, meaning that they behave that way!

That remarkable behavior is not due to any "particle-like" quantization of the electromagnetic field. Quantization of the photon energy is a result of the discrete nature of electron states in atoms.

The movement of an electron in a superposed state couples to another such electron electromagnetically. It is essential that this electromagnetic coupling is bi-directional in space-time to conserve energy in the transaction. The statistical QM formulation needed some mechanism to finalize a transaction and did not recognize the inherent nonlinear positive-feedback that nature built into a pair of coupled wave functions. Therefore, the founders had to put in wave-function collapse "by hand", and it has mystified the field ever since. The NCT formulations *did* understand the inherent nonlinear positive-feedback that nature built into a pair of coupled wave functions, but postulated a unidirectional "Maxwell" treatment of the electromagnetic field that *did not conserve energy*, as we now discuss.

9. Two Atoms at a Distance

We saw that for two atoms to exchange energy, the vector potential A field at atom β must come from atom α, the A field at atom α must come from atom β, and the oscillations must stay in coherent phase, with a particular phase relation during the entire transition. *This phase relation must be maintained even when the two atoms are an arbitrary distance apart.* This is the problem we now address.

To be definite, we consider the case where the two atoms are separated along the x-axis, atom α at $x = 0$ in the excited state and atom β at $x = r$ in its ground state, so their separation r is orthogonal to the z-directed current in the atoms. The "light travel time" from atom α to atom β is thus $\Delta t = r/c$. What is observed is that the energy radiated by atom α at time t is absorbed by atom β at time $t' = t + \Delta t$:

$$\left.\frac{\partial E_\beta(r,t')}{\partial t'}\right|^{t'=t+\Delta t} = -\frac{\partial E_\alpha(0,t)}{\partial t} \tag{48}$$

This behavior is familiar from the behavior of a "particle", which carries its own degrees of freedom with it: It leaves $x = 0$ at time t and arrives at $x = r$ at time $t' = t + \Delta t$ after traveling at velocity c. Thus, Lewis's "photon" became widely accepted as just another particle, with degrees of freedom of its own. We shall see that this assumption violates a wide range of experimental findings.

For atom β, Equation (38) becomes:

$$\frac{\partial E_\beta(r,t')}{\partial t'} = \frac{1}{2\pi}\int_0^{2\pi} -q\mathcal{E}_\alpha(r,t')\frac{\partial\langle z_\beta(r,t')\rangle}{\partial t'}d(t') \qquad \text{where} \qquad t' = t + \Delta t \tag{49}$$

The retarded field from atom α interacts with the motion of the electron in atom β. The only difference from our zero-delay solution is that the energy transfer has its time origin shifted by $\Delta t = r/c$.

This result has required that we choose a *positive* sign for the $\mp r/c$ in Equation (33). By doing that, we are building in an "arrow of time", a preferred time direction, in the otherwise even-handed formulation. In particular, we are assuming that the retarded solution transfers positive energy. So far, everything is familiar and consistent with commonly held Maxwell notions: A retarded solution carrying energy with it. However, we saw that the source atom required a matched vector potential to lose energy.

The standard picture leaves no way for atom α to lose energy to atom β. It does not conserve energy!

When energy is transferred between two atoms spaced apart on the x-axis, the field amplitude must be *"coordinate and symmetrical"* as Lewis described. The field $\mathcal{E}_\alpha(x = r)$ at the second atom due to the current in the first must be exactly equal in magnitude to the the field $\mathcal{E}_\beta(x = 0)$ at the first atom due to the current in the second, but separated in time by Δt: For atom α, Equation (38) becomes

$$\frac{\partial E_\alpha(0,t)}{\partial t} = \int_0^{2\pi} -q\mathcal{E}_\beta(0,t)\frac{\partial\langle z_\alpha\rangle}{\partial t}dt \tag{50}$$

Thus, the field $\mathcal{E}_\beta$ from atom β, which arises from the motion of its electron at time $t' = t + \Delta t$, must arrive at atom α at time t, *earlier* than its motion by Δt. The only field that fulfills this condition is the *advanced* field from atom β, signified by choosing a negative sign for the $\mp r/c$ in Equation (33). That choice uniquely satisfies the requirement for conservation of energy. It also builds complementary "arrows of time" into the formulation—we assume that the advanced solution transfers negative energy to the past and the retarded solution transfers positive energy to the future. These two assumptions create a new non-local "handshake" symmetry that is not expressed in conventional Maxwell E&M.

Once these choices for the $\mp r/c$ in Equation (33) are made, the resulting equations for each of the energy derivatives in Equation (39) are unchanged when $t' = t + \Delta t$ is substituted for t in the expression for $\partial E_\beta/\partial t$. Thus, each transition proceeds in the local time frame of its atom—for all the world (except for amplitude) as if the other atom were local to it. This "locality on the light cone" is the meaning of Lewis' comment:

"In a pure geometry it would surprise us to find that a true theorem becomes false when the page upon which the figure is drawn is turned upside down. A dissymmetry alien to the pure geometry of relativity has been introduced by our notion of causality."

The dissymmetry that concerned Lewis has been eliminated.

This conclusion is completely consistent with the 1909 formulation of Einstein [43], who was critical of the common practice of simply ignoring the advanced solutions for electromagnetic propagation:

"I regard the equations containing retarded functions, in contrast to Mr. Ritz, as merely auxiliary mathematical forms. The reason I see myself compelled to take this view is first of all that those forms do not subsume the energy principle, while I believe that we should adhere to the strict validity of the energy principle until we have found important reasons for renouncing this guiding star."

After defining the retarded solution as f_1, and the advanced solution as f_2, he elaborates:

"Setting $f(x,y,z,t) = f_1$ amounts to calculating the electromagnetic effect at the point x,y,z from those motions and configurations of the electric quantities that took place prior to the time point t.′ "Setting $f(x,y,z,t) = f_2$, one determines the electromagnetic effects from the motions and configurations that take place after the time point t."

"In the first case the electric field is calculated from the totality of the processes producing it, and in the second case from the totality of the processes absorbing it...
Both kinds of representation can always be used, regardless of how distant the absorbing bodies are imagined to be."

The choice of advanced or retarded solution cannot be made a priori: It depends upon the *boundary conditions* of the particular problem at hand. The quantum exchange of energy between two atoms just happens to require one advanced solution carrying negative energy and one retarded solution carrying positive energy to satisfy its boundary conditions at the two atoms, which then guarantees the conservation of energy.

Thus, the even-handed time symmetry of Wheeler–Feynman electrodynamics [17,18] and of the Transactional Interpretation of quantum mechanics [12], as most simply personified in the two-atom photon transaction considered here, arises from the symmetry of the electromagnetic propagation equations, with boundary conditions imposed by the solution of the Schrödinger equation for the electron in each of the two atoms, as foreseen by Schrödinger. We see that the missing ingredients in previous failed attempts, by Schrödinger and others, to derive wave function collapse from the wave mechanics formalism were that advanced waves were not explicitly used as a part of the process.

To keep in touch with experimental reality, we return to our two H atoms spaced a distance r apart. We can estimate the "transition time" τ from Equations (41) and (40):

$$\frac{\partial E_\beta}{\partial t} = P_{\alpha\beta}\, a_\alpha b_\alpha a_\beta b_\beta = \frac{\mu_0 \omega_0^3 d_{12}^2}{8\pi r}\, a_\alpha b_\alpha a_\beta b_\beta. \tag{51}$$

From the green curve in Figure 5, we can estimate the dipole strength, which is q times the "length" between the positive and negative "charge lumps", say $d_{12} \approx 3qa_0$. At the steepest part of the transition, all the a and b terms will be $1/\sqrt{2}$, so

$$\left.\frac{\partial E_\beta}{\partial t}\right|_{\max} \approx \frac{\mu_0 \omega_0^3 (3qa_0)^2}{32\pi r}. \tag{52}$$

From any treatment of the hydrogen spectrum, we obtain, for the 210→100 transition:

$$E_0 = \hbar\omega_0 = \frac{9q^2}{128\pi\epsilon_0 a_0} \quad (53)$$

so the transition time will be:

$$\tau_{12} \approx \frac{E_0}{\left.\frac{\partial E_\beta}{\partial t}\right|_{\max}} \approx \frac{r}{4\,a_0^3\omega_0^3\epsilon_0\mu_0} = \frac{r\,c^2}{4\,a_0^3\omega_0^3} \approx r \times 0.04\,\frac{\text{sec}}{\text{m}}, \quad (54)$$

Thus, if the assumption of the $1/r$ dependence of the vector potential (r dependence of the transition time) were the whole picture, it would take $1/25$ of a second for a transaction to complete if the atoms were suspended one meter apart. Such a long transition time would allow the excited atom's energy to be frittered away by the many possible competitive paths, thus making any modern optical experiment virtually impossible, so *no experiments are done that way!* In real experiments, atoms are coupled by some *optical system*, composed of lenses, mirrors, and the like. That optical system will have some **solid angle** containing paths from one atom to the other. The effect of the optical system is to replace the $1/r$ dependence with Solid Angle$/\lambda$, as described in Section 10. Thus, the 0.04 s transition given by Equation (54) for two isolated atoms 1 m apart becomes 2×10^{-9} s when a 1 steradian optical system is used.

Once again, we caution that the time estimated here is *the time course of the single transaction after a handshake is formed*, which *must not* be confused with the probabilistic time for a transaction to be initiated after excitation of the source atom.

10. Paths of Interaction

We have developed a simple conceptual understanding of how a single quantum $\hbar\omega_0$ of energy is transferred from one isolated atom to another by way of a "photon" transaction. Real experiments with such transactions measure the statistics of many such events as functions of intensity, polarization, time delay, and other variables. Much has been discovered in the process, some results quite surprising, as described for the Freeman–Clauser experiment in Section 13. Thus, the time has come for us to discuss, at a conceptual level, where the probabilities come from. In the wonderful little book *QED* [44], our Caltech colleague the late Richard Feynman gives a synopsis of the method by which light propagating along multiple paths initiates a transaction, which he calls an event:

> *"Grand Principle: The probability of an event is equal to the square of the length of an arrow called the 'probability amplitude.'..."*
>
> *"General Rule for drawing arrows if an event can happen in alternative ways: Draw an arrow for each way, and then combine the arrows ('add' them) by hooking the head of one to the tail of the next."*
>
> *"A 'final arrow' is then drawn from the tail of the first arrow to the head of the last one."*
>
> *"The final arrow is the one whose square gives the probability of the entire event."*

Feynman's "arrow" is familiar to every electrical engineer as a **phasor**, introduced in 1894 by Steinmetz [45,46] as an easy way to visualize and quantify phase relations in alternating-current systems. In physics, the technique is known as the **sum over histories** and led to **Feynman path integrals**. His "probability amplitude" is the amplitude of our vector potential, whose square is the **probability** of a photon.

Feynman then illustrates his Grand Principle with simple examples how a source of light S at one point in space and time influences a receptor of that light P at another point in space and time, as shown in Figure 8. It is somewhat unnerving to many people to learn from these examples that the resultant intensity is dependent on *every possible path* from S to P. We strongly recommend that little book to everyone. That discussion, as well as what follows, details the behavior of highly coherent electromagnetic radiation with a well-defined, highly stable frequency ω and wavelength λ.

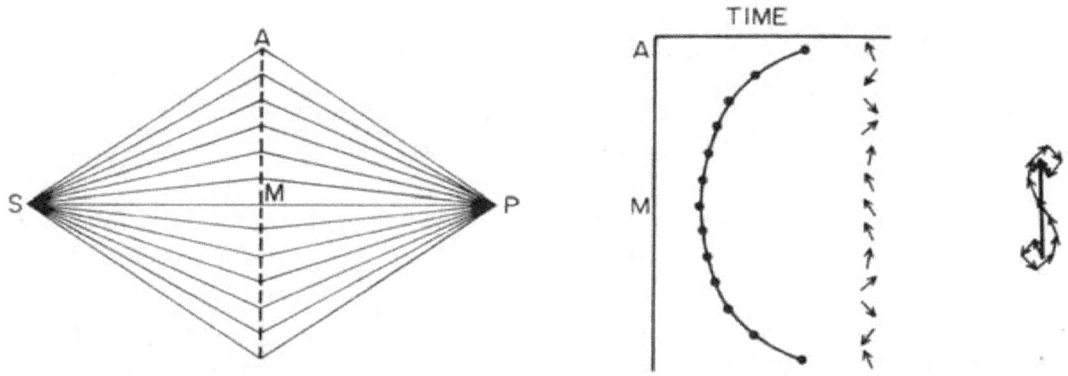

Figure 8. All the paths from coherent light source S to detector P are involved in the transfer of energy. The solid curve on the "TIME" plot shows the propagation time, and hence the accumulated phase, of the corresponding path. Each small arrow on the "TIME" plot is a **phasor** that shows the magnitude (length) and phase (angle) of the contribution of that path to the resultant total vector potential at P. The "sea horse" on the far right shows how these contributions are added to form the total amplitude and phase of the resultant potential. (From Fig. 35 in Feynman's **QED**).

Of course, we have all been taught that light travels in straight lines which spread out as they radiate from the source, and so the resultant intensity decreases as $1/r^2$, where r is the distance from the source S. However, if the light intensity at P depends on *all of the paths*, how can this $1/r^2$ dependence come about? Well, let's follow Feynman's QED logic: [If the reader does not have a copy of QED handy, there is a condensed version in Chapter I-26 of the *Feynman Lectures on Physics.*

We can see from the "seahorse" phasor diagram at the right of Figure 8 that the vast majority of the length of the resultant arrow is contributed by paths very close to the straight line S-M-P. Thus, let's make a rough estimate of how many paths there are near enough to "count." We can see from the diagram that, once the little arrows are plus or minus 90° from the phase of the straight line, additional paths just cause the resultant to wind around in a tighter and tighter circle, making no net progress. Thus, the uppermost and lowermost paths that "count" are about a quarter wavelength longer than the straight line. Let's use r for the straight-line distance S-P, λ for the wavelength, and y for the vertical distance where the path intersects the midline above M. Then, Pythagoras tells us that the length $l/2$ of either segment of the path is

$$\frac{l}{2} = \sqrt{\left(\frac{r}{2}\right)^2 + y^2} \tag{55}$$

Therefore, the entire path length l is

$$l = \sqrt{r^2 + 4y^2} = r\sqrt{1 + 4\frac{y^2}{r^2}} \tag{56}$$

We are particularly interested in atoms at a large distance from each other, and will guess that this means that y is very small compared to r, so all the paths involved are very close to the straight line. We can check that assumption later. Since $y^2/r^2 \ll 1$, we can expand the square root:

$$l \approx r\left(1 + 2\frac{y^2}{r^2}\right) = r + 2\left(\frac{y^2}{r}\right) \tag{57}$$

Thus, the outermost path that contributes is $2y^2/r$ longer than the straight line path. We already decided that maximum extra length of a contributing path would be about a quarter wavelength:

$$2\left(\frac{y^2}{r}\right) \approx \frac{\lambda}{4} \quad \Rightarrow \quad \left(\frac{y}{r}\right)^2 \approx \frac{\lambda}{8r} \tag{58}$$

We can now check our assumption that $y^2/r^2 \ll 1$. If $r = 1$ m and our $210 \to 100$ transition has $\lambda \approx 10^{-7}$ m, then $y^2/r^2 \approx 10^{-8}$, so our assumption is already very good, and gets better rapidly as r gets larger.

How do we estimate the number of paths from S to P? Well, no matter how we choose the path spacing radiating out equally in all directions from S, the number of paths that "count" will be proportional to the solid angle subtended by the outermost such paths. Paths outside that "bundle" will have phases that cancel out as Feynman describes. The angle of the uppermost path is y/r. Paths also radiate out perpendicular to the page to the same extent, so the total number that "count" goes as the solid angle $= \pi(y/r)^2$. Feynman tells us that the resultant amplitude A is proportional to the total number of paths that "count," so we conclude from Equation (58):

$$A \propto \text{Solid Angle} = \pi \left(\frac{y}{r}\right)^2 \approx \frac{\pi\lambda}{8\,r} \tag{59}$$

Thus, this is the fundamental origin of the $1/r$ law for amplitudes.

The intensity is proportional to the square of the amplitude, and therefore goes like $1/r^2$, as we all learned in school. Thus, instead of lines of energy radiating out into space in all directions, Feynman's view of the world encourages us to visualize the source of electromagnetic waves as "connected" to each potential receiver by all the paths that arrive at that receiver in phase. Just to convince us that all this "path" stuff is real, Feynman gives numerous fascinating examples where the $1/r^2$ law doesn't work at all. Our favorite is shown in Figure 9:

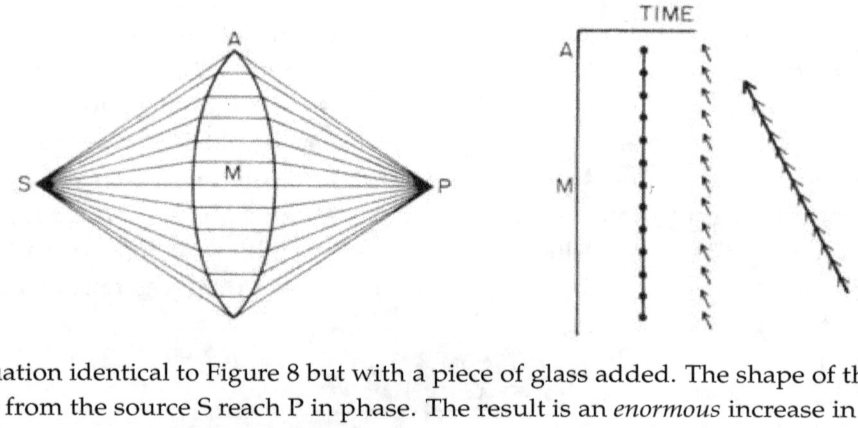

Figure 9. Situation identical to Figure 8 but with a piece of glass added. The shape of the glass is such that all paths from the source S reach P in phase. The result is an *enormous* increase in the amplitude reaching P. (From Fig. 36 in Feynman's **QED**).

The piece of glass does not alter the amplitude of any individual path very much—it might lose a few percent due to reflection at the surfaces. However, it *slows down* the speed of propagation of the light. In addition, the thickness of the glass has been tailored to slow the shorter paths more than the longer paths, so *all paths take exactly the same time.* The net result is that *the oscillating potential propagating along every path reaches P in phase with all the others!* Now, we are adding up all the little phasor arrows *and they all point in the same direction!* The amplitude is enhanced by this little chunk of glass by the factor:

$$\text{EnhancementFactor} = \frac{\text{Lens Solid Angle}}{\text{Solid Angle from Equation (59)}} \approx \frac{8\,r}{\pi\lambda}\text{Lens Solid Angle} \tag{60}$$

For the arrangement shown in Figure 9, at the size it is printed on a normal-sized page, $r \approx 5$ cm, so from From Equation (59), the solid angle of the "bundle" of paths without the glass was of the order of 10^{-7}. The solid angle enclosing the paths through the glass lens is $\approx$1 steradian (sr), so the little piece of glass has increased the potential at P by *seven orders of magnitude*, and the intensity of the light by *fourteen orders of magnitude!* In this and the more general case, we get an important principle:

$$\textbf{Insertion of an optical system replaces the } \frac{\mathbf{1}}{\mathbf{r}} \textbf{ factor by } \frac{8}{\pi\lambda} \times \textbf{Lens Solid Angle} \tag{61}$$

Returning to our two H atoms spaced 1 m apart, we found in Equation (54) that, using the standard $1/r$ potential, the "transition time" for the quantum transaction was $\approx$0.04 s. For the ω_0 wavelength the Rate Enhancement Factor with a 1st optical system is $\approx 2 \times 10^7$, thereby shortening the transition time τ by a factor of $\approx 5 \times 10^{-8}$, making $\tau \approx$ 2 ns.

We have learned an important lesson from Feynman's characterization of propagation phenomena: Changing the configuration of components of the arrangement in what appear to be innocent ways can make *drastic* differences to the resultant potential at certain locations. The reader will find many other eye-opening examples in QED and FLP I-26. We will find in Section 11 that two atoms in a "quantum handshake" form a pattern of paths that greatly increases the potential by which the atoms are coupled, and hence can shorten the transition beyond what is possible with just the optical system.

All the results in statistical QM are probabilities because Heisenberg denied that there was any physics in the transactions. That denial has left the field in a conceptual mess. There is no doubt that statistical QM makes it easy to calculate probabilities of a wide variety of experimental outcomes, and that these predicted outcomes overwhelmingly agree with reality. However that discipline is, by design, powerless to provide reasoning for *how* those outcomes come about. The object of this paper is to *understand the individual transaction, not to calculate probabilities.* Thus, the times quoted above are the times required for the individual event, once initiated, *not* the time constant of some statistical distribution. **We deal with a realm of which statistical QM denies the existence.**

11. Global Field Configuration

We are now in a position to visualize the field configuration for the quantum exchange of energy between two atoms, as analyzed in Section 9, using the locations and coordinated defined there. From Equations (37) and (35), and using $\sin(\phi) = -1$, the total field is composed of the sum of the retarded solution A_α, at distance r_α from atom α and the advanced solution A_β, at distance r_α from atom β:

$$\begin{aligned} A_\alpha &\propto \frac{1}{r_\alpha}\frac{\partial \langle z_\alpha \rangle}{\partial t} = -\frac{1}{r_\alpha}\sin\left(\omega_0(t - r_\alpha/c)\right) \\ A_\beta &\propto \frac{1}{r_\beta}\frac{\partial \langle z_\beta \rangle}{\partial t} \propto \frac{1}{r_\beta}\cos\left(\omega_0(t + \Delta t + r_\beta/c)\right) \end{aligned} \tag{62}$$

Including both x and y coordinates in the distances r_α and r_β from the two atoms, the vector potentials from the two atoms anywhere in the $x - y$ plane are

$$\begin{aligned} A_\alpha(x,y,t) &\propto -\frac{1/\tau}{\sqrt{x^2+y^2}}\sin\left(\omega_0\left(t - \frac{\sqrt{x^2+y^2)}}{c}\right)\right) \\ A_\beta(x,y,t) &\propto \frac{1/\tau}{\sqrt{(x-\Delta x)^2+y^2}}\cos\left(\omega_0\left((t+\Delta t) + \frac{\sqrt{(x-\Delta x)^2+y^2)}}{c}\right)\right) \end{aligned} \tag{63}$$

An example of the total vector potential $A_{\text{tot}} = A_\alpha + A_\beta$ along the x-axis is shown in Figure 10.

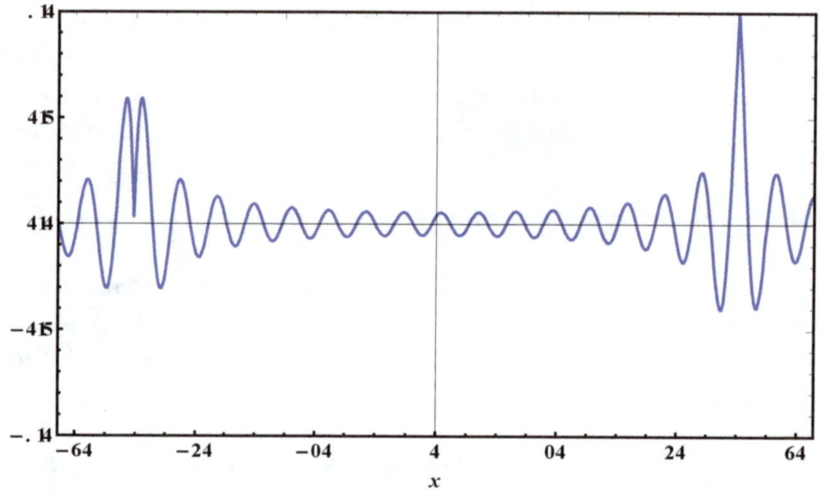

Figure 10. Normalized vector potential along the x-axis (in wavelength/2π) between two atoms in the "quantum handshake" of Equation (63). The wave propagates smoothly from atom α (left) to atom β (right). Animation here [37]. (see Supplementary Materials).

A "snapshot" of the potential of Equation (63) at one particular time for the full $x - y$ plane is shown in Figure 11.

Figure 11. Two atoms in the "quantum handshake" of Equation (63). Animation here [37]. (see Supplementary Materials).

The still image in this figure looks like a typical interference pattern from two sources—a "standing wave." There are high-amplitude regions of constructive interference which appear light blue and yellow on this plot. These are separated from each other by low-amplitude regions of destructive interference, which appear green. In a standing wave, these maxima would oscillate at the transition frequency, with no net motion. The animation, however, shows a totally different story: Instead of oscillating in place as they would in a standing wave, *the maxima of the pattern are moving steadily from the source atom (left) to the receiving atom (right)*. This movement is true, not only of the maxima between the two atoms, but of maxima well above and below the line between the two atoms. These maxima can be thought of as Feynman's paths, each carrying energy along its trajectory from atom α to atom β. For those readers that do not have access to the animations, the same story is illustrated by a stream-plot of the Poynting vector in the x-y plane, shown in Figure 12:

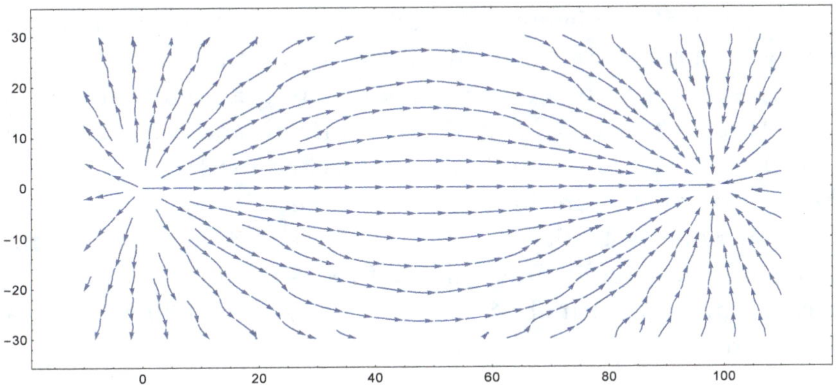

Figure 12. Poynting vector stream lines of the "quantum handshake" of Equation (63).

We can get a more precise idea of the phase relations by looking at the zero crossings of the potential at one particular time, as shown in Figure 13. Paths from atom 1 to atom 2 can be traced through either the high-amplitude regions or the low-amplitude regions. The paths shown in Figure 13 are traced through high-amplitude regions.

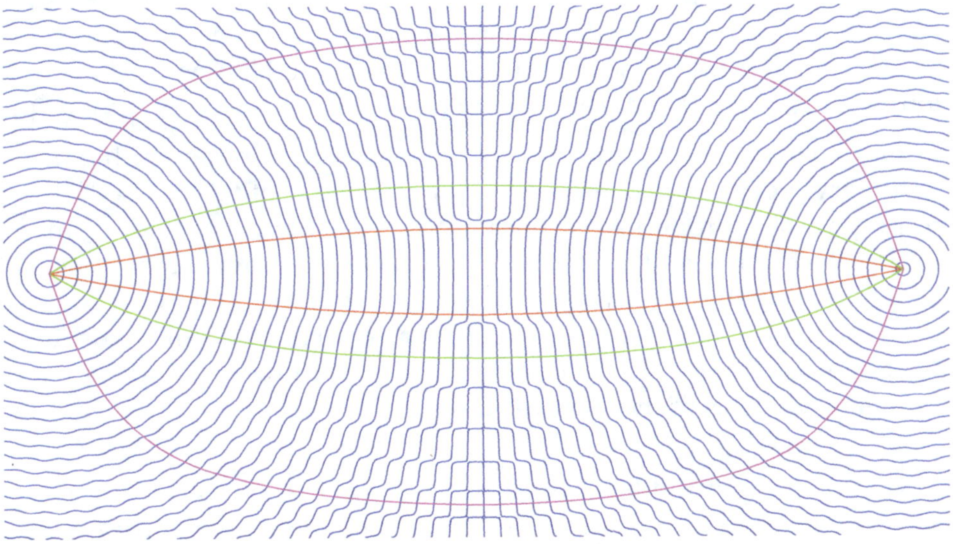

Figure 13. The zero crossings of the handshake vector potential at $t = 0$. Paths near the axis (between the two red lines) are responsible for the conventional $1/r$ dependence of the potential. Paths shown through high-amplitude regions have an even number of zero crossings, and thus the potentials traversing these paths all arrive in phase, thus adding to the central potential.

The central set of paths, delimited by the red lines, are responsible for the conventional $1/r$ dependence of the potential, as described with respect to Equation (59). Working outward from there, each high-amplitude path region is separated from the next by a slim low-amplitude region. It is a remarkable property of this interference pattern that each low-amplitude path has π more phase accumulation along it than the prior high-amplitude path and π less than the next high-amplitude path. The low-amplitude paths are the ones that contribute to the "de-phasing" in this arrangement, but they are very slim and of low amplitude, so they don't de-phase the total signal appreciably. In addition, the phases of the paths through the high-amplitude regions are separated by $2\pi n$, where n is an integer. **All waves propagating from atom α to atom β along high-amplitude paths arrive in phase!**

In Feynman's example shown in Figure 8, there are an equal number of paths of any phase, so every one has an opposite to cancel it out. In Figure 9, the lens makes all paths have *equal time delay*, which then enables them to all arrive with the same phase.

The phase coherence of the advanced-retarded handshake creates a pattern of potentials that has a unique property: It is not like either of Feynman's examples in Figure 8 or Figure 9. Its high-amplitude paths do arrive *in phase*, but by a completely different mechanism. It all starts with the bundle of paths between the red lines, which has the $1/r$ amplitude, just as if there were no quantum mechanism. Then, as the handshake begins to form, additional paths are drawn into the process. The process is self-reinforcing on two levels—the increase in dipole moment and the increase in number of paths that arrive in phase. Paths that formerly would have cancelled the in-phase ones are "squeezed" into extremely narrow regions, all of low amplitude, as can be seen in Figure 13. Thus, a large fraction of the solid angle around the atoms is available for the in-phase high-amplitude paths.

For optical systems with large solid angle, the self-focusing enhancement may still be noticeable in the shape of the transition waveform, but for one-sided systems, like an astronomical telescope, we expect it to be be dominant.

Following Feynman's program has led us to conclude that: The vector potential from all paths sum to make a highly-amplified connection between distant atoms. The advanced-retarded potentials form nature's very own phase-locked loop, which forms nature's own *Giant Lens in the Sky!*

The consequences of this fact are staggering: Once an initial handshake interference pattern is formed between two atoms that have their wave functions synchronized, the strength grows explosively: Not only because the dipole moment of each atom grows exponentially, but, in addition, a substantial fraction of the possible interaction paths between the two atoms propagate through high-amplitude regions, *independent of the distance between them!* Although we have not worked out the difficult second-order dynamics of phase-locking between coupled atoms, we believe that here is the solution to the long-standing mystery of the "collapse of the wave function" of the "photon".

The interaction depends critically on the advanced-retarded potential handshake to keep all paths in phase. Ordinary propagation over very long paths becomes "de-phased" due to the slightest variations of the propagation properties of the medium. By contrast, the advanced and retarted fields are precise negative images of each other on exactly the same light cone, so the phase of high-amplitude handshake paths are always related by an even number of π to the phase of other such paths. Paths having odd numbers of π phase are always of low amplitude, and do not cancel the even-π phase paths as they would in a one-way propagating wave like Feynman used in his illustrations. [A detailed analysis of these properties has not been done. It is a wonderful project for the future.]

The interaction proceeds in the local time frame of each atom because they are linked with the advanced-retarded potential. The waves carrying positive energy from emitter to absorber are retarded waves with positive transit time; they reach the absorber after a single transit time $\Delta t = r/c$. Once they have established a phase-coherent "handshake" connection, Lewis' "*coordinate and symmetrical*" advanced waves with negative transit time are launched toward the emitter, arriving at the precise time and in the precise phase to withdraw energy from the emitter. During the transaction, as long as the "handshake" connection is active, any change in the state of one atom will be directly reflected in the state of the other. Aside from the time-of-flight propagation time to establish the "handshake", there is no additional "round trip" time delay in the quantum-jump process, which proceeds as if the two atoms were local to each other.

Thus, the Transactional Interpretation allows us to conceptualize Niels Bohr's "instantaneous" quantum jump [47] concept that Schrödinger, who expected time-extended classical transitions, found impossible to accept [48].

12. Relevance to the Transactional Interpretation

The calculation that we have presented here, with its even-handed treatment of advanced and retarded four-potentials, was inspired by WFE and the Transactional Interpretation, but it also provides interesting insights that clarify and modify the mechanism by which a transaction forms. Wheeler–Feynman electrodynamics suggests that a retarded wave, arriving at a potential absorber, stimulates that entity to generate a canceling retarded wave accompanied by an advanced wave. The TI suggests that this advanced wave arrives back at the emitter with amplitude $\psi\psi*$, a relation suggesting the Born rule.

However, from our calculation, we see that a slightly different process is described. Both emitter and absorber initially have small admixtures of the opposite eigenstate, giving them dipole moments that oscillate with the *same* difference-frequency ω_0. The two oscillations each have an environment-induced random phase. If the phases have the correct relation [$\sin(\phi) \approx -1$], the dipole moments of both atoms initially increase exponentially, the system becomes a phase-locked loop, and it avalanches to a final state of multi-path energy transfer that satisfies all boundary conditions. In that scenario, the initial confirmation wave is likely to have an amplitude much weaker than WFE would suggest, and the quantity $\psi\psi*$ becomes that used by Schrödinger to provide the electron density function.

How can a linear system generate such nonlinear behavior? While the Schrödinger equation is indeed linear, Equation (43) governing the evolution of transaction formation and wave function collapse is highly nonlinear. Thus, the TI's assertion that the offer wave "stimulates the generation of the confirmation wave" must be modified. Rather, advanced and retarded potentials, boundary conditions at both ends, and a fortuitous matching of phase trigger the nonlinear avalanche in both atoms and brings about the transaction.

We also note that the advanced and retarded waves do not carry "information" in the usual sense in either time direction, but only deliver a pair of oscillating four-potentials to the sites of a pair of oscillating charges, leading dynamically to an initially exponential rise in coupling, a focusing of alternative paths, the formation of a transaction, a transfer of energy, and the enforcement of conservation laws. For the Transactional Interpretation, this phase selection process clarifies the randomizing mechanism by which, in the first stage of transaction formation, the emitter makes a random choice between competing offer waves arriving from many potential absorbers. The offer wave with the best phase is likely to win, even if it comes from far away. [It is sometimes asserted that this handshake situation is only possible in a frozen deterministic four-dimensional "block universe" because of the two-way connections between present and future. We reject this assertion, which is dissected in some detail in Section 9.2 of Ref. [12]. While it is true that the assumption of a block universe would dispel or bypass many of the quantum paradoxes, it would only do so at the terrible price of imposing complete determinism on the universe.].

We saw in the derivation of the coupling of two separated atoms that it was necessary to use the advanced 4-potential *in order to satisfy the law of conservation of energy*. This, not "information transfer", is the role of the quantum handshake in the Transactional Interpretation. The quantum handshakes act to enforce conservation laws and do not form unless all conserved quantities are properly transferred and conserved. This is what is going on in *quantum entanglement*: the separated parts of a quantum system are linked by conservation laws that are enforced by V-shaped three-vertex advanced-retarded quantum handshakes [12] and cannot emerge as a completed transaction unless those conservation laws are satisfied. In this context, we note that the Transactional Interpretation, using such linked advanced-retarded handshakes, is able to explain in detail the behavior of over 26 otherwise paradoxical and mysterious quantum optics experiments and *gedankenexperiments* from the literature. See Chapter 6 of Ref. [12]. If we cannot dismiss the plethora of competing QM interpretations based on their failure in experimental tests, *we should eliminate them when they fail to explain paradoxical quantum optics experiments* (as almost all of them do.)

13. Historic Tests

13.1. The Hanbury–Brown–Twiss Effect and Waves vs. Particles

It is often said that particles in quantum mechanics "travel as waves but arrive as particles". The Hanbury–Brown–Twiss effect [49] (HBT) is an example of this principle. It demonstrates that, in the second-order interference of incoherent wave sources, photons are divisible and are not electromagnetic "billiard balls" that maintain individual identities. The HBT technique was first applied to the measurement of the diameters of nearby stars, e.g., Betelguese, using intensity interferometry with radio waves. The original experiments involved large parabolic radio dishes mounted on rail cars.

A simplified version of an HBT interference measurement is illustrated in Figure 14. Sources 1 and 2 are separated by a distance d_{12}. Both sources emit waves of the same wavelength $\lambda = 2\pi c/\omega$, but are not causally connected. Radiation from the two sources is received by detectors A and B, which are separated by a distance d_{AB}. The line between the sources is parallel to the line between the detectors, and the two lines are separated by a distance L. Hanbury-Brown and Twiss showed that a significant rate of coincident detections—up to a factor of 2 larger than could be ascribed to chance—was observed in this arrangement. For this simple configuration, the probability of a coincident detection in the two detectors, for small d_{AB} and very large L, is proportional to $1 + \cos[2\pi\, d_{AB}\, d_{12}/(\lambda L)]$, which is maximum when $d_{AB} = 0$ and falls off as the detector separation increases, the rate of falloff indicating the value of d_{12}.

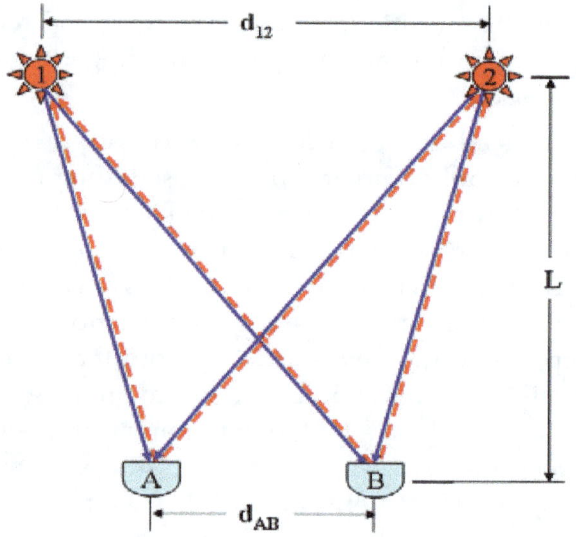

Figure 14. Schematic diagram of the Hanbury–Brown–Twiss effect, with excited atoms 1 and 2 in distant separated sources simultaneously exciting ground state atoms A and B in two separated detectors.

The very fact that coincidences are observed in this experiment reveals a deep truth about electromagnetic coupling: Photons cannot be consistently described as little blobs of mass-energy that travel uniquely from a single source point to a single detector point. In the HBT effect, a whole photon's worth of energy $\hbar\omega$ is assembled at each detector out of fractional energy contributions from each of the two sources.

In the TI description of the HBT event described above, a retarded offer wave is emitted by the source 1 and travels to both detectors A and B. Similarly, a retarded offer wave is emitted by the source 2 and travels to both detectors. Detector A receives a linear superposition of the two offer waves and seeks to absorb the "offered" energy by producing an advanced confirmation wave. If the phases match, as they will in a coincident event, the energy transfer begins with an exponential increase in the dipole moment of each source atom. Atoms in detectors A and B respond similarly, their oscillating dipole moments producing advanced confirmation waves that travel back to the two sources, each of which responds with an increasing dipole moment that enhances the offer waves. A four-atom transaction of the form shown in Figure 14 is formed that removes one photon's worth of energy $\hbar\omega$ from each of the two sources 1 and 2 and delivers one photon's worth of energy $\hbar\omega$ to each of the two detectors A and B.

Neither of the detected "photons" *can be said to have originated uniquely in one of the two sources.* The energy arriving at each detector originated partly in one source and partly in the other. It might be said that each source produced two "fractional photons" and that these fractions from two sources combined at the detector to make a full size "photon". The "particles" transferred have no separate identity that is independent from the satisfaction of the quantum mechanical boundary conditions. The boundary conditions here are those imposed by the HBT geometry, conservation laws, and the detection criteria.

Finally, we note that in many experiments published in the physics literature the HBT effect has been observed and demonstrated not only for photons, but also in the detection of charged π mesons emitted in the ultra-relativistic collision of heavy nuclei. It is observed that the detection probability doubles when the detected particles are close in momentum and position. Furthermore, in nuclear physics experiments with pairs of half-integer-spin neutrons or protons, a Pauli-Exclusion version of the HBT effect has been observed, in which the detection probability drops to zero when the detected particles are close in momentum and position [50]. All of these particle-like entities "travel as waves but arrive as particles".

13.2. Splitting Photons

At the 5th Solvay Conference in 1927, Albert Einstein posed a riddle, sometimes called "Einstein's Bubble Paradox", to the assembled founders of quantum mechanics [12,51]. Einstein's original language was rather convoluted and technical (and in German), but his question can be simply stated as follows:

> *A source emits a single photon isotropically, so that there is no preferred emission direction. According to the quantum formalism, this should produce a spherical wave function Ψ that expands like an inflating bubble centered on the source. At some later time, the photon is detected. Since the photon does not propagate further, its wave function bubble should "pop", disappearing instantaneously from all locations except the position of the detector. In this situation, how do the parts of the wave function away from the detector "know" that they should disappear, and how is it enforced that only a single photon is always detected when only one photon is emitted?*

The implication of his question is that, if a photon is an indivisible particle, it should not be possible to divide one. However, we have already seen in the Hanbury–Brown–Twiss effect that photons are *not* little indivisible billiard balls and that they *can* divide their energy between two receiving atoms. How, then, is it possible that for one photon emitted there is always only one received?

A search for this hypothetical divided-photon behavior is implemented in the setup shown in the left panel in Figure 15, which was enabled once it became possible to build sources of single photons. The idea is that, if a photon is just a short pulse of light, half of it should go through each of the dotted paths, and both halves should be counted at the same time, registering as a coincidence. Of course, the original pulse must have twice the energy required to trigger a detector, so either half by itself would have just enough. However, if the photon was indivisible, as mandated by certain versions of QM, it would make a random decision on which path to take, and no coincidences would be observed. In practice, the number of coincidences is counted for a certain counting period, with the time delay τ between the two detector output pulses as a parameter. Since the photons are generated randomly, the time between successive photons can accidentally range from zero to large, and a plot of the number of correlator outputs vs. time delay τ gives information about the statistics of the source.

Modern versions of the experiment [52] give plots like that shown in the right panel in Figure 15. [Early versions of this kind of experiment suffered from certain defects that made them inconclusive. A definitive version using "heralded photons" was finally accomplished by Clauser in 1973 [53].] Let us look at the experiment from a TI perspective: The source excites one atom, which, due to random coupling, develops a superposition with a tiny presence of ground state and nearly unity excited state. As described for the two-atom photon, the tiny presence in the superposition enables the dipole moment to oscillate, thereby generating a radiating vector potential that propagates along both dotted paths to both detectors. To find a perfectly matched partner atom is rare, but, when one matches up, a quantum handshake grows up connecting them.

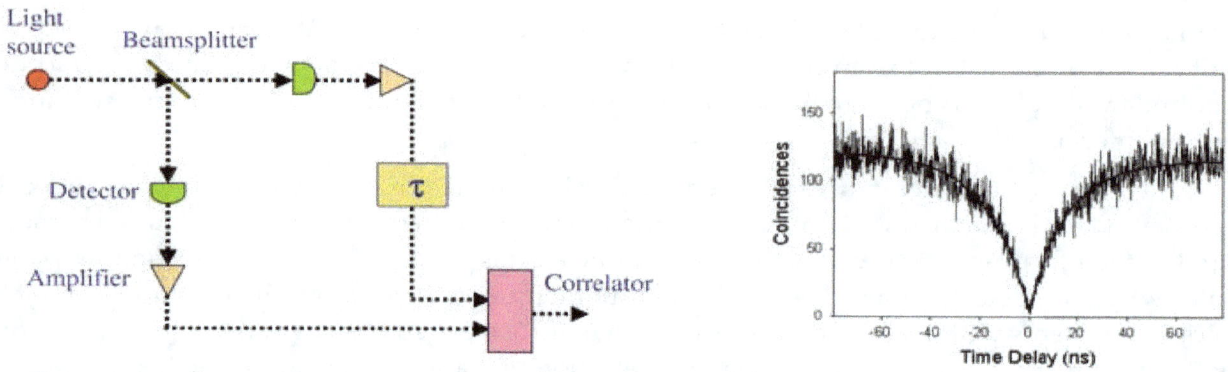

Figure 15. Left: Schematic of the photon-splitting experiment. **Right**: Plot of the number of coincidences vs. time delay between the arriving pulses. The source on average generates a photon per ≈12 nsec. The finite number counted at zero delay is consistent with the accidental presence of two excited atoms.

There are then three possible outcomes:

1. The handshake goes to completion and the partner atom is in one of the detectors, in which case only the chosen detector registers an output.
2. The handshake goes to completion, but the partner atom is not in one of the detectors. In this case, no output is registered from either detector.
3. The initial stages of a handshake begins in *two* partner atoms, one in each detector. When the source atom has de-excited, both of the detector atoms are left in mixed states with roughly equal components of ground-state and excited state wave functions, as was illustrated in Figure 7. This is not a stable configuration because both of the detector atoms have oscillating dipole moments sending strong "unrequited" advanced confirmation waves. These waves are in phase at and focused on the source, and they are likely to find another well-phased excited-state atom there or nearby that will complete the four-way transaction. Thus, a four-atom HBT event should be created, in which there are two emissions and two detections. Such an unlikely event would register as an "accidental" case of two simultaneous emissions in the same time window. There will never be an event with a single emission and two detections.

Thus, we see that the outcomes of the experiment predicted by TI and QM are essentially identical. Certainly, no solid conclusion can be drawn from this experiment as to whether quantization occurs in the field or in the transaction.

13.3. Freedman–Clauser Experiment

In our introductory discussion of Schrödinger's visualization of his newly-invented Wave Mechanics in Section 3, we described how Clauser and his colleagues, through experiments that were heroic at the time, were able to show that no "local, realistic theory" was compatible with their results [6–11]. We now describe the earliest conclusive version of these EPR experiments and show how our TI approach gives a simple and natural explanation of the otherwise mysterious outcome. A sketch of the arrangement is shown in Figure 16.

The atomic configuration used was the three-level system of the Ca atom, shown at the right of the figure. The atomic wave functions were an upper $4p^2\ S_0$ state Ψ_3 of frequency ω_3, a middle $4p4s\ P_1$ state Ψ_2 of frequency ω_2, and a $4s^2\ S_0$ ground state Ψ_1 of frequency ω_1, so that the cascade starts and ends in an S_0 state of zero angular momentum.

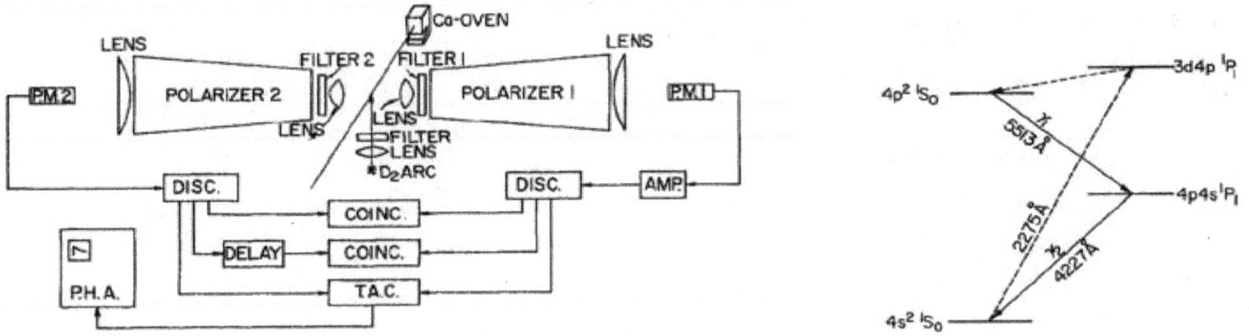

Figure 16. Left: Schematic of the polarization-correlation experiment. Right: Energy levels of the Ca atoms used in this experiment. From Freedman and Clauser 1972 [6].

The state wave functions are of the same form as those given in Equation (2):

$$\Psi_3 = R_3(r)\, e^{-i\omega_3 t} \qquad \Psi_2 = f_2(r)\cos(\theta)\, e^{-i\omega_2 t} = R_2(r,\theta)\, e^{-i\omega_2 t} \qquad \Psi_1 = R_1(r)\, e^{-i\omega_1 t} \tag{64}$$

The transition state is a superposition of these three states: The analysis is a simple extension of that for the two-level system, starting with Equation (10):

$$\Psi = ae^{i\phi_a} R_1 e^{-i\omega_1 t} + be^{i\phi_b} R_2 e^{-i\omega_2 t} + ce^{i\phi_c} R_3 e^{-i\omega_3 t}, \tag{65}$$

The charge density of the mixed state is then:

$$\begin{aligned}
\rho &= q\Psi^* \Psi \\
\frac{\rho}{q} &= \left(ae^{-i\phi_a} R_1 e^{i\omega_1 t} + be^{-i\phi_b} R_2 e^{i\omega_2 t} + ce^{-i\phi_c} R_3 e^{i\omega_3 t}\right)\left(ae^{i\phi_a} R_1 e^{-i\omega_1 t} + be^{i\phi_b} R_2 e^{-i\omega_2 t} + ce^{i\phi_c} R_3 e^{-i\omega_3 t}\right) \\
&= a^2 R_1^2 + b^2 R_2^2 + c^2 R_3^2 \\
&+ 2abR_1 R_2 \cos((\omega_2 - \omega_1)t + (\phi_b - \phi_a)) \\
&+ 2acR_1 R_3 \cos((\omega_3 - \omega_1)t + (\phi_a - \phi_c)) \\
&+ 2bcR_2 R_3 \cos((\omega_3 - \omega_2)t + (\phi_b - \phi_c))
\end{aligned} \tag{66}$$

The dipole srength d_{ij} for the three terms is given by straightforward extension of Equation (18)

$$d_{12} = 2q \int R_1 R_2 z \qquad d_{23} = 2q \int R_3 R_2 z \qquad d_{13} = 2q \int R_1 R_3 z = 0 \tag{67}$$

By symmetry around the ($+z$ axis) of the spherical coordinate system, both d_{12} and d_{23} are in the $\vec{z}$ direction determined by the direction of the Ψ_2 wave function. In general, that direction will shift around in space depending on the coupling of the atom to others. However, in any given situation, there is only one z-axis that defines the "North pole" direction, and both dipole moments are oscillating in that direction. Thus, it is the Ψ_2 state, shared by both transactions of the 3-state "cascade" that aligns the linear polarizations of the two interlocking transactions.

13.3.1. Dynamics of the Transaction

The cascade atom is emitted from the oven shown at the top of the figure. It is is initially "pumped" by the D_2 arc when it is centered between the two lenses. The excitation is along the 2275Å dashed path on the energy diagram, from which it relaxes into the $4p^2\, S_0$ excited state where $a \approx 0$, $b \approx 0$, $c \approx 1$, leaving a small residual admixture of the middle and ground states. By Equation (66) and the obvious generalization of Equation (19), the superposition begins to oscillate with dipole moment:

$$\text{dipole moment} = q\,\langle z \rangle = d_{12}ab\,\cos(\omega_{12}t + \phi_{ab}) + d_{23}bc\,\cos(\omega_{23}t + \phi_{bc}) \tag{68}$$

where we have included only non-zero time-varying terms terms and have used:

$$\omega_{12} = (\omega_2 - \omega_1) \qquad \omega_{23} = (\omega_3 - \omega_2) \qquad \phi_{ab} = (\phi_b - \phi_a) \qquad \phi_{bc} = (\phi_c - \phi_b) \tag{69}$$

Because $\omega_{12} \neq \omega_{23}$, the two terms in the dipole moment do not interact over periods of many cycles, and each term can couple to a separate atom to form its own quantum handshake. The analysis has already been done starting with Equation (35) and ending with Equation (41), and applies directly to the upper (ω_{23}) transaction with atom we will call α, which has level spacing exactly equal to E_{23}.

Since c and a_α both start almost equal to 1 and have the same derivative, we can set $c = a_\alpha$.

$$\frac{\partial E_\alpha}{\partial t} = -E_{23}\frac{\partial c^2}{\partial t} = -E_{23}\frac{\partial a_\alpha^2}{\partial t} = -P_\alpha\, c\, b\, a_\alpha b_\alpha \sin(\phi_\alpha) = -P_\alpha\, c^2\, b\,\sqrt{1-c^2}\,\sin(\phi_\alpha) \tag{70}$$

2

Similarly, we describe the fraction a of the superposition in the ground state due to the lower 5227Å (ω_{12}) transaction with atom we will call β, which has level spacing exactly equal to E_{12}.

Since a and b_β both start almost equal to 0 and have the same derivative, we can set $a = b_\beta$.

$$\frac{\partial E_\beta}{\partial t} = E_{12}\frac{\partial a^2}{\partial t} = E_{12}\frac{\partial b_\beta^2}{\partial t} = P_\beta\, a\, b\, a_\beta b_\beta \sin(\phi_\beta) = P_\beta\, a^2\, b\,\sqrt{1-a^2}\,\sin(\phi_\beta) \tag{71}$$

These relations may then be expressed more compactly as:

$$\frac{\partial c^2}{\partial t} = -\frac{1}{\tau_\alpha}\,c^2\, b\,\sqrt{1-c^2} \qquad \frac{\partial a^2}{\partial t} = \frac{1}{\tau_\beta}\,a^2\, b\,\sqrt{1-a^2} \qquad a^2+b^2+c^2=1 \tag{72}$$

where $1/\tau_\beta = P_\beta\,\sin(\phi_\beta)/E_{12}$ and $1/\tau_\alpha = P_\alpha\,\sin(\phi_\alpha)/E_{23}$ express the respective strength of coupling to each atom, and the last relation constrains the superposition to contain exactly one electron. Equation (72) is identical in form to Equation (43) with the exception of the shared state amplitude b occurring in both derivatives.

The behavior of this arrangement, shown in Figure 17, is very instructive. To review in brief: The cascade atom is prepared by providing the energy to promote the electron to the upper state: $c \approx 1$, $a \approx 0$, $b \approx 0$. The preparation is never perfect, so there is always a small residual of b and a components in the initial superposition. The small admixture of b and c, by Equation (68), creates an oscillating dipole moment at frequency ω_{23}, the amplitude of which is shown as the red curve in the right panel of Figure 17. That oscillating dipole moment initiates a vector potential "offer wave" which propagates outward according to Equation (37). When that vector potential couples to the wave function of another atom of the same frequency, its wave function oscillates with the vector potential in the correct phase to withdraw energy according to Equation (50). The vector potential from that oscillation propagates *backwards in time*, so it is "felt" by the cascading atom as if it had been there all along.

The Transactional Interpretation is applied to the Freedman–Clauser experiment as shown in Figure 18. Two-way handshakes between the source and the two polarimeters are joined at the source and must satisfy the boundary condition, based on conservation of angular momentum in a system that begins and ends in a state of zero angular momentum that the polarizations must match for the two offer waves emitted back-to-back. This V-shaped transaction holds the key to understanding the mechanism behind quantum nonlocality in EPR expertiments.

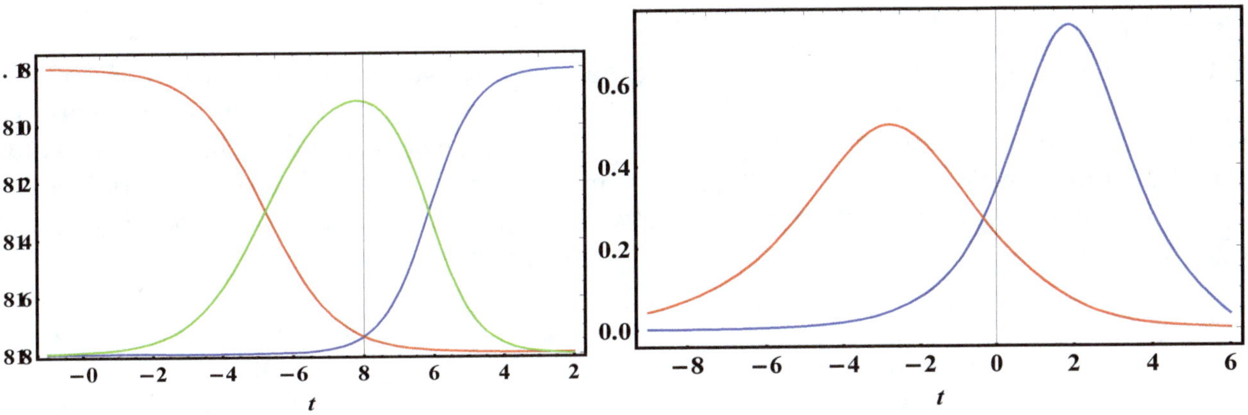

Figure 17. Left: Superposition contributions of the upper state c^2 (red), shared middle level b^2 (green) and ground state a^2 (blue). **Right**: Amplitude of dipole oscillations due to upper transition at ω_{23} (red) and lower transition at ω_{12} (blue), from. The horizontal axis in both plots is time in units of $\tau_\alpha = 1.5\,\tau_\beta$.

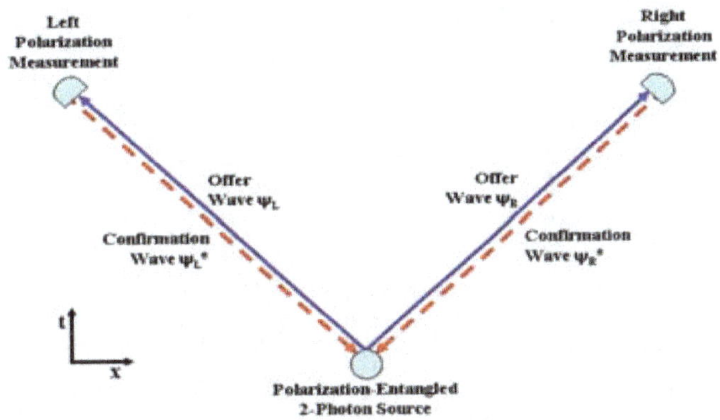

Figure 18. Three-vertex transaction formed by a detection event in the Freedman–Clauser experiment. Linked transactions between the source and the two polarimeters cannot form unless the source boundary condition of matching polarization states is met.

It was pointed out by one of our reviewers that the kind of description just given might appeal to readers who are not accustomed to standard relativistic space-time diagrams, on which events on the "light cone" are local in the sense that $r^2 - c^2t^2 = 0$, and that we had, in some measure, ignored G.N. Lewis' chiding quoted earlier "A dissymmetry alien to the pure geometry of relativity has been introduced by our notion of causality." Thus, let's try it again:

A quantum handshake is an antisymmetric bidirectional electromagnetic connection between two atoms on a light cone, whose direction of time is the direction of positive energy transfer.

Either way we look at it, the cascade atom and atom α are locked in phase and amplitude at frequency ω_{23}, and the locked amplitude of oscillation of the wave functions of the two atoms is growing with time. In the process, the z-axes of both atoms becomes better and better aligned. Meanwhile, the small superposition amplitudes b and a are growing, thus developing a growing oscillation at ω_{12}, shown by the blue curve in the right panel of Figure 17. Since both the upper and lower levels are S states, they have no effect on the orientation of the oscillation, which is determined by the shared middle level, which is a P state that has a definite direction in space. That direction determines the direction of oscillation of any superposition involving that P state. The vector potential from the nascent ω_{12} oscillation recruits a willing atom β whose level spacing is precisely match to ω_{12}, and forms an embryonic quantum handshake of its own whose z-axis is already determined by the fully developed ω_{23} oscillation. From there, both transactions were completed, with z axes aligned, in very much the same way we have described for a single photon.

13.3.2. The NCT-Killer Result

The unique aspect of the Freedman–Clauser experiment was separating the back-to-back paths of the two wave propagations with filters that selected the ω_{23} path to the left and the ω_{12} path to the right. Each path was equipped with its own glass-plate-stack linear polarizer, whose angle could be adjusted. The number of coincidences of signals from the left and right photomultipliers P.M.1 and P.M.2 were plotted as a function of the angle ϕ between the polarization axes of the polarizers, and is shown in Figure 19.

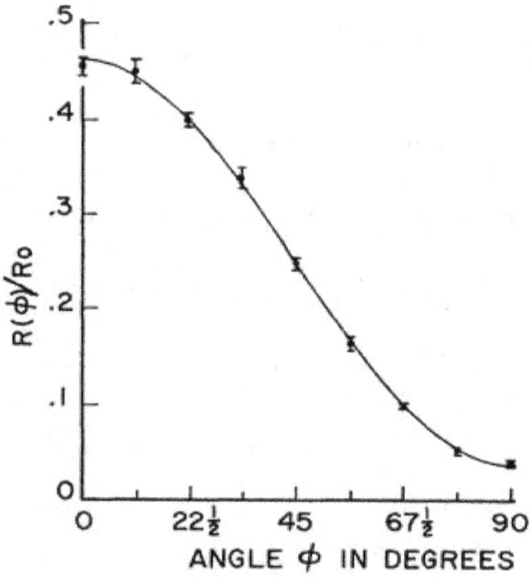

Figure 19. Coincidence rate vs. angle ϕ between the polarizers, divided by the rate with both polarizers removed. The solid line is the prediction by quantum mechanics, calculated using the measured efficiencies of the polarizers and solid angles of the experiment.

The key aspect of Figure 19 is that at $\phi = 90°$ the coincidence rate drops to essentially zero. The graph does show a small non-zero coincidence rate at 90°, but this is because of the imperfect linear polarization discriminations of the glass-plate-stack polarizing filters and the finite solid angles of the detection paths. Jaynes' NCT approach failed to reproduce [32] this result. That failure, in the view of most of the field, "falsified" the NCT approach to quantum phenomena and caused it to be subsequently ignored.

We can understand this result very simply from our TI perspective, reasoning directly from Figure 17: By the time the blue ω_{12} transition is just getting started, the red ω_{23} transition is well along, and has connected to a partner in P.M.2 through polarizer 2. Since the polarizer only transmits a vector potential aligned with its axis of polarization, the z axes of both the cascade atom and atom α will be well aligned with polarizer 2. Now, as the blue oscillation just begins to build, its axis, as part of the same P wave function, will also be aligned with polarizer 2, at rotation angle θ_2.

The way that these glass-plate-stack polarizers work is that they only pass the component of propagating vector potential along their axis of polarization, the orthogonal component being reflected out of the direction of propagation. Thus, the fraction of θ_2 polarized vector potential that can pass through a perfect θ_1 oriented polarizer 1 is just $\cos(\theta_2 - \theta_1)$. Thus, the amplitude of the nascent ω_{12}, θ_2 polarized "offer wave" propagating outward from the cascade atom through polarizer 1 will be proportional to $\cos(\theta_2 - \theta_1)$. By Feynman's *Grand Principle*, the probability amplitude of an "offer wave" actually forming a transaction is proportional to the amplitude of the vector potential. Since, in this example, the ω_{23} transaction in P.M.2 has already formed, the slightly later formation of an ω_{12} transaction in P.M.1 will count as a coincidence. The probability of coincidence counts will therefore be proportional to the square of the probably amplitude which, for this case, is $\cos^2(\theta_2 - \theta_1)$ which, when corrected as noted, gives the solid line in Figure 19 and is zero when $\phi = 90°$. Once again, the bi-directional non-local nature of the quantum handshake predicts the observed outcome of this historic experiment.

14. Conclusions

The development of our understanding of quantum systems began with a physical insight of deBroglie: Momentum was the wave vector of a propagating wave of some sort. Schrödinger is well known for developing a sophisticated mathematical structure around that central idea, only a shadow of which remains in current practice. What is less well known is that Schrödinger also developed a deep understanding of the *physical meaning* of the mathematical quantities in his formalism. That physical understanding enabled him to *see* the mechanism responsible for the otherwise mysterious quantum behavior. Meanwhile, Heisenberg, dismissing visual pictures of quantum processes, had developed a matrix formulation that dealt with only the probabilities of transitions—what he called "measurables." It looked for a while as if we had two competing quantum theories, until Schrödinger and Dirac showed that they gave the same answers. However, the stark contrast between the two approaches was highlighted by the ongoing disagreement in which Bohr and Heisenberg maintained that the transitions were events with no internal structure, and therefore there was nothing left to be understood, while Einstein and Schrödinger believed that the statistical formulation was only a stopgap and that a deeper understanding was possible and was urgently needed. This argument still rages on.

Popular accounts of these ongoing arguments, unfortunately, usually focus on the 1930 Solvay Conference confrontation between Bohr and Einstein that was centered around Einstein's clock paradox, a clever attempted refutation of the uncertainty principle [51]. Einstein is generally considered to have lost to Bohr because he was "stuck in classical thinking." However, as detailed in ***The Quantum Handshake*** [12], Einstein's effort was doomed from the start and was also beside the point. The uncertainty principle is simply a Fourier-algebra property of *any system described by waves*. Both parties to the Solvay argument lacked any real clarity as to how to handle the intrinsic wave nature of matter. In the introduction, we quoted Einstein's deepest concern with statistical QM:

There must be a deeper structure to the quantum transition.

Back in 1926, the field was faced with a choice: Schrödinger's wave function in three-dimensional space, or Heisenberg and Born's matrices, in as many dimensions as you like. The choice was put forth clearly by Hendrik Lorentz [54] in a letter to Schrödinger in May, 1926:

> *"If I had to choose now between your wave mechanics and the matrix mechanics, I would give the preference to the former because of its greater intuitive clarity, so long as one only has to deal with the three coordinates x,y,z. If, however, there are more degrees of freedom, then I cannot interpret the waves and vibrations physically, and I must therefore decide in favor of matrix mechanics. However, your way of thinking has the advantage for this case too that it brings us closer to the real solution of the equations; the eigenvalue problem is the same in principle for a higher dimensional q-space as it is for a three-dimensional space.*
>
> *"There is another point in addition where your methods seem to me to be superior. Experiment acquaints us with situations in which an atom persists in one of its stationary states for a certain time, and we often have to deal with quite definite transitions from one such state to another. Therefore, we need to be able to represent these stationary states, every individual one of them, and to investigate them theoretically. Now a matrix is the summary of all possible transitions and it cannot at all be analyzed into pieces. In your theory, on the other hand, in each of the states corresponding to the various eigenvalues, E plays its own role."*

Thus, the real choice was between the intuitive clarity of Schrödinger's wave function and the ability of Heisenberg–Born matrix mechanics to handle more degrees of freedom. That ability was immediately put to the test when Heisenberg [55] worked out the energy levels of the helium atom, in which two electrons shared the same orbital state and their correlations could not be captured by wave functions with only three spatial degrees of freedom. That amazing success set the field on the path of eschewing Schrödinger's views and moving into multi-dimensional Hilbert space, which was further ossified by Dirac and von Neumann. Schrödinger's equation had been demoted to a bare matrix equation, engendering none of the *intuitive clarity*, the ability to *interpret the waves and vibrations physically* so treasured by Lorentz.

The matrix formulation of statistical QM, as now universally taught in physics classes, saves us the "tedious" process of analyzing the details of the transaction process. That's the good news. The bad news is that *it actively prevents us from learning anything about the transaction process, even if we want to!*

What has been left out is, as Einstein said, any *"description of the individual system"*.

Thus, it was left to the more practical-minded electrical engineers and applied scientists to resurrect, each in their own way, Schrödinger's way of thinking because they needed a *"description of the individual system"* to make progress. Electrons in conductors were paired into standing waves, which could carry current when the propagation vector of one partner was increased and that of the other partner decreased. Energy gaps resulted from the interaction of the electron wave functions with the periodic crystal lattice. Those same electron wave functions can "tunnel" through an energy gap in which they decay exponentially with distance. The electromagnetic interaction of the collective wave functions in superconducting wires leads to a new formulation of the laws of electromagnetism without the need for Maxwell's equations [16]. The field of Quantum Optics was born. Conservation of momentum became the matching of wavelengths of waves such that interaction can proceed. When one such wave is the wave function of an electron in the conduction band and the other is the wave function of a hole in the valence band of a semiconductor, matching of the wavelengths of electron, hole, and photon leads to light emission near the band-gap energy.

When that emission intensity is sufficient, the radiation becomes coherent—a semiconductor laser. These insights, and many more like them, have made possible our modern electronic technology, which has transformed the entire world around us.

Each of them requires that, as Lorentz put it: *we... represent these stationary states, every individual one of them, and to investigate them theoretically.*

Each of them also requires that we analyze the transaction involved very much the way we have done in this paper.

What we have presented is a detailed analysis of the most elementary form of quantum transition, indicating that the simplest properties of solutions of Schrödinger's equation for single-electron atomic states, the conservation of energy, and a symmetry property of relativistic laws of electromagnetic propagation, together with Feynman's insight that all paths should be counted, give a unique form to the photon transaction between two atoms.

We have extended this approach to experiments involving three atoms. The reason we can treat situations with more than one electron using a wave-mechanics Schrödinger equation that only works for one electron is that the non-local bi-directional electromagnetic coupling between wave functions can be factored into a retarded wave propagating forward in time and an advanced wave propagating backward in time, the vector potential of each partner in a photon transaction being incorporated in the opposite partner's one-electron Schrödinger equation.

These calculations are, of course, not general proofs that in every system the offer/confirmation exchange always triggers the formation of a transaction. They do, however, represent demonstrations of that behavior in tractable cases and constitute prototypes of more general transactional behavior. They further demonstrate that the transaction model is implicit in and consistent with the Schrödinger wave mechanics formalism, and they demonstrate how transactions, as a space-time standing waves connecting emitter to absorber, can form.

We see that the missing ingredients in previous failed attempts by others to derive wave function collapse from the standard quantum formalism were:

1. Advanced waves were not explicitly used as part of the process.
2. The "focusing" property of the advanced-retarded radiation pattern had not been identified.

Although many complications are avoided by the simplicity of these two-atom and three-atom systems, they clearly illustrate that *there is internal structure to quantum transitions* and that *this structure is amenable to physical understanding*. Each of them is an example of Einstein's "*description of the individual system*". Through the Transactional Interpretation, the standard quantum formalism is seen as an ingenious shorthand for arriving at probabilities without wading through the underlying details that Schrödinger described as "tedious".

Although the internal mechanism detailed above is of the simplest form, it describes the most mysterious behavior of quantum systems coupled at a distance, as detailed in [12]. All of these behaviors can be exhibited by single-electron quantum systems coupled electromagnetically. The only thing "mysterious" about our development is our unorthodox use of the advanced-retarded electromagnetic solution to conserve energy and speed up the transition. Therefore, we have learned some interesting things by analyzing these simple transactions!

This experience brings us face to face with the obvious question: **What if Einstein was right?**

If there is internal structure to these simple quantum transitions, there must also be internal structure to the more complex questions involving more than one electron, which cannot be so simply factored!

In this case, we should find a way to look for it. That would require that we effectively time-travel back to 1926 and *grok* the questions those incredibly talented scientists were grappling with at the time.

To face into the *conceptual* details of questions involving an overlapping multi-electron system is a daunting task that has defeated every attempt thus far. We strongly suspect that the success achieved by the matrix approach—adding three more space dimensions and one spin dimension for each additional electron—came at the cost of being "lost in multi-dimensional Hilbert space." Heisenberg's triumph with the helium atom led into a rather short tunnel that narrowed rapidly in the second row of the periodic table.

Quantum chemists work with complex quantum systems that share many electrons in close proximity, and thus must represent many overlapping degrees of freedom. Their primary goal is to find the ground state of such systems. Lorentz's hope—that the intuitive insights of Schrödinger's wave function in three dimensions would *bring us closer to the real solution* in systems with more than one electron—actually helped in the early days of quantum chemistry: Linus Pauling visualized chemical bonds that way, and made a lot of progress with that approach. It is quite clear that the covalent bond has a wave function in three dimensions, even if we don't yet have a fully "quantum" way of handling it in three dimensions. The Hohenberg–Kohn theorems [56] demonstrate that the ground-state of a many-electron system is uniquely determined by its electron density, which depends on only three spatial coordinates. Thus, the chemists have a three-dimensional wave function for many electrons! They use various approaches to minimize the total energy, which then gives the best estimate of the true ground state.

These approaches have evolved into **Density Functional Theory** (DFT), and are responsible for amazingly successful analyses of an enormous range of complex chemical problems. The original Thomas–Fermi–Hohenberg–Kohn idea was to make the Schrödinger equation just about the 3d density. The practical implementations do not come close to the original motivation because half-integer spin, Pauli exclusion, and 3N dimensions are still hiding there. DFT, as it stands today, is a practical tool for generating numbers rather than a fundamental way of thinking. Although it seems unlikely at present that a more intuitive view of the multi-electron wave function will emerge from DFT, the right discovery of how to adapt 3D thinking to the properties of electron pairs could be a major first step in that direction.

When we look at even the simplest two-electron problems, we see that our present understanding uses totally ad hoc rules to eliminate configurations that are otherwise sensible: The most outrageous of these is the **Pauli Exclusion Principle**, most commonly stated as: *Two electrons can only occupy the same orbital state if their spins are anti-parallel.*

It is the reason we have the periodic table, chemical compounds, solid materials, and electrical conductors. It is just a rule, with no underlying *physical* understanding. We have only mathematics to cover our ignorance of *why it is true physically.* [The matrix formulation of QM has a much fancier mathematical way of enforcing this rule. The associated quantum field theory axiom is *not* a physical understanding.]

There is no shame in this—John Archibald Wheeler said it well:

"We live on an island surrounded by a sea of ignorance.

As our island of knowledge grows, so does the shore of our ignorance."

The founders made amazing steps forward in 1926.
Any forward step in science *always* opens new questions that we could not express previously. However, we need to make it absolutely clear *what it is that we do not yet understand:*

- We do not yet understand the mechanism that gives the 3D wave function its "identity", which causes it to be normalized.
- We do not yet have a *physical* picture of how the electron's wave function can be endowed with half-integer "spin", which why it requires a full 720° (twice around) rotation to bring the electron's wave function back to the same state, why both matter and electron antimatter states exist, and why the two have opposite parity.
- We do not yet have a *physical* understanding of how two electron wave functions interact to enforce Pauli's Exclusion Principle.

However, our analysis *has* allowed us to understand *conceptually* several things that have been hidden under the statistics: We saw that the Bose–Einstein property of photons can be understood as arising from the symmetry of electromagnetic coupling together with the movement of electron charge density of a superposed state. There was nothing "particle-like" about the electromagnetic coupling. Indeed, the two-way space-time symmetry of the photon transaction cannot really be viewed as the one-way symmetry of the flight of a "photon particle." Thus, looking at the mechanism of the "wave-function collapse" gives us a deeper view of the "boson" behavior of the photon: The rate of growth of oscillation of the superposed state is, by Equation (26), proportional to the oscillating electromagnetic field. When the oscillating currents of all the atoms are in phase, the amplitudes of their source contributions add, and any new atom is correspondingly more likely to synchronize its contribution at that same phase.

Thus, the "magic" bosonic properties of photons, including the quantization of energy $\hbar\omega$ and tendency to fall into the same state, are simply properties of single-electron systems coupled electromagnetically: Their two-way space-time symmetry is in no way "particle-like." It seems as though there is, after all, a fundamental conceptual difference between "matter" and "coupling".

Perhaps, it is the stubborn determination of theoretical physicists to make everything into particles residing in a multi-dimensional Hilbert space that has delayed for so long Lorentz's *greater intuitive clarity*—a deeper *conceptual, physical understanding* of simple quantum systems.

Thanks to modern quantum optics, we are experimentally standing on the shoulders of giants: We can now routinely realize radio techniques, such as phase-locked loops, at optical frequencies. The old argument that "everything is just counter-clicks" just doesn't cut it in the modern world!

Given the amazing repertoire of these increasingly sophisticated experiments with coherent optical-frequency quantum systems, many of the "mysterious" quantum behaviors seem more and more physically transparent when viewed as arising from the transactional symmetry of the interaction, rather than from the historic "photon-as-particle" view. The bottom line is that Schrödinger wave mechanics can easily deal with issues of quantum entanglement and nonlocality in atomic systems coupled by matched advanced/retarded 4-potentials. It remains to be seen whether this wave-based approach can be extended to systems involving the emission/detection of quark-composites or leptons.

Our Caltech colleague Richard Feynman left a legacy of many priceless quotations; a great one is:

"However, the *Real Glory* of *Science*
is that we can *Find* a *Way* of *Thinking*
such that the *Law* is *Evident!*"

What he was describing is **Conceptual Physics**.

From our new technological vantage point, it is possible to develop Quantum Science in this direction, and make it accessible to beginning students.

We urge new generations of talented researchers to take this one on.

Be Fearless—as they were in 1926!

Author Contributions: Both authors contributed to this work in roughly equal amounts. It started as a short internal report written by J.G.C. discussing Mathematica calculations based on C.A.M.'s book. Following this, C.A.M. made major contributions in expansion to the present length and produced most of the discussion of equations and formalism. All authors have read and agreed to the published version of the manuscript.

Acknowledgments: The authors are grateful to Jamil Tahir-Kheli for sharing his unmatched knowledge and understanding of physics history, and his deeply insightful critique of the approach developed here, during many thoughtful discussions down through the years. We thank Ruth Kastner and Gerald Miller for helpful comments and for rubbing our noses in the probabilistic approach to QM. We are particularly grateful to Nick Herbert for asking about transition time in our calculations, which led us to important new insights. David Feinstein, Glenn Keller, Ed Kelm, and Lloyd Watts caught many bugs and offered helpful suggestions. We thank Jordan Maclay for asking us to write this paper for a special issue of *Symmetry* and for making useful comments and suggestons. Finally, we thank our four anonymous reviewers, whose penetrating comments and questions led to major improvements in this paper.

Abbreviations

The following abbreviations are used in this manuscript:

DFT	density functional theory
EPR experiment	Einstein, Podolsky, and Rosen experiment demonstrating nonlocality
NCT	neoclassical theory, i.e., Schrödinger's wave mechanics plus Maxwell's equations
QM	quantum mechanics
TI	the Transactional Interpretation of quantum mechanics [12]
WFE	Wheeler–Feynman electrodynamics

References

1. Einstein, A. Reply to Bohr. In *The Tests of Time: Readings in the Development of Physical Theory*; Lisa M.D., Arthur, F.G., Glenn, N.S., Eds.; Princeton University Press, Princeton, NJ, USA, 2003; ISBN 978-0691090849.
2. Einstein, A.; Podolsky, B.; Rosen, N. Can Quantum-Mechanical Description of Physical Reality Be Considered Complete? *Phys. Rev.* **1935**, *47*, 777–785. [CrossRef]
3. Bell, J.S. On the Einstein Podolsky Rosen paradox. *Physics* **1964**, *1*, 195. [CrossRef]
4. Bell, J.S. On the Problem of Hidden Variables in Quantum Mechanics. *Rev. Mod. Phys.* **1966**, *38*, 447. [CrossRef]
5. Camhy-Val, C.; Dumont, A.M. Absolute Measurement of Mean Lifetimes of Excited States Using the Method of Correlated Photons in a Cascade. *Astron. Astrophys.* **1970**, *6*, 27.
6. Freedman, S.J.; Clauser, J.F. Experimental Test of Local Hidden-Variable Theories. *Phys. Rev. Lett.* **1972**, *28*, 938. [CrossRef]
7. Fry, E.S.; Thompson, R.C. Experimental Test of Local Hidden-Variable Theories. *Phys. Rev. Lett.* **1976**, *37*, 465. [CrossRef]
8. Aspect, A.; Grangier, P.; Roger, G. Experimental Investigation of a Polarization Correlation Anomaly. *Phys. Rev. Lett.* **1981**, *47*, 460. [CrossRef]

9. Clauser, J.F. Experimental Investigation of a Polarization Correlation Anomaly. *Phys. Rev. Lett.* **1976**, *36*, 1223–1226. [CrossRef]
10. Aspect, A.; Grangier, P.; Roger, G. Experimental Realization of Einstein-Podolsky-Rosen-Bohm Gedankenexperiment: A New Violation of Bell's Inequalities. *Phys. Rev. Lett.* **1982**, *49*, 91. [CrossRef]
11. Aspect, A.; Dalibard, J.; Roger, G. Experimental Test of Bell's Inequalities Using Time-Varying Analyzers. *Phys. Rev. Lett.* **1982**, *49*, 1804. [CrossRef]
12. Cramer, J.G., *The Quantum Handshake—Entanglement, Nonlocality and Transactions;* Springer: Berlin/Heidelberg, Germany, 2016; ISBN 978-3-319-24640-6.
13. Kastner, R.E. *The Transactional Interpretation of Quantum Mechanics: The Reality Of Possibility*; Cambridge University Press: London, UK, 2012; ISBN 978-1108407212.
14. Kastner, R.E.; Cramer, J.G. Quantifying Absorption in the Transactional Interpretation. *Int. J. Quantum Found.* **2018**, *4*, 210–222.
15. Kastner, R.E. Decoherence and the Transactional Interpretation. *Int. J. Quantum Found.* **2020**, *6*, 24–39.
16. Mead, C. *Collective Electrodynamics: Quantum Foundations of Electromagnetism*; The MIT Press: Cambridge, MA, USA, 2000; ISBN 0-262-13378-4.
17. Wheeler, J.A.; Feynman, R.P. Interaction with the Absorber as the Mechanism of Radiation. *Rev. Mod. Phy.* **1945**, *17*, 157. [CrossRef]
18. Wheeler, J.A.; Feynman, R.P. Classical Electrodynamics in Terms of Direct Interparticle Action. *Rev. Mod. Phys.* **1949**, *21*, 425. [CrossRef]
19. Cramer, J.G. The Arrow of Electromagnetic Time and Generalized Absorber Theory. *Found. Phys.* **1983**, *13*, 887. [CrossRef]
20. Lewis, G.N. The Nature of Light. *Proc. Natl. Acad. Sci.* **1926**, *12*, 22–29. [CrossRef]
21. Lewis, G.N. The conservation of photons. *Nature* **1926**, *118*, 874–875. [CrossRef]
22. Erwin Schrödinger. An Undulatory Theory of the Mechanics of Atoms and Molecules. *Phys. Rev.* **1926**, *28* 1049–1070. [CrossRef]
23. Jaynes, E.T. *Some Aspects of Maser Theory*; C Microwave Laboratory Report No. 502; Stanford University: Stanford, CA, USA, 1958; Available online: http://faculty.washington.edu/jcramer/TI/Jaynes_1958.pdf (accessed on 17 August 2020).
24. Jaynes, E.T.; Cummings, F.W. Comparison of Quantum and Semiclassical Radiation Theory with Application to the Beam Maser. *Proc. IEEE* **1963**, *51*, 89–109. [CrossRef]
25. Lamb, W.E., Jr.; Scully, M.O. The photoelectric effect without photons. In *Polarization: Matiere et Rayonnement*; Bederson, H., Ed.; Presses universitaires de France: Paris, France, 1969.
26. Crisp, M.D.; Jaynes, E.T. Radiative Effects in Semiclassical Theory. *Phys. Rev.* **1969**, *179*, 1253.[CrossRef]
27. Stroud, C.R., Jr; Jaynes, E.T. Long-Term Solutions in Semiclassical Radiation Theory. *Phys. Rev. A* **1970**, *1*, 106. [CrossRef]
28. Leiter, D. Comment on the Characteristic Time of Spontaneous Decay in Jaynes's Semiclassical Radiation Theory. *Phys. Rev A* **1970**, *2*, 259. [CrossRef]
29. Jaynes, E.T. Reply to Leiter's Comment. *Phys. Rev A* **1970**, *2*, 260. [CrossRef]
30. Jaynes, E.T. Survey of the Present Status of Semiclassical Radiation Theories. In *Coherence and Quantum Optics: Proceedings of the Third Rochester Conference on Coherence and Quantum Optics held at the University of Rochester, June 21–23, 1972*, Mandel, L., Wolf, E., Eds.; Springer: Berlin/Heidelberg, Germany, 1973.
31. Mandel, L.; Wolf, E. (Eds.) *Coherence and Quantum Optics: Proceedings of the Third Rochester g on Coherence and Quantum Optics held at the University of Rochester, June 21–23, 1972*; Springer: Berlin/Heidelberg, Germany, 1973.
32. Clauser, J.F. Experimental Limitations to the Validity of Semiclassical Radiation Theories. In *Coherence and Quantum Optics: Proceedings of the Third Rochester Conference on Coherence and Quantum Optics held at the University of Rochester, June 21–23, 1972*, Mandel, L.; Wolf, E. Eds.; Springer: Berlin/Heidelberg, Germany, 1973.
33. Cramer, J.G. Generalized Absorber Theory and the Einstein Podolsky Rosen Paradox. *Phys. Rev. D* **1980**, *22*, 362–376. [CrossRef]
34. Cramer, J.G. The Transactional Interpretation of Quantum Mechanics. *Rev. Mod. Phys.* **1986**, *58*, 647–687. [CrossRef]

35. Erwin Schrödinger. *Collected Papers on Wave Mechanics*, 3rd ed.; AMS Chelsea Publishing: Providence, RI, USA, 1982; p. 220.
36. Millikan, R.A. *The Electron*; University of Chicago Press: Chicago, IL, USA, 1917.
37. The Three .mov Animation Files Need to be Downloaded and Placed in the Same Folder as This .pdf File. Set them to "Repeat". Available online: https://1drv.ms/u/s!Ap3rYYlMocgZgZxju0KRgd5mTjXK9A?e=qR7mQy (accessed on 17 August 2020).
38. Erwin Schrödinger. *Collected Papers on Wave Mechanics*; Blackie & Son Ltd., London, UK, 1928; p. 58.
39. Erwin Schrödinger. *Four Lectures on Wave Mechanics Delivered at the Royal Institution, London, on 5th, 7th, 12th, and 14th March, 1928*; Blackie & Son Limited: London/Scotland, UK, 1928.
40. Jackson, J.D. *Classical Electrodynamics*, 3rd ed.; John Wiley & Sons: New York, NY, USA, 1999; ISBN 978-0471309321.
41. Ruth E. Kastner, *Understanding Our Unseen Reality: Solving Quantum Riddles*; Imperial College Press: London, UK, 2015; ISBN 978-1-787326-695-1.
42. Dirac, P.A.M. The Quantum Theory of Emission and Absorption of Radiation. *Proc. R. Soc.* **1927**, *114*, 243–265.
43. Ritz, W.; Einstein, A. Zur Aufklärung. *Phys. Z.* **1909**, *10*, 323–324.
44. Feynman, R.P. *QED: The Strange Theory of Light and Matter*; Princeton University Press: Princeton, NJ, USA, 1985; ISBN 978-0691024172.
45. Steinmetz, C.P.; Berg, E.G. Complex Quantities and Their Use in Electrical Engineering. In Proceedings of the International Electrical Congress, Chicago, IL, USA, 21–25 August 1893; American Institute of Electrical Engineers: New York, NY, USA, 1894; pp. 33–74.
46. Steinmetz, C.P. *Theory and Calculation of Alternating Current Phenomena*, 3rd. ed.; rev. and enl.; Electrical World and Engineer, Inc.: New York, NY, USA, 1900.
47. Heisenberg, W. Reminiscences from 1926 and 1927. In *Niels Bohr, A Centenary Volume*; French, A.P., Kennedy, P.J., Eds.; Harvard University Press: Cambridge, MA, USA, 1985; ISBN 978-0674624153.
48. Schrödinger, E. Are there quantum jumps? *Br. J. Phil. Sci.* **1952**, *3*, 109–123; 233–242. [CrossRef]
49. Brown, R.H.; Twiss, R.Q. A new type of interferometer for use in radio astronomy. *Philos. Mag.* **1954**, *45*, 663–682. [CrossRef]
50. Padula, S.S. HBT Interferometry: Historical Perspective. *Braz. J. Phys.* **2005**, doi:10.1590/S0103-97332005000100005.
51. Jammer, M. *The Philosophy of Quantum Mechanics: The Interpretations of Quantum Mechanics in Historical Perspective*; John Wiley & Sons.: New York, NY, USA, 1974; ISBN 978-0471439585.
52. Lounis, B.; Bechtel, H.A.; Gerion, D.; Alivisatos, P.; Moerner, W.E. Photon antibunching in single CdSe/ZnS quantum dot fluorescence. *Chem. Phys. Lett.* **2000**, *329*, 399–404.[CrossRef]
53. Clauser, J.F. Experimental distinction between the quantum and classical field-theoretic predictions for the photoelectric effect. *Phys. Rev. D* **1974**, *9*, 853–860. [CrossRef]
54. Lorentz, H.A. *Letter to Schrödinger, 27 May 1926*; Letters on Wave Mechanics: 76; Philosophical Library/Open Road: New York, NY, USA, 2011.
55. Heisenberg, W. Mehrkörperproblem und Resonanz in der Quantenmechanik. *Z. für Physik* **1926**, *38*, 411. [CrossRef]
56. Hohenberg, P.; Kohn, W. Inhomogeneous electron gas. *Phys. Rev.* **1964**, *136*, B864–B871.[CrossRef]

8

Supersymmetry of Relativistic Hamiltonians for Arbitrary Spin

Georg Junker [1,2]

[1] European Southern Observatory, Karl-Schwarzschild-Straße 2, D-85748 Garching, Germany; gjunker@eso.org

[2] Institut für Theoretische Physik I, Universität Erlangen-Nürnberg, Staudtstraße 7, D-91058 Erlangen, Germany

Abstract: Hamiltonians describing the relativistic quantum dynamics of a particle with an arbitrary but fixed spin are shown to exhibit a supersymmetric structure when the even and odd elements of the Hamiltonian commute. Here, the supercharges transform between energy eigenstates of positive and negative energy. For such supersymmetric Hamiltonians, an exact Foldy–Wouthuysen transformation exists which brings it into a block-diagonal form separating the positive and negative energy subspaces. The relativistic dynamics of a charged particle in a magnetic field are considered for the case of a scalar (spin-zero) boson obeying the Klein–Gordon equation, a Dirac (spin one-half) fermion and a vector (spin-one) boson characterised by the Proca equation. In the latter case, supersymmetry implies for the Landé g-factor $g = 2$.

Keywords: relativistic wave equation; Klein–Gordon equation; Dirac equation; Proca equation; supersymmetry

1. Introduction

Soon after the formulation of non-relativistic quantum mechanics by Heisenberg, Born, Jordan and Schrödinger in 1925 and 1926, Klein [1] and Gordon [2] made first attempts to develop a relativistic quantum wave formalism. This Klein–Gordon equation is known to have certain deficits for its quantum mechanical interpretation but nowadays is well accepted as being the correct quantum wave formalism for spin-zero particles. To overcome the problems of the Klein–Gordon equation, Dirac in 1928 [3,4] made an ansatz for a wave equation being linear in the time derivative and thus found his famous equation describing the relativistic quantum dynamics of spin one-half fermions. The relativistic spin-one equation, also known as Proca equation, has been developed by Proca [5] in 1936. In the same year Dirac [6] and later Fierz and Pauli [7,8] investigated relativistic wave equations for arbitrary spin. See also the later work by Bhabha [9]. A group theoretical discussion of such wave equations was then given by Bargmann and Wigner in 1948 [10].

In their fundamental work in 1950, Foldy and Wouthuysen [11] constructed a unitary transformation which separates the positive and negative energy states; that is, the Dirac Hamiltonian became block-diagonal. This work has triggered the study of the Hamiltonian form for the other wave equations. For example, Foldy [12] investigated the Klein–Gordon equation and Feshbach and Villars [13] made a unified approach to a Hamiltonian form of the Klein–Gordon and Dirac equation. The Hamiltonian form for the Proca equation has been studied by various authors including Duffin [14], Kemmer [15], Yukawa, Sakata and Taketani [16,17], Corben and Schwinger [18], as well as by Schrödinger [19]. The problem of restoring a block-diagonal Hamiltonian via a so-called exact Foldy–Wouthuysen (FW) transformation for a particle with an arbitrary spin is still attracting many researchers. See, for example, the recent work by Silenko [20] and Simulik [21] and the references therein.

The aim of the present work was to investigate another aspect of such relativistic Hamiltonians related to an underlying supersymmetric structure. Here, it shall be emphasised that this kind of supersymmetry is not related to the supersymmetry (SUSY) known from quantum field theory where the supercharges transform a bosonic state into a fermionic state and vice versa. Here, SUSY is to be understood in the context of supersymmetric quantum mechanics, where the supercharges transform between states of positive and negative Witten parity. Despite the fact that supersymmetric quantum mechanics was originally introduced by Nicolai [22] as the $(0+1)$-dimensional limit of SUSY quantum field theories, it is rather independent of the latter. Supersymmetric quantum mechanics became rather popular with the model introduced by Witten [23], being in essence a one-dimensional non-relativistic quantum system, which still finds many application in various areas of physics [24,25]. The first extension of supersymmetric quantum mechanics to the relativistic Dirac Hamiltonian is due to Jackiw [26] and Ui [27] and has found many applications, for example, in the analysis of the electronic properties of topological superconductors and graphene [28].

The main purpose of the present work was to show that such supersymmetric structure may also be established for other relativistic systems going beyond that of the Dirac Hamiltonians. That is, we will show under the requirement that the odd and even part of a relativistic Hamiltonian commute with each other it is possible to establish a SUSY structure similar to what is know in the Dirac case. To the best of our knowledge, such an extension for the Klein–Gordon case was only briefly discussed by Thaller in Section 5.5.3 of his book [29] and more explicit by Znojil [30]. Here, we will present a general approach of supersymmetric quantum mechanics for relativistic Hamiltonians for arbitrary but fixed spin. The explicit discussion will be limited to the scalar or spin-zero case, the Dirac case for spin-$\frac{1}{2}$ and the vector boson case, i.e., spin-one. In all three cases we consider the well-known problem of a charged particle in the presence of a magnetic field but now from the point of view of supersymmetric quantum mechanics. It turns out that all three models, in essence, are closely related to their non-relativistic counterparts in essentially the same way. In addition, SUSY requires for vector bosons a Landé g-factor $g = 2$.

The paper is organised as follows. In the next section, the basic structure of relativistic Hamiltonians for an arbitrary spin are recalled. In Section 3, we then show that whenever the odd part commutes with the even part of such a Hamiltonian it is possible to construct an $N = 2$ SUSY structure very similar to what is known for supersymmetric Dirac Hamiltonians [29,31]. It is also recalled that there exists an exact FW transformation bringing the Hamiltonian into a block-diagonal form. In Section 4, we explicitly discuss the cases of a charged spin-zero, spin one-half and spin-one particle in an external magnetic field. In all three cases, which cover all the currently known charged elementary particles, we find that the eigenvalue problem of the relativistic Hamiltonian can indeed be reduced to that of a non-relativistic one. Section 5 discusses the resolvent of supersymmetric relativistic Hamiltonians and again shows that for the three cases under consideration the Green's function, in essence, may be reduced to that of the associated non-relativistic Hamiltonian. Finally, in Section 6, we present a short conclusion and an outlook for possible further investigations and in the Appendix A we collect some useful relations for the spin $s = 1$ case which are not that commonly known.

2. Relativistic Hamiltonians for Arbitrary Spin

In the Hamiltonian form of relativistic quantum mechanics, one puts the wave equation into a Schrödinger-like form

$$i\hbar\partial_t\Psi = H\Psi\,. \tag{1}$$

The Hamiltonian in the above equation in general is of the form

$$H = \beta m + \mathcal{E} + \mathcal{O}\,, \tag{2}$$

where $\beta^2 = 1$ acts as a grading operator and m standard for the particle's mass. In addition to the mass term βm, the operator $\mathcal{E}$ represents the remaining even part of the Hamiltonian; that is, it commutes with the grading operator, $[\beta, \mathcal{E}] = 0$. The operator $\mathcal{O}$ denotes the odd part of H and obeys the anticommutation relation $\{\beta, \mathcal{O}\} = 0$. For a particle with spin s, $s = 0, \frac{1}{2}, 1, \frac{3}{2}, \ldots$, the Hilbert space $\mathcal{H}$ on which H acts is given by

$$\mathcal{H} = L^2(\mathbb{R}^3) \otimes \mathbb{C}^{2(2s+1)}\,, \tag{3}$$

that is, the wave function Ψ in (1) is a spinor with $2(2s+1)$ components [20]. Let us note that we can decompose the Hilbert space $\mathcal{H}$ into a direct sum of the two eigenspaces of the grading operator β with eigenvalue $+1$ and -1, respectively,

$$\mathcal{H} = \mathcal{H}_+ \oplus \mathcal{H}_-\,, \qquad \mathcal{H}_\pm := L^2(\mathbb{R}^3) \otimes \mathbb{C}^{2s+1}\,. \tag{4}$$

Obviously, $\mathcal{H}_\pm$ are simultaneously the subspaces where eigenvalues of H are positive and negative, respectively.

For simplicity, let us put the relativistic Hamiltonian (2) into the form

$$H = \beta\mathcal{M} + \mathcal{O}\,, \tag{5}$$

where we have absorbed the mass m in the even mass operator $\mathcal{M} := m + \beta\mathcal{E}$ with $[\beta, \mathcal{M}] = 0$. Let us note here that above Hamiltonian is self-adjoint, i.e., $H = H^\dagger$, only for the case of fermions where $s = \frac{1}{2}, \frac{3}{2}, \ldots$ is a half-odd integer. For bosons, where s takes integer values, the Hamiltonian is pseudo-hermitian, that is, $H = \beta H^\dagger \beta$.

Choosing a representation where β takes the diagonal form

$$\beta = \begin{pmatrix} 1 & 0 \\ 0 & -1 \end{pmatrix}, \tag{6}$$

where in the above the 1 denotes a $(2s+1)$-dimensional unit matrix. The even and odd operators obeying $[\beta, \mathcal{M}] = 0$ and $\{\beta, \mathcal{O}\} = 0$ are necessarily of the form

$$\mathcal{M} = \begin{pmatrix} M_+ & 0 \\ 0 & M_- \end{pmatrix}, \qquad \mathcal{O} = \begin{pmatrix} 0 & A \\ (-1)^{2s+1}A^\dagger & 0 \end{pmatrix}, \tag{7}$$

where $M_\pm : \mathcal{H}_\pm \mapsto \mathcal{H}_\pm$ with $M_\pm^\dagger = M_\pm$ is an operator mapping positive and negative energy states into positive and negative energy states, respectively. Whereas $A : \mathcal{H}_- \mapsto \mathcal{H}_+$ maps a negative energy state into a positive energy state and $A^\dagger : \mathcal{H}_+ \mapsto \mathcal{H}_-$ vice versa. With the above representation, the general relativistic spin-s Hamiltonian then takes the form

$$H = \begin{pmatrix} M_+ & A \\ (-1)^{2s+1}A^\dagger & -M_- \end{pmatrix}. \tag{8}$$

In the following section, we will show that under the assumption that the even and odd parts of the Hamiltonian (8) commute, i.e., $[\mathcal{M}, \mathcal{O}] = 0$, it is possible to establish an $N = 2$ supersymmetric structure being well-studied in supersymmetric quantum mechanics. This condition, which we will call the SUSY condition, also allows for an exact Foldy–Wouthuysen transformation.

3. Supersymmetric Relativistic Hamiltonians for Arbitrary Spin

As stipulated above, let us assume that the even mass operator $\mathcal{M}$ and the odd operator $\mathcal{O}$ commute. This SUSY condition implies that

$$M_+A = AM_-\,, \qquad A^\dagger M_+ = M_-A^\dagger\,. \tag{9}$$

As a consequence of this, the squared Hamiltonian (8) becomes block diagonal

$$H^2 = \begin{pmatrix} M_+^2 + (-1)^{2s+1} AA^\dagger & 0 \\ 0 & M_-^2 + (-1)^{2s+1} A^\dagger A \end{pmatrix} . \tag{10}$$

Inspired by the construction of a SUSY structure for supersymmetric Dirac Hamiltonians [31], let us introduce the following SUSY Hamiltonian

$$H_{\mathrm{SUSY}} := \frac{(-1)^{2s+1}}{2mc^2}(H^2 - \mathcal{M}^2) = \frac{1}{2mc^2}\begin{pmatrix} AA^\dagger & 0 \\ 0 & A^\dagger A \end{pmatrix} \geq 0 \tag{11}$$

and the complex supercharges

$$Q := \frac{1}{\sqrt{2mc^2}}\begin{pmatrix} 0 & A \\ 0 & 0 \end{pmatrix}, \qquad Q^\dagger = \frac{1}{\sqrt{2mc^2}}\begin{pmatrix} 0 & 0 \\ A^\dagger & 0 \end{pmatrix}. \tag{12}$$

Here, $m > 0$ is an arbitrary mass-like parameter, representing, for example, the mass of the relativistic particle in (2). It is obvious that these operators generate a transformation between positive and negative energy states. A straightforward calculation shows that these operators together with the Witten parity operator $W := \beta$ form an $N = 2$ SUSY system; that is,

$$H_{\mathrm{SUSY}} = \{Q, Q^\dagger\}, \qquad \{Q, W\} = 0, \qquad Q^2 = 0 = (Q^\dagger)^2, \qquad [W, H_{\mathrm{SUSY}}] = 0, \qquad W^2 = 1. \tag{13}$$

Let us note that $\mathcal{M}$ under condition (9) commutes with all operators of above algebra and thus constitutes a centre of the SUSY algebra (13). Hence, a relativistic arbitrary-spin Hamiltonian (8) obeying the SUSY condition (9) may be called a *supersymmetric relativistic arbitrary-spin Hamiltonian.*

Let us also note that for a supersymmetric relativistic arbitrary-spin Hamiltonian, there exists an exact Foldy–Wouthuysen transformation U which brings (8) into the block-diagonal form [20]

$$\begin{aligned} H_{\mathrm{FW}} := \; & UHU^\dagger = \beta\sqrt{H^2} \\ = \; & \begin{pmatrix} \sqrt{M_+^2 + (-1)^{2s+1} 2mc^2 H_+} & 0 \\ 0 & -\sqrt{M_-^2 + (-1)^{2s+1} 2mc^2 H_-} \end{pmatrix}, \end{aligned} \tag{14}$$

where the partner Hamiltonians $H_\pm$ are defined by

$$H_+ := \frac{1}{2mc^2} AA^\dagger , \qquad H_- := \frac{1}{2mc^2} A^\dagger A . \tag{15}$$

In fact, it is known that under condition $[\mathcal{M}, \mathcal{O}] = 0$, the exact Foldy–Wouthuysen transformation is explicitly given by [20,32]

$$U := \frac{|H| + \beta H}{\sqrt{2H^2 + 2\mathcal{M}|H|}} = \frac{1 + \beta \,\mathrm{sgn}\, H}{\sqrt{2 + \{\mathrm{sgn}\, H, \beta\}}}, \qquad \mathrm{sgn}\, H := \frac{H}{\sqrt{H^2}}. \tag{16}$$

As a side remark, let us mention that the four projections operators

$$P^\pm := \tfrac{1}{2}\,[1 \pm W] , \qquad \Lambda^\pm := \tfrac{1}{2}\,[1 \pm \mathrm{sgn}\, H] , \tag{17}$$

projecting onto the subspaces of positive/negative Witten parity and positive/negative eigenvalues of H, respectively, are related to each other via the same unitary transformation as H and H_{FW}

$$P^\pm = U\Lambda^\pm U^\dagger . \tag{18}$$

That is, the positive and negative energy eigenspaces are transformed via U into spaces of positive and negative Witten parity. In fact, one may verify that U may be represented in terms of these projection operators as follows

$$U = \frac{P^+\Lambda^+ + P^-\Lambda^-}{\sqrt{(P^+\Lambda^+ + P^-\Lambda^-)(\Lambda^+P^+ + \Lambda^-P^-)}} \,. \tag{19}$$

The non-negative partner Hamiltonians $H_\pm \geq 0$ are essential isospectral which means that their strictly positive eigenvalues are identical. The corresponding eigenstates are related to each other via a SUSY transformation. To be more explicit, let us assume these are given by

$$H_\pm \phi_\varepsilon^\pm = \varepsilon \phi_\varepsilon^\pm \,, \qquad \phi_\varepsilon^\pm \in \mathcal{H}_\pm \,, \qquad \varepsilon > 0 \,, \tag{20}$$

then the SUSY transformation reads [31]

$$\phi_\varepsilon^+ = \frac{1}{\sqrt{2mc^2\varepsilon}} A \phi_\varepsilon^- \,, \qquad \phi_\varepsilon^- = \frac{1}{\sqrt{2mc^2\varepsilon}} A^\dagger \phi_\varepsilon^+ \,. \tag{21}$$

Note that the energy eigenvalue ε may be degenerate and above relations are valid for each of these energy eigenstates. We omit an additional index in $\phi_\varepsilon^\pm$ enumerating such a degeneracy. In addition, both partner Hamiltonians $H_\pm$ may have a non-trivial kernel; that is, there may exist one or several eigenstates with

$$H_\pm \phi_0^\pm = 0 \,. \tag{22}$$

In this case, SUSY is said to be unbroken [31]. For these ground states, again we omit the index for a possible degeneracy. There exists no SUSY transformation relating ϕ_0^+ and ϕ_0^-. The breaking of SUSY can be studied via the so-called Witten index Δ [33], which in the current context is identical to the Fredholm index of A, if it is a Fredholm operator, that is,

$$\Delta \equiv \operatorname{ind} A := \dim\ker A - \dim\ker A^\dagger = \dim\ker H_- - \dim\ker H_+ \,. \tag{23}$$

Obviously a non-vanishing Witten index indicates that SUSY is unbroken. In connection with [29]

$$\dim\ker\,(Q + Q^\dagger) = \dim\ker A + \dim\ker A^\dagger = \dim\ker H_- + \dim\ker H_+ \tag{24}$$

the kernels of $H_\pm$; that is, the number of zero-energy states of $H_\pm$ are known. In general, however, the operator A is not Fredholm and hence some regularized indices are studied [29,31].

Due to the SUSY condition (9), the mass operators commute with the associated partner Hamiltonians, $[M_\pm, H_\pm] = 0$, and therefore have an identical set of eigenstates. Let us denote the corresponding eigenvalues of $M_\pm$ by $m_\pm c^2$; that is,

$$M_\pm \phi_\varepsilon^\pm = m_\pm c^2 \phi_\varepsilon^\pm \,, \qquad \varepsilon \geq 0 \,, \tag{25}$$

then obviously the eigenvalues and eigenstates of (14),

$$H_{\rm FW} \psi_\varepsilon^\pm = E_\pm \psi_\varepsilon^\pm \,, \tag{26}$$

are given by

$$E_\pm = \pm\sqrt{m_\pm^2 c^4 + (-1)^{2s+1} 2mc^2\varepsilon} \,, \qquad \psi_\varepsilon^+ = \begin{pmatrix} \phi_\varepsilon^+ \\ 0 \end{pmatrix} , \qquad \psi_\varepsilon^- = \begin{pmatrix} 0 \\ \phi_\varepsilon^- \end{pmatrix} . \tag{27}$$

Here, let us note that the mass eigenvalues may depend on the energy eigenvalues, $m_\pm = m_\pm(\varepsilon)$. Using the relation (9) in combination with the SUSY relation (21) and the above eigenvalue

Equation (25), one may verify that $m_+(\varepsilon) = m_-(\varepsilon)$ for all $\varepsilon > 0$. In essence, this means that the spectrum of a supersymmetric relativistic Hamiltonian is symmetric about zero with a possible exception at $\pm m_\pm(0)c^2$; that is, if $\varepsilon = 0$, which may only occur in the case of unbroken SUSY.

The eigenstates of the original Hamiltonian (8) are then easily found via the unitary transformation $\Psi_\varepsilon^\pm = U^\dagger \psi_\varepsilon^\pm$ having the same eigenvalues $E_\pm$ given above. Hence, the eigenvalue problem of a supersymmetric relativistic spin-s Hamiltonian can be reduced to the simultaneous eigenvalue problems for $M_\pm$ and $H_\pm$ on $\mathcal{H}_\pm$.

It will turn out in the examples to be discussed below that the partner Hamiltonians $H_\pm$ and the mass operators $M_\pm$ are in essence represented by a non-relativistic Schrödinger-like Hamiltonian $H_{\rm NR}$ and/or some constant operator. To be more precise, we will show for all three cases—$s = 0, \frac{1}{2}$ and 1—discussed below that the FW-transformed relativistic Hamiltonian takes the form

$$H_{\rm FW} = \beta mc^2\sqrt{1+\frac{2H_{\rm NR}}{mc^2}} \tag{28}$$

with $H_{\rm NR}$ representing the associate non-relativistic Hamiltonian as

$$\lim_{c\to\infty}\left(P^\pm H_{\rm FW} P^\pm \mp mc^2\right) = \pm \lim_{c\to\infty}\left(mc^2\sqrt{1+2H_{\rm NR}/mc^2} - mc^2\right) = \pm H_{\rm NR}\,. \tag{29}$$

4. Examples

In the following subsections we will consider a relativistic charged particle with charge e and mass m in an external magnetic field $\vec{B} := \vec{\nabla}\times\vec{A}$ characterised by a vector potential $\vec{A}$. The symbol $\vec{\pi}$ stands for the kinetic momentum; that is, $\vec{\pi} := \vec{p} - e\vec{A}/c$, where c denotes the speed of light. We will consider the case of a scalar particle with spin $s = 0$, a Dirac particle with $s = \frac{1}{2}$ and a vector boson having spin $s = 1$.

4.1. The Klein–Gordon Hamiltonian with Magnetic Field

The Schrödinger form of the Klein–Gordon equation as been considered by Feshbach and Villars in [13] where they have shown that the Klein–Gordon equation for a charged particle in a magnetic field can be put into the Schrödinger form (1) where the pseudo-Hermitian Hamiltonian is given by

$$H = \begin{pmatrix} mc^2 + \frac{\vec{\pi}^2}{2m} & \frac{\vec{\pi}^2}{2m} \\ -\frac{\vec{\pi}^2}{2m} & -mc^2 - \frac{\vec{\pi}^2}{2m} \end{pmatrix}, \qquad \mathcal{H} = L^2(\mathbb{R}^3)\otimes\mathbb{C}^2\,. \tag{30}$$

Comparing this with the general form (8), we may identify the operators $M_\pm$ and A as follows:

$$M_\pm = \frac{1}{2m}\vec{\pi}^2 + mc^2\,, \qquad A = \frac{1}{2m}\vec{\pi}^2\,. \tag{31}$$

Obviously, these operators are identical up to an additional constant $M_\pm = A + mc^2$ and hence the condition (9); that is, $[M_\pm, A] = 0$ is trivially fulfilled. In other words, the Klein–Gordon Hamiltonian (30) for a charged scalar particle in the presence of an arbitrary magnetic field represents a supersymmetric relativistic spin-zero Hamiltonian. Let us note that the two operators (31) in essence are given by the Landau Hamiltonian $H_{\rm L} := (\vec{p} - e\vec{A}/c)^2/2m = \vec{\pi}^2/2m$ of a non-relativistic spinless charged particle of mass m in a magnetic field; that is, $M_\pm = H_{\rm L} + mc^2$ and $A = H_{\rm L}$. As a consequence, the eigenvalue problem for (30) is reduced to that of $H_{\rm L}$. The FW transformed Hamiltonian explicitly reads

$$H_{\rm FW} = \begin{pmatrix} \sqrt{(H_{\rm L}+mc^2)^2 - H_{\rm L}^2} & 0 \\ 0 & -\sqrt{(H_{\rm L}+mc^2)^2 - H_{\rm L}^2} \end{pmatrix} = \beta mc^2\sqrt{1+\frac{2H_{\rm L}}{mc^2}}\,. \tag{32}$$

Let us denote the eigenvalues of $H_{\rm L}$ by ϵ, then the eigenvalues of $M_\pm$ and $H_\pm = H_{\rm L}^2/2mc^2$ read

$$m_\pm c^2 = mc^2 + \epsilon\,, \qquad \varepsilon = \frac{\epsilon^2}{2mc^2} \tag{33}$$

and via relation (27) we find the eigenvalues of the Klein–Gordon Hamilonian

$$E_\pm = \pm\sqrt{m_\pm^2 c^4 - 2mc^2\varepsilon} = \pm\sqrt{(mc^2+\epsilon)^2 - \epsilon^2} = \pm mc^2\sqrt{1+\frac{2\epsilon}{mc^2}}\,, \tag{34}$$

which is a result expected from relation (32). For a non-vanishing constant magnetic field, say in the z-direction $\vec{B} = B\vec{e}_z$ with $B \neq 0$, the eigenvalues of $H_{\rm L}$ are the well-known Landau levels [34,35]

$$\epsilon = \hbar\omega_c\left(n+\frac{1}{2}\right) + \frac{\hbar^2 k_z^2}{2m}\,, \qquad n \in \mathbb{N}_0\,, \qquad k_z \in \mathbb{R}\,, \qquad \omega_c := \frac{|eB|}{mc}\,. \tag{35}$$

As $\epsilon \geq \hbar\omega_c/2 > 0$ so is $\varepsilon > 0$ and hence $\dim\ker H_- = \dim\ker H_- = 0$. In other words, the Witten index (23) vanishes and SUSY is broken for the Klein–Gordon Hamiltonian in a constant magnetic field.

4.2. The Dirac Hamiltonian with Magnetic Field

The Dirac equation representing the relativistic dynamics of spin-$\frac{1}{2}$ fermions has intensively been studied since its introduction. See, for example, the excellent book by Thaller [29]. For a charged particle in an arbitrary external magnetic field, the Dirac Hamiltonian reads

$$H = \begin{pmatrix} mc^2 & c\vec{\sigma}\cdot\vec{\pi} \\ c\vec{\sigma}\cdot\vec{\pi} & -mc^2 \end{pmatrix}, \tag{36}$$

where $\vec{\sigma} = (\sigma_1, \sigma_2, \sigma_3)^T$ stands for a three-dimensional vector who's components are given by the Pauli matrices acting on $\mathbb{C}^2$, thus representing the spin-$\frac{1}{2}$ degree of freedom. Comparing this with the general form (8), we may identify the operators $M_\pm = mc^2$ and $A = c\sigma\cdot\vec{\pi} = A^\dagger$ and note that condition (9) is trivially fulfilled. Hence, the Dirac Hamiltonian (36) is indeed supersymmetric and its FW transformed form is known [32] to be expressible in terms of the non-relativistic Pauli Hamiltonian for a spin-$\frac{1}{2}$ particle with Landé g-factor $g = 2$.

$$H_{\rm P} := \frac{1}{2m}(\vec{\sigma}\cdot\vec{\pi})^2 = \frac{1}{2m}(\vec{p} - e\vec{A}/c)^2 - \frac{e\hbar}{mc}\vec{B}\cdot\vec{\sigma}\,. \tag{37}$$

Obviously, the partner Hamiltonians can be identified with the Pauli Hamiltonian $H_\pm = A^2/2mc^2 = H_{\rm P}$ and we find for the FW transformed Dirac Hamiltonian

$$H_{\rm FW} = \begin{pmatrix} \sqrt{m^2c^4 + 2mc^2 H_{\rm P}} & 0 \\ 0 & -\sqrt{m^2c^4 + 2mc^2 H_{\rm P}} \end{pmatrix} = \beta mc^2\sqrt{1+\frac{2H_{\rm P}}{mc^2}}\,. \tag{38}$$

As shown by Aharonov and Casher [36], for a magnetic field having only a z-component, $\vec{B} = B(x,y)\vec{e}_z$, the ground-state energy of $H_{\rm P}$ is zero and the Witten index (23) in essence is given by the flux $F = \int_{\mathbb{R}^2} {\rm d}x{\rm d}y B(x,y)$ measured in units of the magnetic flux quantum $\Phi_0 := 2\pi\hbar c/|e|$. Hence, SUSY is unbroken in this case. For details, see for example [31], where it is shown that the Pauli Hamiltonian for this magnetic field provides an additional SUSY structure within both subspaces $\mathcal{H}_\pm$.

Finally, for an electron ($e < 0$) in a constant magnetic field, the eigenvalues of $H_{\rm P}$ are well-known

$$\varepsilon = \hbar\omega_c\left(n+\frac{1}{2}+s_z\right) + \frac{\hbar^2 k_z^2}{2m}\,, \qquad n \in \mathbb{N}_0\,, \qquad k_z \in \mathbb{R}\,, \qquad s_z \in \{-\tfrac{1}{2}, \tfrac{1}{2}\}\,, \tag{39}$$

and the eigenvalues for (36) determined via (27) are the relativistic Landau levels first obtained by Rabi in 1928 [37]

$$E_{\pm} = \pm\sqrt{m^2c^2 + \hbar^2c^2k_z^2 + 2mc^2\hbar\omega_c(n + 1/2 + s_z)}\,. \tag{40}$$

The degeneracy for each set of quantum numbers (n, s_z, k_z) is given by the largest integer which is strictly less than $|F|/\Phi_0$ and is only finite in case the magnetic field has a compact support.

4.3. The Spin-1 Hamiltonian with Magnetic Field

Initiated by Proca's work [5], several authors have studied relativistic spin-one wave equations. Let us mention here the work by Duffin [14], by Kemmer [15] and by Yukawa, Sakata and Taketani [16,17]. An early study of the eigenvalue problem is due to Corben and Schwinger [18]. A relativistic Hamiltonian was studied, for example, by Young and Bludman [38], by Krase et al. [39] and by Tsai and coworkers [40,41]. The later work by Daicic and Frankel [42] presents an alternative solution to eigenvalue problem of the spin-one Hamiltonian in an external magnetic field. A recent treatment via FW transformation can be found in ref. [43].

The Hamiltonian of a charged spin-one particle with a priori arbitrary g-factor is given by, see for example [38,42,43],

$$H = \begin{pmatrix} mc^2 + \frac{\vec{\pi}^2}{2m} - \frac{ge\hbar}{2mc}(\vec{S}\cdot\vec{B}) & \frac{\vec{\pi}^2}{2m} - \frac{1}{m}(\vec{S}\cdot\vec{\pi})^2 + \frac{(g-2)e\hbar}{2mc}(\vec{S}\cdot\vec{B}) \\ -\frac{\vec{\pi}^2}{2m} + \frac{1}{m}(\vec{S}\cdot\vec{\pi})^2 - \frac{(g-2)e\hbar}{2mc}(\vec{S}\cdot\vec{B}) & -mc^2 - \frac{\vec{\pi}^2}{2m} + \frac{ge\hbar}{2mc}(\vec{S}\cdot\vec{B}) \end{pmatrix}, \tag{41}$$

where $\vec{S} = (S_1, S_2, S_3)^T$ is a vector who's components are 3×3 matrices acting on $\mathbb{C}^3$ and obeying the $SO(3)$ algebra $[S_i, S_j] = \mathrm{i}\varepsilon_{ijk}S_k$ representing the spin-one-degree of freedom of the particle. Again, we may identify the operators

$$M_{\pm} := mc^2 + \frac{\vec{\pi}^2}{2m} - \frac{ge\hbar}{2mc}(\vec{S}\cdot\vec{B})\,, \qquad A := \frac{\vec{\pi}^2}{2m} - \frac{1}{m}(\vec{S}\cdot\vec{\pi})^2 + \frac{(g-2)e\hbar}{2mc}(\vec{S}\cdot\vec{B}) = A^\dagger\,. \tag{42}$$

From now on, let us assume that the magnetic field $\vec{B}$ is constant, i.e., $\vec{A} = \frac{1}{2}\vec{B}\times\vec{r}$. Under this condition, one may verify that

$$[M_{\pm}, A] = (g-2)\frac{e\hbar}{2m^2c}\left[(\vec{S}\cdot\vec{B}), (\vec{S}\cdot\vec{\pi})^2\right]\,. \tag{43}$$

Hence, the SUSY condition (9) is fulfilled if and only if $g = 2$. In other words, the relativistic spin-one Hamiltonian (41) is a supersymmetric Hamiltonian if the gyromagnetic factor is given by $g = 2$. For a detailed discussion, we refer to the recent paper [44]. Here, we remark that the "Vector Boson" Hamiltonian

$$H_{\mathrm{V}} := \frac{\vec{\pi}^2}{2m} - \frac{e\hbar}{mc}(\vec{S}\cdot\vec{B}) \tag{44}$$

represents the non-relativistic Hamiltonian of a charged spin-one particle in a magnetic field with gyromagnetic factor $g = 2$. Note that (44) is related to the quantity α introduced by Weaver [45] by $H_V = \frac{\alpha}{2m}$. That this is indeed the non-relativistic version of (41) was already mentioned in ref. [39]. With this we have $M_{\pm} = H_{\mathrm{V}} + mc^2$ and with relation (3.18) and (3.19) from ref. [42], see also the Appendix A, a straightforward calculation shows that $A^2 = H_{\mathrm{V}}^2$. That is, the partner Hamiltonians read $H_{\pm} = H_{\mathrm{V}}^2/2mc^2$ and the transformed FW Hamiltonian takes the form

$$H_{\mathrm{FW}} = \begin{pmatrix} \sqrt{(H_{\mathrm{V}} + mc^2)^2 - H_{\mathrm{V}}^2} & 0 \\ 0 & -\sqrt{(H_{\mathrm{V}} + mc^2)^2 - H_{\mathrm{V}}^2} \end{pmatrix} = \beta mc^2\sqrt{1 + \frac{2H_{\mathrm{V}}}{mc^2}} \tag{45}$$

The eigenvalues of (44) are given by (we assume $e < 0$)

$$\epsilon = \hbar\omega_c \left(n + \frac{1}{2} + s_z\right) + \frac{\hbar^2 k_z^2}{2m}, \qquad n \in \mathbb{N}_0, \qquad k_z \in \mathbb{R}, \qquad s_z \in \{-1, 0, 1\}. \tag{46}$$

Hence, the spectrum of the partner Hamiltonians $H_\pm$ is given by $\varepsilon = \epsilon^2/2mc^2$ and SUSY is unbroken as $\epsilon = 0$ for $n = 0$, $s_z = -1$ and $k_z = \pm 1/\lambda_L$ with $\lambda_L := \sqrt{\hbar/m\omega_c} = \sqrt{\hbar c/|eB|}$ being the Lamor length [42]. That is, SUSY is unbroken for a spin-1 particle in a homogeneous magnetic field but the Witten index remains zero as $H_+ = H_-$ and therefore $\dim\ker H_+ = \dim\ker H_-$. The corresponding eigenvalues of (41) are given by

$$E_\pm = \pm\sqrt{m^2c^2 + \hbar^2c^2k_z^2 + 2mc^2\hbar\omega_c(n + 1/2 + s_z)}, \tag{47}$$

which is identical in form to the Dirac case (40) but s_z now taking the integer values as given in (46). In fact, for $k_z = 0$, $n = 0$ and $s_z = -1$, the above eigenvalues would become complex if $|B| > m^2c^3/|e|\hbar$. Such large magnetic fields would imply $\lambda_L < \lambda_C := \hbar/mc$;that is, the Lamor wavelength being smaller than the Compton wavelength of the vector boson. Note that confining a quantum particle to a region of the order of its Compton wavelength $\Delta x \sim \lambda_C$ implies by the uncertainty relation a momentum fluctuation $\Delta p \sim mc$ and thus a single particle description is no longer appropriate. In other words for such large magnetic fields a description via quantum field theory must be applied.

5. The Resolvent of Supersymmetric Relativistic Arbitrary-Spin Hamiltonians

In this section, we want to study the resolvent or Green's function of supersymmetric relativistic arbitrary-spin Hamiltonians defined as

$$G(z) := \frac{1}{H - z}, \qquad z \in \mathbb{C}\backslash \operatorname{spec} H. \tag{48}$$

For this, is it convenient to first look at the iterated resolvent which is given by

$$g(\zeta) := \frac{1}{H^2 - \zeta}, \qquad \zeta \in \mathbb{C}\backslash \operatorname{spec} H^2 \tag{49}$$

and is related with (48) via the obvious relation

$$G(z) = (H + z)\, g(z^2). \tag{50}$$

H^2 is block-diagonal and so is g; hence, it can be put into the form

$$g(\zeta) := \begin{pmatrix} g^+(\zeta) & 0 \\ 0 & g^-(\zeta) \end{pmatrix} \quad \text{with} \quad g^\pm(\zeta) := \frac{1}{M_\pm^2 + (-1)^{2s+1} 2mc^2 H_\pm - \zeta}. \tag{51}$$

As a result, the resolvent (48) can be expressed in terms of (51) as follows:

$$G(z) = \begin{pmatrix} (z + M_+)g^+(z^2) & Ag^-(z^2) \\ (-1)^{2s+1}A^\dagger g^+(z^2) & (z - M_-)g^-(z^2) \end{pmatrix}. \tag{52}$$

In the following subsections, we will explicitly consider the three cases discussed in the previous section. It will turn out that for these three cases, the diagonal elements $g^\pm$ of the iterated Green's function can be expressed in terms of the Green's function of the corresponding non-relativistic Hamiltonian $H_{\rm NR}$; that is,

$$g^{\pm}(\zeta) = \frac{1}{2mc^2}\, G_{\rm NR}\left(\tfrac{\zeta}{2mc^2} - \tfrac{mc^2}{2}\right), \qquad G_{\rm NR}(\xi) := \frac{1}{H_{\rm NR} - \xi}, \qquad \xi \in \mathbb{C}\backslash \operatorname{spec} H_{\rm NR}. \tag{53}$$

Note that the relation $\xi = z^2/2mc^2 - mc^2/2$, which can be put into the form $z = \pm mc^2\sqrt{1 + 2\xi/mc^2}$, in essence reflects the relation (28).

5.1. *The Resolvent of the Klein–Gordon Hamiltonian with Magnetic Field*

Following the discussion of the first example of Section 4, we may express all relevant operators in terms of the Landau Hamiltonian $H_{\rm L} = \vec{\pi}^2/2m$. Explicitly, we have

$$M_{\pm} = H_{\rm L} + mc^2\,, \qquad H_{\pm} = H_{\rm L}^2/2mc^2\,, \qquad A = H_{\rm L}\dagger\,, \tag{54}$$

which results in the iterated resolvents

$$g^{\pm}(\zeta) := \frac{1}{(H_{\rm L} + mc^2)^2 - H_{\rm L}^2 - \zeta} = \frac{1}{2mc^2} G_{\rm L}\left(\tfrac{\zeta}{2mc^2} - \tfrac{mc^2}{2}\right), \tag{55}$$

where $G_{\rm L}$ stands for the Green function of the Landau Hamiltonian in terms of which the Klein–Gordon Hamiltonian reads

$$H = \begin{pmatrix} mc^2 + H_{\rm L} & H_{\rm L} \\ -H_{\rm L} & -mc^2 - H_{\rm L} \end{pmatrix}. \tag{56}$$

The Green's function then reads in terms of the Landau Hamiltonian

$$G(z) = \frac{1}{2mc^2}\begin{pmatrix} z + mc^2 + H_{\rm L} & H_{\rm L} \\ -H_{\rm L} & z - mc^2 - H_{\rm L} \end{pmatrix} G_{\rm L}\left(\tfrac{z^2}{2mc^2} - \tfrac{mc^2}{2}\right). \tag{57}$$

5.2. *The Resolvent of the Dirac Particle in a Magnetic Field*

As in the above discussion, let us first recall the observations made in Section 4.2; that is,

$$M_{\pm} = mc^2\,, \qquad H_{\pm} = A^2/2mc^2 = H_{\rm P}\,, \qquad A = c\vec{\sigma}\cdot\vec{\pi}\,, \tag{58}$$

which provide us with the components of the iterated kernel

$$g^{\pm}(\zeta) := \frac{1}{2mc^2 H_{\rm P} + m^2c^4 - \zeta} = \frac{1}{2mc^2} G_{\rm P}\left(\tfrac{\zeta}{2mc^2} - \tfrac{mc^2}{2}\right), \tag{59}$$

where $G_{\rm P}(\epsilon) := (H_{\rm P} - \epsilon)^{-1}$ is the resolvent of the non-relativistic Pauli Hamiltonian. In terms of this Pauli Green's function and the spin projection operator A, the Dirac Green's function can be put into the form

$$G(z) = \frac{1}{2mc^2}\begin{pmatrix} z + mc^2 & A \\ A & z - mc^2 \end{pmatrix} G_{\rm P}\left(\tfrac{z^2}{2mc^2} - \tfrac{mc^2}{2}\right). \tag{60}$$

Some explicit examples have been worked out in ref. [32].

5.3. *The Resolvent of a Vector Boson in a Magnetic Field*

From Section 4.3, let us recall the relevant operators as follows:

$$M_{\pm} = H_{\rm V} + mc^2\,, \qquad H_{\pm} = H_{\rm V}^2/2mc^2\,, \qquad A = \frac{\vec{\pi}^2}{2m} - \frac{1}{m}(\vec{S}\cdot\vec{\pi})^2\,, \tag{61}$$

where the vector Hamilton H_V is given in Equation (44). Recalling that $A^2 = H_V^2$, we find for the iterated Green's functions

$$g^{\pm}(\zeta) := \frac{1}{(H_V + mc^2)^2 - H_V^2 - \zeta} = \frac{1}{2mc^2} G_V\left(\frac{\zeta}{2mc^2} - \frac{mc^2}{2}\right) \tag{62}$$

with $G_V(\epsilon) := (H_V - \epsilon)^{-1}$. The relativistic spin-one Hamiltonian explicitly reads

$$H = \begin{pmatrix} mc^2 + H_V & A \\ -A & -mc^2 - H_V \end{pmatrix} \tag{63}$$

and leads us to the Green's function

$$G(z) = \frac{1}{2mc^2} \begin{pmatrix} z + mc^2 + H_V & A \\ -A & z - mc^2 - H_V \end{pmatrix} G_V\left(\frac{z^2}{2mc^2} - \frac{mc^2}{2}\right). \tag{64}$$

6. Summary and Outlook

In this work we have considered relativistic one-particle Hamiltonians for an arbitrary but fixed spin s and have shown that under the condition, that its even part commutes with its odd part, a SUSY structure can be established. Here, the SUSY transformations map states of negative energy to those of positive energy and vice versa. This is different to the usual SUSY concepts in quantum field theory where those charges transform bosonic into fermionic states and vice versa. As examples, we have chosen the physically most relevant cases of a massive charged particle in a magnetic field for the cases of a scalar particle ($s = 0$), a Dirac fermion ($s = 1/2$) and a vector boson ($s = 1$). In the case of a constant magnetic field, SUSY is broken for $s = 0$ but remains unbroken for $s = 1/2$ and $s = 1$. The Witten index is only non-zero in the Dirac case but vanishes for the bosonic cases discussed. However, all three cases have resulted in the notable observation (28) that the FW-transformed Hamiltonian H_{FW} is entirely expressible in terms of a corresponding non-relativistic Hamiltonian H_{NR}. As $H_{FW}^2 = H^2$, the relativistic energy–momentum relation can be put into the form

$$H^2 = m^2c^4 + 2mc^2 H_{NR}, \tag{65}$$

which allows us to relate H_{NR} with the SUSY Hamiltonian (11).

There naturally arises the desire to also study the higher-spin cases $s \geq 3/2$. The corresponding free-particle Hamiltonians have been constructed, for example, by Guertin [46] in a unified way. However, as Guertin mentions, only for the cases discussed here, i.e., $s = 0, 1/2$ and 1, the corresponding Hamiltonians are local operators.

Another route for further investigation would be to consider more exotic magnetic fields. For example, choosing an imaginary vector potential such that the kinetic momentum takes the form $\vec{\pi} = \vec{p} + im\omega\vec{r}$ in essence leads for $s = 1/2$ to the so-called Dirac oscillator, which is know to exhibit such a SUSY structure [32]. To the best of our knowledge, the corresponding Klein–Gordon and vector boson oscillators have not yet been studied in the context of SUSY. Similarly, following the discussion of ref. [32] on the Dirac case, one may extend these discussion on a path-integral representation of the iterated Green's functions to the bosonic cases $s = 0$ and $s = 1$.

Acknowledgments: The author has enjoyed enlightening discussions with Simone Warzel for which I am very thankful. The comments made by the referees are very much appreciated.

Appendix A. Some Useful Relations for the Spin-One Case

In this Appendix A we present a few relations which provide some additional steps used in Section 4.3. For an arbitrary magnetic field let us recall that the components of the kinetic momentum given by $\pi_j = p_j - (e/c)A_j$ obey the commutation relation

$$[\pi_k, \pi_l] = (\mathrm{i}\hbar e/c)\, \varepsilon_{klm} B_m \tag{A1}$$

where we use Einstein's summation convention for repeated indices. From this relation one may derive the commutator $[\pi_k, \vec{S}\cdot\vec{\pi}] = (\mathrm{i}\hbar e/c)\,\varepsilon_{klm} S_l B_m$ which in turn leads us to

$$\left[\vec{\pi}^2, \vec{S}\cdot\vec{\pi}\right] = (e\hbar/c)[\vec{S}\cdot\vec{B}, \vec{S}\cdot\vec{\pi}] + (e\hbar/c) S_k S_l \left(\pi_l B_k - B_l \pi_k\right) . \tag{A2}$$

For an arbitrary magnetic field the components of the kinetic momentum do not commute with the components of the magnetic field. However, if we now assume that the magnetic field is constant one may commute in the last term these components. That is, under the assumption that $\vec{B} = const.$ we arrive at

$$\left[\vec{\pi}^2, \vec{S}\cdot\vec{\pi}\right] = (2e\hbar/c)[\vec{S}\cdot\vec{B}, \vec{S}\cdot\vec{\pi}] , \tag{A3}$$

which in turn results in

$$\left[\vec{\pi}^2, (\vec{S}\cdot\vec{\pi})^2\right] = \frac{2e\hbar}{c}\left[(\vec{S}\cdot\vec{B}), (\vec{S}\cdot\vec{\pi})^2\right] . \tag{A4}$$

Note that relation (A3) was already given in Equation (3.17) of ref. [42]. With the help of (A4) it is easy to calculate the commutator

$$[M_\pm, A] = (g-2)\frac{e\hbar}{2m^2c}\left[(\vec{S}\cdot\vec{B}), (\vec{S}\cdot\vec{\pi})^2\right] + (2g-2)\frac{e\hbar}{2m^2c}\left[\vec{\pi}^2, (\vec{S}\cdot\vec{B})\right] . \tag{A5}$$

Noting that we have derived this under the assumption of a constant magnetic field the last commutator in above expression vanishes and hence we arrive at Equation (43). Note that $[S_k, S_l] = \mathrm{i}\varepsilon_{klm} S_m$ and therefore the first term on the right-hand-side above even for a constant magnetic field only vanishes when $g = 2$.

With the assumption that the magnetic field is constant and utilising below properties of the spin-one matrices

$$S_i S_j S_k + S_k S_j S_i = \delta_{ij} S_k + \delta_{jk} S_i , \qquad \varepsilon_{ijk} S_i S_j B_k = \mathrm{i}\vec{S}\cdot\vec{B} \tag{A6}$$

one may verify the relations (see Equations (3.18) and (3.19) in ref. [42])

$$\begin{aligned} (\vec{S}\cdot\vec{\pi})^4 &= \left(\vec{\pi}^2 - \tfrac{2e\hbar}{c}(\vec{S}\cdot\vec{B})\right)(\vec{S}\cdot\vec{\pi})^2 + \tfrac{e\hbar}{c}(\vec{B}\cdot\vec{\pi})(\vec{S}\cdot\vec{\pi}) , \\ \left\{(\vec{S}\cdot\vec{B}), (\vec{S}\cdot\vec{\pi})^2\right\} &= \left(\vec{\pi}^2 - \tfrac{e\hbar}{c}(\vec{S}\cdot\vec{B})\right)(\vec{S}\cdot\vec{B}) + (\vec{B}\cdot\vec{\pi})(\vec{S}\cdot\vec{\pi}) . \end{aligned} \tag{A7}$$

Noting that for $g = 2$ we have

$$H_{\mathrm{V}} := \frac{\vec{\pi}^2}{2m} - \frac{e\hbar}{mc}(\vec{S}\cdot\vec{B}) , \qquad A = \frac{\vec{\pi}^2}{2m} - \frac{1}{m}(\vec{S}\cdot\vec{\pi})^2 \tag{A8}$$

and with above relations (A7) immediately follows that $A^2 = H_{\mathrm{V}}^2$ as claimed in the main text. Finally, let us mention the explicit form of the energy eigenfunctions can also be found in ref. [42].

References

1. Klein, O. Quantentheorie und fünfdimensionale Relativitätstheorie. *Z. Phys.* **1926**, *37*, 895–906.
2. Gordon, W. Der Comptoneffekt nach der Schrödingerschen Theorie. *Z. Phys.* **1926**, *40*, 117–133.
3. Dirac, P.A.M. The Quantum Theory of the Electron. *Proc. R. Soc. A* **1928**, *117*, 610–624.

4. Dirac, P.A.M. The Quantum Theory of the Electron Part II. *Proc. R. Soc. A* **1928**, *118*, 351–361.
5. Proca, A. Sur la théorie ondulatoire des électrons positifs et négatifs. *J. Phys. Radium* **1936** , *7*, 347–353.
6. Dirac, P.A.M. Relativistic Wave Equations. *Proc. R. Soc. A* **1936**, *155*, 447–459.
7. Fierz, M. Über die relativistische Theorie kräftefreier Teilchen mit beliebigem Spin. *Helv. Phys. Acta* **1939**, *12*, 3–37.
8. Fierz, M.; Pauli, W. On Relativistic Wave Equations for Particles of Arbitrary Spin in an Electromagnetic Field. *Proc. R. Soc. A* **1939**, *173*, 211–232.
9. Bhabha, H.J. Relativistic wave equations for the elementary particles. *Rev. Mod. Phys.* **1945**, *17*, 200–216.
10. Bargmann, V.; Wigner, E.P. Group Theoretical Discussion of Relativistic Wave Equations. *Proc. Natl. Acad. Sci. USA* **1948**, *34*, 211–223.
11. Foldy, L.L.; Wouthuysen, S.A. On the Dirac Theory of Spin 1/2 Particles and Its Non-Relativistic Limit. *Phys. Rev.* **1950**, *78*, 29–36.
12. Foldy, L.L. Synthesis of Covariant Particle Equations. *Phys. Rev.* **1956**, *102*, 568–581.
13. Feshbach, H.; Villars, F. Elementary Relativistic Wave Mechanics of Spin 0 and Spin 1/2 Particles. *Rev. Mod. Phys.* **1958**, *30*, 24–45.
14. Duffin, R.J. On The Characteristic Matrices of Covariant Systems. *Phys. Rev.* **1938**, *54*, 1114.
15. Kemmer, N. The particle aspect of meson theory. *Proc. R. Soc. A* **1939**, *173*, 91–116.
16. Yukawa, H.; Sakata, S.; Taketani, M. On the Interaction of Elementary Particles. III. *Proc. Phys.-Math. Soc. Jpn.* **1938**, *20*, 319–340.
17. Sakata, S.; Taketani, M. On the Wave Equation of Meson. *Proc. Phys.-Math. Soc. Jpn.* **1940**, *22*, 757–770.
18. Corben, H.C.; Schwinger, J. The Electromagnetic Properties of Mesotrons. *Phys. Rev.* **1940**, *58*, 953–968.
19. Schrödinger, E. The wave equation for spin 1 in Hamiltonian form. *Proc. R. Soc. A* **1955**, *229*, 39–43.
20. Silenko, A. Exact form of the exponential Foldy-Wouthuysen transformation operator for an arbitrary-spin particle. *Phys. Rev. A* **2016**, *94*, 032104.
21. Simulik, V.M. Relativistic Equations for arbitray Spin, especially for the Spin $s = 2$. *Ukr. J. Phys.* **2019**, *64*, 1064–1068.
22. Nicolai, H. Supersymmetry and spin systems. *J. Phys. A* **1976**, *9* 1497–1506.
23. Witten, E. Dynamical breaking of supersymmetry. *Nucl. Phys. B* **1981** , *188* 513–554.
24. Cooper, F.; Khare, A.; Sukhatme, U. Supersymmetry and quantum mechanics. *Phys. Rep.* **1995**, *251*, 267–385.
25. Junker, G. *Supersymmetric Methods in Quantum and Statistical Physics*, 1st ed.; Springer: Berlin, Germany, 1996.
26. Jackiw, R. Fractional charge and zero modes for planar systems in a magnetic field. *Phys. Rev. D* **1984**, *29* 2375–2377.
27. Ui, H. Supersymmetric quantum mechanics and fermion in a gauge field of $(1+2)$ dimensions. *Prog. Theor. Phys.* **1984**, *72*, 192–193.
28. Sarma, S.D.; Adam, S.; Hwang, E.H.; Rossi, E. Electronic transport in two-dimensional graphene. *Rev. Mod. Phys.* **2011**, *83* 407–470.
29. Thaller, B. *The Dirac Equation*; Springer: Berlin, Germany, 1992.
30. Znoil, M. Relativistic supersymmetric quantum mechanics based on Klein-Gordon equation. *J. Phys. A* **2004**, *37*, 9557.
31. Junker, G. *Supersymmetric Methods in Quantum, Statistical and Solid State Physics, Enlarged and Revised Edition*; IOP Publishing: Bristol, UK, 2019.
32. Junker, G.; Inomata, A. Path integral and spectral representations for supersymmetric Dirac-Hamiltonians *J. Math. Phys.* **2018**, *59*, 052301.
33. Witten, E. Constraints on supersymmetry breaking. *Nucl. Phys. B* **1982**, *202* 253–316.
34. Fock, V. Bemerkung zur Quantelung des harmonischen Oszillators im Magnetfeld. *Z. Phys.* **1928**, *47* 446–448.
35. Landau, L. Diamagnetismus der Metalle. *Z. Phys.* **1930**, *64*, 629–637.
36. Aharonov, Y.; Casher, A. Ground state of a spin-$\frac{1}{2}$ charged particle in a two-dimensional magnetic field. *Phys. Rev. A* **1978**, *19*, 2461–2462.
37. Rabi, I.I. Das freie Elektron im homogenen Magnetfeld nach der Diracschen Theorie. *Z. Phys.* **1928**, *49*, 507–511.
38. Young, J.A.; Bludman, S.A. Electromagnetic Properties of a Charged Vector Meson. *Phys. Rev.* **1963**, *131*, 2326–2334.

39. Krase, L.D.; Lu, P.; Good, R.H., Jr. Stationary States of a Spin-1 Particle in a Constant Magnetic Field. *Phys. Rev. D* **1971**, *3*, 1275–12798.
40. Tsai, W.-Y.; Yildiz, A. Motion of Charged Particles in a Homogeneous Magnetic Field. *Phys. Rev. D* **1971** *4*, 3643–3648.
41. Goldman, T.; Tsai, W.-Y. Motion of Charged Particles in a Homogeneous Magnetic Field II. *Phys. Rev. D* **1971**, *4*, 3648–3651.
42. Daicic, J.; Frankel, N.E. Relativistic spin-1 bosons in a magnetic field. *J. Phys. A Math. Gen.* **1993**, *26*, 1397–1408.
43. Silenko, A.J. High precision description and new properties of a spin-1 particle in a magnetic field. *Phys. Rev. D* **2014**, *89*, 121701.
44. Junker, G. Supersymmetric quantum mechanics requires $g = 2$ for vector bosons. *Eur. Phys. Lett.* **2020**, *130*, 30003.
45. Weaver, D.L. Application of a single-particle-theory calculation of $g - 2$ to spin one. *Phys. Rev. D* **1976**, *14*, 2824–2825.
46. Guertin, R.F. Relativistic Hamiltonian Equations for Any Spin. *Ann. Phys.* **1974**, *88*, 504–553.

9

The Heisenberg Uncertainty Principle as an Endogenous Equilibrium Property of Stochastic Optimal Control Systems in Quantum Mechanics

Jussi Lindgren [1,*] and Jukka Liukkonen [2]

1 Department of Mathematics and Systems Analysis, Aalto University, 02150 Espoo, Finland
2 Nuclear and Radiation Safety Authority, STUK, 00880 Helsinki, Finland; jukka.liukkonen@stuk.fi
* Correspondence: jussi.lindgren@aalto.fi

Abstract: We provide a natural derivation and interpretation for the uncertainty principle in quantum mechanics from the stochastic optimal control approach. We show that, in particular, the stochastic approach to quantum mechanics allows one to understand the uncertainty principle through the "thermodynamic equilibrium". A stochastic process with a gradient structure is key in terms of understanding the uncertainty principle and such a framework comes naturally from the stochastic optimal control approach to quantum mechanics. The symmetry of the system is manifested in certain non-vanishing and invariant covariances between the four-position and the four-momentum. In terms of interpretation, the results allow one to understand the uncertainty principle through the lens of scientific realism, in accordance with empirical evidence, contesting the original interpretation given by Heisenberg.

Keywords: stochastic optimal control; stochastic mechanics; thermodynamic equilibrium; uncertainty principle

---

1. Introduction

This research article aims to provide an intuitive explanation for the quantum phenomenon of measurement uncertainty and it also explains why the Born rule should hold, i.e., why the probability of finding a test particle should be proportional to the square of the wave function. The article builds on the basic framework presented in the authors' recent article [1] and it improves the framework there. The basic framework is built on the assumption that quantum mechanics should be seen through the framework of stochastic optimal control theory; stochastic dynamic optimization in a coordinate-invariant manner on the Minkowski spacetime. In particular, the algebraic structure including the imaginary units can be understood through this framework. However, even though we have shown how relativistic and non-relativistic wave equations of quantum mechanics are the special cases of the corresponding Hamilton–Jacobi–Bellman transport equation for the value function, little consideration was given to the Born rule and the Heisenberg uncertainty principle. In this article, we consider these issues in more detail as an extension and follow-up.

The Heisenberg uncertainty principle is a concept in quantum mechanics that is as elusive as it is important, as there seems to be no unanimous agreement in the literature on its real meaning, content and implications. Originally, Werner Heisenberg linked the principle with some external interference stemming from the measurement experiment [2]. More recently, Ozawa has linked it more abstractly with measurement errors and disturbances, see [3]. There is also a line of research, which relates the uncertainty principle more in line with the approach of the present authors, see, e.g., [4]. On the other hand, Popper [5] and Ballentine [6] already promoted over 50 years ago an interpretation where the uncertainty principle should be seen merely as a statistical scatter relationship. The approach by Bohm [7] is also in line with a statistical ensemble interpretation.

We argue in this article that indeed a more straightforward statistical interpretation for the uncertainty principle is possible when one considers the conjugate variables as random vectors with some possible linear covariance structure. In this respect, in particular, what seems to be especially fruitful is to consider the optimal spacetime diffusions and the respective Fokker–Planck equations or Kolmogorov forward equations for the stochastic optimal control model. The Fokker–Planck equations determine the transport behavior of the transition probability density, as the test particle undergoes the optimal spacetime diffusion, which minimizes the expected action.

In [1] it was shown that the optimal four-velocity is essentially a four-gradient of the value function J, which is typical for such linear–quadratic (stochastic) optimal control problems. This optimal behavior of the test particle is a random motion, which has a drift into the optimal direction and is disturbed by a Brownian noise. We have argued that the diffusion model might be just a phenomenological model in the sense that it is the spacetime itself that is fluctuating [8] and this can be modeled by vector diffusions on the Minkowski spacetime.

We use the Einstein summation convention throughout the article so that upper indices and lower indices represent contravariant and covariant objects, which are generally different. We use slightly different index notation compared to [1] in order to avoid misunderstandings. We use the normalization that the reduced Planck's constant is unity, $\hbar = 1$.

2. The Evolution of the Transition Probability Density and the Stationary Distribution

As was shown in [1] by the authors, the key equations of quantum mechanics can be obtained as optimality equations of certain coordinate invariant stochastic optimization programs on Minkowski spacetimes. Here, as in that model, the test particle undergoes a diffusion process on the Minkowski spacetime; we can analyze the transition probabilities and possible stationary distributions as well.

We argue that the stationary distribution is indeed the relevant distribution in terms of falsifiable experiments, as in order to test any predictions about observables or expectations, the distribution must be sampled many times, by which time the ensemble distribution is the relevant probability distribution. This is due to the property that a stochastic system with a *confining potential* (and other mild technical requirements) yields a stochastic process, which is *ergodic* and the transition probability converges to the Gibbs distribution *exponentially fast*, see [9]. A confining potential means a potential which loosely guarantees that the test particle is "confined" due to the gradient drift of the stochastic process. In this sense, the stochastic optimal control approach to quantum mechanics is actually quite close conceptually to non-equilibrium and equilibrium statistical mechanics.

The basic framework of the stochastic optimal control problem is similar to the set-up in [1], we have a spacetime diffusion for the test particle:

$$dX^{\mu} = u^{\mu} ds + \sigma^{\mu} dW^{\mu} \tag{1}$$

where X^{μ} is the random four-position, the contravariant index $\mu = 0,1,2,3$ represents the spacetime coordinates in the (flat) Minkowski spacetime and s is the proper time. There are four independent standard Brownian motions dW^{μ} with a scale $\sigma^{\mu\mu} = -\frac{1}{2im}$. We use the notation $\sigma^{\mu\nu} = \sigma^{\mu}\sigma^{\nu}$, $\mu = 0,1,2,3$, $\nu = 0,1,2,3$ for the diffusion matrix, so it is a tensor product of two vectors. Note that in the denominator, we have the imaginary unit i. The mass of the test particle is m. A capital letter X refers to the random variable itself, whereas a small x refers to the realization coordinate of the random variable. For convenience, we have used a normalization in which the speed of light is unity, $c = 1$. The four-velocity is represented by u^{μ}. We use slightly different scales (same for all $\mu = 0,1,2,3$) for the Brownian motions compared to [1], as we believe it makes the set-up even more clear and simple. The diffusion matrix should be chosen in such a way that the Hamilton–Jacobi–Bellman equation can be linearized in the simplest possible manner using the Hopf–Cole transformation, as in [1]. Utilizing Occam's razor, the simplest such structure comes from the option:

$$\sigma^{\mu\mu} = -\frac{1}{2im}, \text{ otherwise zero (independent Brownian motions)} \tag{2}$$

The functional to be optimized is given by:

$$S = \int \int_{\tau}^{T} (\frac{1}{2} m g_{\mu\nu} u^{\mu} u^{\nu} - V(x)) ds p(x,s) dV. \tag{3}$$

With the invariant volume form $dV = \sqrt{g} dcx_0 dx_1 dx_2 dx_3$, with the metric determinant g. The metric tensor on the Minkowski spacetime is given by $g_{\mu\nu}$. The transition probability density to the four-position x at proper time s is given by $p(x,s)$. The Lagrangian is the four-dimensional generalization of the usual Lagrangian in classical mechanics in a stochastic environment.

We note that the model, as in [1], has a gradient structure, as the optimal drift (necessary condition for the optimal control problem) is given by:

$$u^{\mu} = -\frac{1}{im} \nabla^{\mu} J, \tag{4}$$

where J is the value function.

The resulting nonlinear Hamilton–Jacobi–Bellman optimality equation for this system, as in [1], is given by:

$$\frac{\partial J}{\partial \tau} - iV(x) - \frac{1}{2im} \nabla_{\mu} J \nabla^{\mu} J + \frac{1}{2} \sigma^2 \nabla_{\mu} \nabla^{\mu} J = 0, \tag{5}$$

where $\sigma^2 = \sigma^{\mu}_{\mu} = -\frac{1}{im}$, which is the sum of the diagonal elements of the diffusion matrix. Note that the Minkowski metric implies that the trace includes the minus sign when summing the elements together. The Hamilton-Jacobi-Bellman equation can then be linearized to obtain the Stueckelberg and Schrödinger wave equations, as in [1], when we use the Hopf–Cole transformation $J = \log \varphi$, where φ is the wave function. This implies that the wave function is connected to the optimal expected action.

We then expect the respective Fokker–Planck equation for the optimal stochastic process to yield a stationary solution, which is the well-known Gibbs distribution from equilibrium statistical mechanics and thermodynamics. As such, we can analyze the properties of this diffusion process by indeed considering the respective Fokker–Planck equation, which is the partial differential equation governing the transition probability $p(x,s)$ density, where x is the four-position. The Fokker–Planck equation for the spacetime diffusion is given by:

$$\frac{\partial p}{\partial s} = \nabla_{\mu} \Big(-u^{\mu} p + \frac{1}{2} \nabla_{\nu} (\sigma^{\mu\nu} p) \Big), \tag{6}$$

where we have used the Einstein summation convention. Although in general we cannot solve explicitly the transition probability density for all proper times s, we can, however, in this case solve the stationary distribution, if the value function J has some mild properties, see [9]. The stationary distribution is obtained by setting the expression in the brackets to be zero. The more general family of stationary distributions can be obtained by setting only the partial derivative of the transition density with respect to proper time to zero. For our purposes, setting the expression in the brackets to zero suffices. As the symmetric diffusion matrix $\big(\sigma^{\mu\mu} = -\frac{1}{2im}, \textit{ otherwise zero}\big)$ is constant (no correlated noise), we must have: $\frac{1}{2} \nabla^{\mu} p = \nabla^{\mu} J p$, where on the right side the gradient operates only on J, from which we can directly deduce that the stationary distribution is indeed the Gibbs distribution:

$$p = \frac{1}{Z} e^{2J}, \tag{7}$$

where Z is a normalization constant (the partition function in statistical mechanics). Note that the value function could be complex-valued in general and therefore we need to require additional properties from the system as we want the stationary probability density to be real. We allow the four-position of the test particle to be complex-valued, but we require that the probability density is real. A complex four-position might also have interesting links with the algebra of octonions.

2.1. The Requirement that the Stationary Density is Real Leads to Born's Rule

Remember that, as in [1], we assume a simple relationship (Hopf–Cole transformation) between the value function J and the wave function φ: $J = \log \varphi$. As it holds that the complex conjugates also have the relationship: $J^* = \log \varphi^*$, it is straightforward to see that if the value function is of the form $J = a + ib$, then:

$$2a = \log \varphi\varphi^*. \tag{8}$$

Note now that as we need to require that the stationary distribution is purely real, then the Gibbs distribution implies that $= \frac{1}{Z}\, \varphi\varphi^*$, i.e., we can deduce that the Born rule must hold. We cannot infer from this that the Born rule holds at all times, but we can guarantee that when sampling from the stationary ensemble, the Born rule holds. This constructive demonstration has an interesting parable with the quantum equilibrium hypothesis in Bohmian mechanics [7].

2.2. The Heisenberg Uncertainty Principle

In this subsection, we show how the stochastic optimal control approach to quantum mechanics can help to understand the meaning of Heisenberg's uncertainty principle. In terms of abstract notions, this principle is usually derived by the assumption that the momentum and position operators do not commute. There is no generally agreed common understanding of the principle beyond of the assumption that conjugate variables such as position and momentum cannot be measured simultaneously to an arbitrary degree of certainty.

If one considers the stochastic approach toquantum mechanics, the uncertainty principle can be understood in an intuitive manner by treating the four-position and four-momentum as random vectors. The uncertainty principle is then a statement about the covariance of these random vectors. As such, it does not necessarily then say anything about how experiments are limited or affected by the experimenter, but instead the principle only observes that the covariance does not vanish between these two random vectors. In its usual form, the uncertainty principle is given as (we have used the normalization $\hbar = 1$):

$$\sigma_X \sigma_P \geq \frac{1}{2}, \tag{9}$$

where the sigmas are the standard deviations of the four-position X and four-momentum P, when using the Born rule in the stationary ensemble. The key is to observe that we can use the Cauchy–Schwartz inequality:

$$\sigma_X \sigma_P \geq \left|Cov(X, P)\right|. \tag{10}$$

So that the uncertainty principle holds, we need to have a non-vanishing absolute covariance of the four-position and the four-velocity (up to a scale). As covariance is defined by

$$Cov(X, P) = E[XP] - E[X]E[P], \tag{11}$$

where $E[[]$ is the mathematical expectation operator, we should be able to show that $E[XP] \neq E[X]E[P]$ where X is the four-position and P is the four-momentum. As we can center the random variables without any loss of generality, it suffices to show that

$$E[XP] \neq 0. \tag{12}$$

Consider now again the stationary distribution. As testing the prediction for an expected value of a random variable makes sense only if we can sample the distribution many times and measure the average (law of large numbers), we argue that in terms of any proper experiment, the uncertainty principle concerns only the stationary ensemble averages. Therefore, we calculate the covariance based on the thermodynamical ensemble distribution, which we showed is the Gibbs distribution. This means that all expectations from now on are calculated with respect to the stationary Gibbs distribution.

Consider first the expectation $E[XP]$. Remembering that $P^\mu = mu^\mu = i\nabla^\mu J$, we can consider the integral:

$$E[XP] = i \int X^\mu \nabla^\mu J p \sqrt{g} d^4x, \tag{13}$$

where we integrate over the invariant volume form $\sqrt{g}d^4x$. Using the implicit definition of the stationary distribution (stationary solution for the Fokker–Planck equation) $\frac{1}{2}\nabla^\mu p = \nabla^\mu Jp$, we have:

$$E[XP] = -\int X^\mu \nabla^\mu J p d^4x = -\frac{1}{2}\int X^\mu \nabla^\mu p d^4x. \tag{14}$$

By integration by parts, we have (assuming that the probability flux vanishes at the boundary):

$$-\tfrac{1}{2}\int X^\mu \nabla^\mu p d^4x = \tfrac{1}{2}\int \nabla^\mu X^\mu p d^4x = \\ \tfrac{1}{2}\nabla^\mu X^\mu \int p d^4x = \tfrac{1}{2}\nabla^\mu X^\mu \neq 0.$$

where in the last part we have used the fact that the total probability mass is unity and that the $\mu : th$ component derivative of the $\mu : th$ position vector component spits out constants. This shows that the covariance does not vanish and that Heisenberg's uncertainty principle holds (up to a scale). Note that the expectation is invariant as it is a constant.

3. Discussion and Interpretation of the Result

The uncertainty principle obtained is presented in an objective manner through the properties of the stationary distribution and the gradient structure of the underlying stochastic process. The epistemological limit for measuring conjugate variables, i.e., four-position and four-momentum, can be understood intuitively through the stochastic optimal control framework. Intuitively, the stochastic optimal control approach describes an ensemble of test particles with the optimal gradient drift and undergoing a diffusion, which seeks to find a stationary point for the value function. If one prepares such a system and measures the position and momentum repeatedly in order to test the predictions about observables, one will find that the statistical spreads of the position and momentum of the test particle have an inverse relationship. The existence of the lower limit for the product of standard deviations for the position and momentum does not in this framework, however, preclude the possibility of measuring the position and momentum simultaneously. Based on the present result, it only says that when measurements are repeated, the statistical spread obeys the uncertainty principle. Importantly, the constructive proof in this study does not require any external interference from the measurement apparatus, merely the structure of the gradient drift and the underlying diffusion imply the uncertainty principle.

Indeed, the linear–quadratic structure of the stochastic optimal control problem implies that the optimal four-velocity of the test particle has a gradient structure. The gradient structure is somewhat similar to the guidance equation of the pilot wave in the de Broglie–Bohm approach [7]. We therefore argue that Bohmian mechanics is essentially very close to the present stochastic optimal control approach, and that the *guidance equation* is close to the optimal drift velocity of the present model. Unfortunately, when David Bohm developed his model, stochastic optimal control theory had not been developed yet. Moreover, the *quantum equilibrium* in Bohmian mechanics seems to be quite close to the stationary distribution derived in this study. Indeed, it is shown here that the stochastic process under the optimal gradient drift converges exponentially fast to a stationary distribution (thermodynamical equilibrium), where one has a non-vanishing covariance between the four-position and the four-momentum.

Whereas in Bohmian mechanics the ontology of the theory is rather complicated with *pilot waves*, the present approach is more minimalistic: as the *spacetime has fluctuations* at small scales, the test particle obeys a random walk with a drift in such a way that the drift is a "gradient search", which tries to find a dynamically optimal route for the test particle in order to make the expected action stationary. This system results in an equilibrium distribution for the random walk, and this "thermodynamic ensemble" is what we believe corresponds to the quantum equilibrium in Bohmian mechanics. The model is realistic and objective in line with the inclinations of Karl Popper and the statistical/ensemble interpretation of quantum mechanics. The reason why nature seems to obey variational principles is nevertheless a deep question. Why the drift is a gradient search for the value function is no more or less deep a question than why nature prefers various optimal programs, be it in classical mechanics, electromagnetism or general relativity. In the present model, the test particle has a drift into the direction, in which the value function decreases rapidly. These teleological questions are equally deep and equally valid in the Hamilton–Jacobi formulation of classical mechanics. In this sense, the particle just "knows" the direction in which the expected action decreases most rapidly. Therefore, we do not need any "hidden variables" in quantum mechanics any more than we need such "hidden variables" in classical or statistical mechanics. In the Hamilton–Jacobi approach of classical mechanics, the velocity of the test particle is equivalently linked to the gradient of the value function.

Intuitively, the content of the uncertainty principle is, in the present model, that the product of the standard deviations has a lower bound in terms of the stationary distribution due to the covariance. If one would like to have a small standard deviation for the position, one would need to have a rather steep value function in order to constrain and confine the particle within a small location in terms of the stationary distribution (due to the exponential structure of the Gibbs distribution). A steep and a localized value function J would then imply a large standard deviation for the gradient of the value function. There is then a natural tradeoff between the localizations of four-position and four-momentum (Figure 1). The key is the gradient structure of the stochastic process, where the test particle tries to seek a minimum for the value function as in a "gradient search".

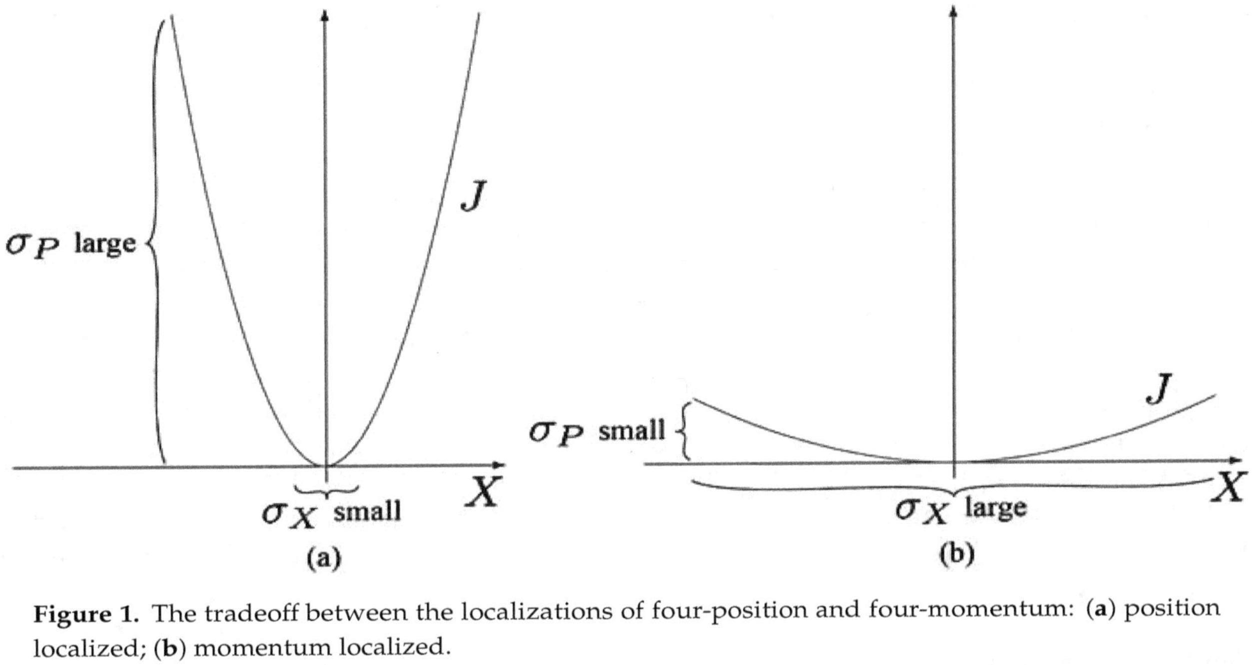

Figure 1. The tradeoff between the localizations of four-position and four-momentum: (**a**) position localized; (**b**) momentum localized.

Therefore, if one seeks an equilibrium system which has a very deep potential well and thus good localization and confinement properties, one needs to accept the large variability in the gradient of the potential (= value) function. This understanding of the uncertainty principle then implies that the lower bound for the product of standard deviations has nothing to do with experiments interfering with the set-up, but it is merely a natural trade-off implied by the linear–quadratic structure of the stochastic optimal control program. Indeed, one can see the framework through the lens of thermodynamics, where the system relaxes towards a thermodynamic equilibrium, which is just the stationary state for the respective Fokker–Planck equation. The results in this paper seem to therefore strongly support the statistical or the ensemble interpretation of quantum mechanics, put forward by, for example, Ballentine [6].

4. Conclusions

We believe that the Heisenberg uncertainty principle can be understood in a fruitful manner by considering it as a stationarity property of stochastic or statistical mechanics. The ontological and epistemological problems related to the uncertainty principle can be mitigated with such an interpretation as in the stochastic optimal control framework; the tradeoff of standard deviations between the four-position and the four-momentum can be understood intuitively through the confinement properties of the value function. The sharp localization, and thus small standard deviation of the test particle four-position, requires a rather steep value function locally, which naturally implies large variability for the gradient of the value function and thus for the four-momentum. Within this interpretation, it still holds that we cannot measure momentum and position simultaneously to an arbitrary degree of precision repetitively, but the uncertainty principle has a new, more intuitive meaning based on scientific realism and objectivism. The interpretation of the uncertainty principle, in which the measurement process itself causes the interference and the uncertainty limit, should be therefore finally abandoned.

The present model allows scientific realism and objectivism in the sense that the test particle obeys an optimal diffusion irrespective of the observer and its position and momentum are uniquely defined and measurable in principle at each point in spacetime, but as we repetitively measure something such as electrons from a beam, the fluctuations of the spacetime induce such randomness to the system that on average we can only "see" the ensemble in the "thermodynamic" equilibrium. In this sense, we argue that this study, together with [1], is very much a constructive model of an objective interpretation of quantum mechanics put forward by Karl Popper [5], among others. Moreover, the interpretation given in this article implies that the uncertainty principle does not logically rule out determinism as such.

There is, however, a conceptual connection to Bohmian mechanics, due to the gradient drift structure and the quantum equilibrium, but as we have argued, the diffusion is a phenomenological model for the fluctuation of the spacetime itself. In a sense, the "hidden variables" are not needed, because the medium (spacetime) is stochastic. It is a modeling paradigm, from which one cannot, however, deduce whether the universe is deterministic or not. The optimization procedure, which nature seems to obey, just implies that certain random variables (due to the fluctuations of spacetime) must have a non-vanishing correlation. These conclusions are also in line with rather recent experimental evidence; see [10,11]. The present result seems to also support the paradigm that such uncertainty principles are general features of specific stochastic systems, as has been indicated, for example, in [4]. The symmetries of the present framework are manifested clearly in the property that the covariance of the four-position and the four-momentum is non-vanishing and invariant.

Author Contributions: J.L. (Jussi Lindgren) invented the link between the stationary distribution and the uncertainty principle through the linear–quadratic optimization program; J.L. (Jukka Liukkonen) edited the article and wrote the discussion and conclusions. Both authors reviewed and approved the article. All authors have read and agreed to the published version of the manuscript

References

1. Lindgren, J.; Liukkonen, J. Quantum mechanics can be understood through stochastic optimization on spacetimes. *Sci. Rep.* **2019**, *9*, 1–8. [CrossRef]
2. Heisenberg, W. Über den anschaulichen inhalt der quantentheoretischen kinematik und mechanik. In *Original Scientific Papers Wissenschaftliche Originalarbeiten*; Springer: Berlin/Heidelberg, Germany, 1985; pp. 478–504.
3. Ozawa, M. Physical content of Heisenberg's uncertainty relation: Limitation and reformulation. *Phys. Lett. A* **2003**, *318*, 21–29. [CrossRef]
4. Koide, T.; Kodama, T. Generalization of uncertainty relation for quantum and stochastic systems. *Phys. Lett. A* **2018**, *382*, 1472–1480. [CrossRef]
5. Popper, K. *Quantum Theory and the Schism in Physics*; Psychology Press: Hove, UK, 1992.
6. Ballentine, L.E. The statistical interpretation of quantum mechanics. *Rev. Mod. Phys.* **1970**, *42*, 358–381. [CrossRef]
7. Bohm, D. A suggested interpretation of the quantum theory in terms of "hidden" variables. I. *Phys. Rev.* **1952**, *85*, 166. [CrossRef]
8. Frederick, C. Stochastic space-time and quantum theory. *Phys. Rev. D* **1976**, *13*, 3183–3191. [CrossRef]
9. Pavliotis, G.A. *Stochastic Processes and Applications, Diffusion Processes, the Fokker-Planck and Langevin Equations*; Springer: New York, NY, USA, 2014.
10. Rozema, L.A.; Darabi, A.; Mahler, D.H.; Hayat, A.; Soudagar, Y.; Steinberg, A.M. Violation of Heisenberg's measurement-disturbance relationship by weak measurements. *Phys. Rev. Lett.* **2012**, *109*, 100404. [CrossRef]
11. Erhart, J.; Sponar, S.; Sulyok, G.; Badurek, G.; Ozawa, M.; Hasegawa, Y. Experimental demonstration of a universally valid error–disturbance uncertainty relation in spin measurements. *Nat. Phys.* **2012**, *8*, 185–189. [CrossRef]

Permissions

All chapters in this book were first published in MDPI; hereby published with permission under the Creative Commons Attribution License or equivalent. Every chapter published in this book has been scrutinized by our experts. Their significance has been extensively debated. The topics covered herein carry significant findings which will fuel the growth of the discipline. They may even be implemented as practical applications or may be referred to as a beginning point for another development.

The contributors of this book come from diverse backgrounds, making this book a truly international effort. This book will bring forth new frontiers with its revolutionizing research information and detailed analysis of the nascent developments around the world.

We would like to thank all the contributing authors for lending their expertise to make the book truly unique. They have played a crucial role in the development of this book. Without their invaluable contributions this book wouldn't have been possible. They have made vital efforts to compile up to date information on the varied aspects of this subject to make this book a valuable addition to the collection of many professionals and students.

This book was conceptualized with the vision of imparting up-to-date information and advanced data in this field. To ensure the same, a matchless editorial board was set up. Every individual on the board went through rigorous rounds of assessment to prove their worth. After which they invested a large part of their time researching and compiling the most relevant data for our readers.

The editorial board has been involved in producing this book since its inception. They have spent rigorous hours researching and exploring the diverse topics which have resulted in the successful publishing of this book. They have passed on their knowledge of decades through this book. To expedite this challenging task, the publisher supported the team at every step. A small team of assistant editors was also appointed to further simplify the editing procedure and attain best results for the readers.

Apart from the editorial board, the designing team has also invested a significant amount of their time in understanding the subject and creating the most relevant covers. They scrutinized every image to scout for the most suitable representation of the subject and create an appropriate cover for the book.

The publishing team has been an ardent support to the editorial, designing and production team. Their endless efforts to recruit the best for this project, has resulted in the accomplishment of this book. They are a veteran in the field of academics and their pool of knowledge is as vast as their experience in printing. Their expertise and guidance has proved useful at every step. Their uncompromising quality standards have made this book an exceptional effort. Their encouragement from time to time has been an inspiration for everyone.

The publisher and the editorial board hope that this book will prove to be a valuable piece of knowledge for researchers, students, practitioners and scholars across the globe.

List of Contributors

Marisol Bermúdez-Montaña
Instituto de Ciencias Nucleares, Universidad Nacional Autónoma de México, Apartado Postal 70-543, CDMX 04510, Mexico
Facultad de Química, Universidad Nacional Autónoma de México, Apartado Postal 70-543, CDMX 04510, Mexico

Marisol Rodríguez-Arcos and Renato Lemus
Instituto de Ciencias Nucleares, Universidad Nacional Autónoma de México, Apartado Postal 70-543, CDMX 04510, Mexico

José M. Arias
Departamento de Física Atómica, Molecular y Nuclear, Facultad de Física, Universidad de Sevilla, Apartado 1065, 41080 Sevilla, Spain
Instituto Carlos I (iCI) de Física Teórica y Computacional, Universidad de Sevilla, Apartado 1065, 41080 Sevilla, Spain

Joaquín Gómez-Camacho
Departamento de Física Atómica, Molecular y Nuclear, Facultad de Física, Universidad de Sevilla, Apartado 1065, 41080 Sevilla, Spain
CN de Aceleradores (U. Sevilla, J. Andalucía, CSIC), 41092 Sevilla, Spain

Emilio Orgaz
Facultad de Química, Universidad Nacional Autónoma de México, Apartado Postal 70-543, CDMX 04510, Mexico

Elena V. Orlenko
Higher School of Engineering Physics, Institute of Physics, Nanotechnology and Telecommunication, Peter the Great Saint Petersburg Polytechnic University, 195251 St. Petersburg, Russia

Fedor E. Orlenko
Scientific and Educational Center for Biophysical Research in the Field of Pharmaceuticals, Saint Petersburg State Chemical Pharmaceutical University (SPCPA), 197376 St. Petersburg, Russia

Miloslav Znojil
The Czech Academy of Sciences, Nuclear Physics Institute, Hlavní 130, 25068 Řež, Czech Republic
Department of Physics, Faculty of Science, University of Hradec Králové, Rokitanského 62, 50003 Hradec Králové, Czech Republic

G. Jordan Maclay
Quantum Fields LLC, St. Charles, IL 60174, USA

Dalia Saha, Susmita Sanyal and Abhik Kumar Sanyal
Department of Physics, Jangipur College, Murshidabad 742213, West Bengal, India

Christiane Quesne
Physique Nucléaire Théorique et Physique Mathématique, Université Libre de Bruxelles, Campus de la Plaine CP229, Boulevard du Triomphe, B-1050 Brussels, Belgium

John Gleason Cramer
Department of Physics, University of Washington, Seattle, WA 98195, USA

Carver Andress Mead
California Institute of Technology, Pasadena, CA 91125, USA

Georg Junker
European Southern Observatory, Karl-Schwarzschild-Straße 2, D-85748 Garching, Germany
Institut für Theoretische Physik I, Universität Erlangen-Nürnberg, Staudtstraße 7, D-91058 Erlangen, Germany

Jussi Lindgren
Department of Mathematics and Systems Analysis, Aalto University, 02150 Espoo, Finland

Jukka Liukkonen
Nuclear and Radiation Safety Authority, STUK, 00880 Helsinki, Finland

Index

www.ingramcontent.com/pod-product-compliance
Lightning Source LLC
LaVergne TN
LVHW081018210926
848485LV00004B/894
* 9 7 8 1 6 4 7 2 5 4 6 6 7 *